AF553806

SILK CULTURE :
A BIOCHEMICAL APPROACH

By

P. N. Panday

S.K. Sharan **P.K. Mishra**

A.P.H. Publishing Corporation

New Delhi 110002

Published by

S.B. Nangia
APH Publishing Corporation
4435-36/7, Ansari Road, Darya Ganj
New Delhi-110 002
PH. : 23274050

Email : aphbooks@gmail.com

2021

₹.3995/-

Typesetting at

Rawat Computers
Gandhi Nagar, Delhi

Printed at

Balaji Offset
Navin Shahdara, Delhi-32

This Book is dedicated to
the parents of
Dr. S.K. Sharan,
Late Vaidehi Ballave Sharan
and Radhika Devi

Acknowledgements

The first author express his deep sense of gratitude to Late Dr. B. R. R. P. Sinha, Director, central Tasar Research and Training institute, central Silk Board, Piska Nagari, Ranchi, India who allowed him to carry out this study on behalf of the Member Secretary, Central Silk Board, Madivalla, Hosur Road, Bangalore-560068, India.

His sincere thanks are also to his dear friend Shri Bishwa Mohan Kumar Singh who helped him in every stance during writing of this book. He is thankful to his wife Smt. Rita Sinha who was present at every stage during course of completing this book and allowed him to lead an uninterrupted life.

Preface

Antheraea mylitta Drury is a sericigenous insect belonging to family Saturniidae of order Lepidoptera. The species produces specific silk known as Tasar and growing of Tasar cocoons is known as Tasar culture. Tasar culture is a traditional occupation of the tribes of tropical India since time immemorial. Its origin and history are lost in antiquity. Lord Rama's nuptial gift to Sita includes "tussar" silk, as we learn from Ramayana, which vividly testifies the ancient nature and origin of tasar silk in India. However, since ages, tasar silk production remained as a traditional craft of the poor tribal folk. Incidentally, nature also has gifted the region with dense forest having rich distribution of tasar silkworm food plants. Combination of salubrious climate, dense population of food plants in the deciduous and semi-deciduous forests and the tribal folk constitute the essential inputs for tasar silk production.

The sericulture activity is essentially an agro-based cottage industry having both agricultural and industrial activities intertwined together and provides employment to masses in rural India. Sericulture covers cultivation of silkworms host plants, production of silkworm eggs, silkworm rearing cocoon production, reeling, spinning weaving and utilization of silk waste and other bye-products and finally the production of silk.

For long, Tasar production remained un-organised with low productivity. However, with gradual scientific and organisational development, the culture is moving towards systematisation and increased productivity and taking a shape of agro-industry. Higher productivity and systematisation is the future of industry and is achievable through scientific investigations and back up. There are many inherent causes of low productivity and these include polyphagous nature of feeding, distribution to area having varying biotic and abiotic factors, weak voltinism and disease.

A. mylitta is highly polyphagous and it shows wide range of adaptation. Most harmoniously, it has adapted to feed and survive on *Terminalia tomentosa* (W & A) and *Terminalia arjuna* (Bedd.), both in semi-domesticated and wild conditions. Substantial share of Tasar cocoon collections are from trees of *Shorea robusta* (Roxburghii.), *Anogeissus latifolia* (Wall), *Lagerstroemia spp.*, etc. in the forests of tropical India and are called "wild tasar cocoons." These semi-domesticated and wild varieties available in different pockets of tropical India are termed as different ecoraces, eco-types or biotypes which not only differ among themselves based on their host plants, but in their niche, behaviour, cocoon colours and commercial characteristics. Examples of such semi-domesticated biotypes of *A. mylitta* are Daba, Sukinda, Laria, Raily, Sarihan, Modal, Tira, Bhandara, etc. Survey made so far reveals that major economically important ecoraces are distributed mainly in Central India within 24-16°N and 80-88°E, whereas occurrence has been reported within 12-30°N and 72-96°E. Though there is a notion that each biotype is specific to a particular region, overlapping of habitat at the edge zone can not be ruled out. Maximum production of Tasar cocoons comes from the regions of moist deciduous forests.

Insects as well as other organisms do not live in a constant environment optimum for growth and survival. With the changes in environmental conditions, insects are able to adapt to and survive unfavourable seasons or conditions that do not support their continued development. Such behaviour of insects is termed as quiescence, diapause, hibernation or aestivation depending upon the insects' programming to skip unfavourable environmental conditions. In case of *A. mylitta,* diapause has been observed to be facultative. Based on nature of voltinism, they have been placed under three categories: univoltine, bivoltine and trivoltine. Normally, it behaves as bivoltine completing two generations in a year. However, in warmer climates, part of the population turns tri-voltine. The average larval period during first, second and third crops are around 30, 45 and 65 days, respectively. The winter crop harvested cocoons enter into diapause and emerge as moths giving rise to first crop during next year monsoon, which coincides with the sprouting of leaves in host trees. Commercially exploited ecotypes of *a. mylitta* are either bivoltine (BV) or trivoltine (TV). The diapause duration of bivoltine and trivoltine races were observed as 205-241 days and 141-175 days, respectively. The life cycle passes through four stage egg, larva, pupa and adult. Voltinism in *A. mylitta* is not very strict and their behaviour changes based on the population raised at varying degrees of latitude, longitude, altitude and temperature (Sharan *et. al.,* 1994). During favourable season, humidity helps in the triggering of moth emergence in *A. mylitta* thereby humidity helps in avoiding pupil mortality due to desiccation of diapausing pupae. The voltinism has been found to be stable for a particular zone but on shifting the stock from one place having one regimen of day's length and temperature cycle to another, voltinism gets affected. Authors, while handling the large population of *A. mylitta,* came across such instances that the entire population in a particular area was lost due to vagaries of the nature or due to incidence of various natures of bacterial, viral or fungal diseases. Reasons for crop failures of, especially those related with diseases and irregularity in voltinism, may be known in a better way if the haemolymph physiology of this important insect is studied.

Insect haemolymph is best described as the extra cellular circulating fluid that fills the body cavity or haemocoel. It is physically isolated from direct contact with body tissues by a thin permeable membrane, which lines the haemocoel. In insects the haemolymph, like the blood of higher animals, is comprised of two main components: the plasma and the corpuscles or haemocytes. Quantitative methods have been applied to record the population of these haemocytes such as total and differential haemocytes count, absolute number based on haemolymph volume, which depicts upon the stage of development, physiological state and sexual differences in insects. The population of haemocytes are reported to change during course of ontogenetic development either due to normal division of haemocytes or due to haemocytopoietic differentiation.

The haemocytes play a major role in synthesis, storage and transportation of nutritional material required for insect survival and they play a role in defence mechanism by way of phagocytosis and nodule formation neutralizing the harmful effects of pathogen. The localisation of general proteins, bound lipids, nucleic acids, glycogen and phosphatases in the haemocytes cytoplasm and nucleus provide a glimpse of the physiological state of an insect. The composition of haemolymph tends to vary in response to various conditions or activities such as ontogenetic effects and distribution pattern. Physical properties, such as pH, specific gravity and osmotic pressure are always maintained at their optimum level during insects development and survival state. Substances available in dissolved form in haemolymph are proteins, amino acids, uric acids, carbohydrates, lipids and inorganic components *etc.* Haemolymph cations plays a major role in maintaining the osmotic pressure of insect blood and in Endopterygotes (Lepidoptera, Hymenoptera and Coleoptera), the cations, such as Na^+, K^+, Ca^{++} and Mg^{++} are the major osmoeffectors. Carbohydrates are an important constituent in insects haemolymph and are found in high concentration in the form of trehalose which plays a major role during moulting, metamorphosis and diapause in insects.

Nitrogenous compounds, like proteins, amino acids, uric acid, *etc.* in insect haemolymph and their variation in different physiological state of an insect has been a subject of great interest. The nitrogenous substances vary in its occurrence depending upon developmental stages; sex-specificity; season and dietary conditions; moulting; metamorphosis and diapause among insects. Uric acid is a major nitrogenous end product the level of which is influenced by feeding and development state of an insect. Diapause associated proteins have been described in insects which appear in haemolymph during diapause and disappear from haemolymph once diapause stage is over. Cholesterol is one of the main energy sources, transported through haemolymph plasma predominantly in a non-esterified form, and generally plays an important role in general histogenesis and development of estrogenic tissues and it helps in maturation of spermatocytes.

Considering the importance of haemolymph in sects' life, the structure of haemocytes, their population variation and physiology of haemolymph of *A. mylitta* in non-diapausing and diapausing generations' ontogenetic stages have been reported in this book. The book has in all five Chapters: Introduction, Materials and methods, Observations, Results and Discussion, Conclusion; and Bibliography. Various haematological parameters, such as types of haemocytes, their total and differential counts in relation to body weight and haemolymph volume of larvae, pupae and adults of have been reported. Through histochemical tests, the presence of general proteins, DNA, RNA, PA/S substances, glycogen, bound lipids in cytoplasm and nucleus of different haemocytes has been confirmed Physical properties, such as pH and specific gravity of haemolymph and the presence of important osmolytes, such as Na^+, K^+, Ca^{++} and Mg^{++} are depicted. Important organic constituents of haemolymph trehalose; quantitative total proteins, total amino acids, qualitative proteins through disc gel electrophoresis, uric acid and lipids in the form of cholesterol have been described.

The findings enumerated in this Book may help in furthering the physiological and pathological studies.

—**Authors**

List of abbreviations

Abbreviations	Explanation
μ	Micron
ANOVA	Analysis of Variance
Ca	Calcium
CD	Critical difference in ANOVA
Cm (Cms)	Centimetre (Centimetres)
CV%	Co-efficient of variance (%)
Deg.	Degenerated cells
DHC (DHCs)	Differential haemocyte count (Differential haemocyte counts)
DNA	Deoxyribonucleic acid
Eq	Equivalent weight
FAA (FAAs)	Free amino acid (Free amino acids)
GR (GRs)	Granulocyte (Granulocytes)
h (hs)	Hour (Hours)
Hec (Hecs)	Hectare (Hectares)
K	Potassium
l	Litre
M	Minutes
MC (MCs)	Mitotic cell (Mitotic cells)
Mg	Magnesium
mm (mms)	Millimetre (Millimetres)
MSS	Mean sum of squares
Mtr (mtrs)	Meter (Meters)
Na	Sodium
NS	Non-significant

OD	Optical Density
OE (OEs)	Oenocytoid (Oenocytoids)
PL (PLs)	Plasmatocyte (Plasmatocytes)
PLF (PLFs)	Fusiform Plasmatocyte (Fusiform Plasmatocytes)
PR (PRs)	Prohaemocyte (Prohaemocytes)
RNA	Ribonucleic acid
SE	Standard Error
SP (SPs)	Spherulocyte (Spherulocytes)
SS	Sum of squares
THC (THCs)	Total haemocyte count (Total haemocyte counts)

Contents

1

Introduction

India is a developing country, masses living in rural areas. Agriculture is the main source of gainful employment to the majority of those living in rural areas of India. In order to achieve the objective of alleviating rural poverty and to arrest migration of rural folks to cities, there is a need of diversification of agriculture sector to other important avocations. One of such avocations is sericulture, which generates employment on a large scale.

The sericulture activity is essentially an agro-based cottage industry having both agricultural and industrial activities intertwined together. Primarily, it covers cultivation of host plants, production of silkworm eggs, silkworm rearing cocoon production, reeling, spinning weaving and utilization of silk waste and other by-products and finally the production of silk.

In comparison with other natural textile-fibre in use worldwide, silk occupies only a minuscule slice of world production, consumption and demand. In pure volume terms, the share of silk in the global fibre market is thus dwarfed by the two other major categories viz., cotton and wool. Apart from cotton and wool man made synthetic fibres further pull down the ratio of utilization of silk fibre in the world. Global output of the major textile fibre of different categories between 1985 and 1999 were summarized in the International Trade Centre Silk Review, Geneva (2001). (Table 1.1).

Table 1.1. Global output of major textile fibres.

(Unit: Thousand Metric Tonnes)

Category/year	1985	1991	1995	1999
Cotton	17,540	20,830	19,200	19,200
Synthetics	12,515	16,440	20,200	28,200
Cellulose Fibres	2,999	2,860	3,000	2,700
Wool	1,673	1,940	1,600	1,400
Silk	59	75	100	76

Source: International Trade Centre - Silk Review - 2001

Silk always ranks comparably higher than any other textile fibre known in human history or in use in present times because of its appreciation in terms of consumer preference, unit values, fashion significance, employment generation, income earning opportunities in producer communities and the intangible value related to the preservation of mankind's cultural heritage.

However, out of the total world raw silk production of mulberry silk of 89,656 metric tonnes in 2002, India's share is to the tune of 14,617 metric tonnes (Anonymous, 2004). The production of silk is dominated by China and India. The production by other countries is very limited or negligible (Table 1.2).

Table 1.2. World raw silk production (mulberry)

Unit: Metric Tonnes

Country	1995	1996	1997	1998	1999	2000	2001	2002
China	64,613	59,000	52,700	49,430	55,990	60,000	62,560	64,100
India	12,884	129,954	14,048	14,260	13,944	14,432	15,842	14,617
Japan	3,240	2,580	1,920	1,080	650	557	431	394
Brazil	2,468	2,270	2,120	1,821	1,554	1,389	1,485	1,607
Korea Rep.	946	506	272	210	200	165	157	154
Uzbekistan	1,320	2,500	2,000	1,500	923	1,100	1,260	1,260
Thailand	1,313	1,144	1,039	900	1,000	955	1,510	1,510
Vietnam	2,100	1,500	1,000	862	780	780	2,035	2,200
Others	2,967	2,766	2,117	1,684	1,250	1,952	1,692	3,814
Total	91,851	85,220	77,216	71,747	76,291	81,330	86,972	89,656

Sericulture in India is as old as Indian culture. According to the Indian legend, sericulture was introduced about 2,000 years back; the wild silk moth producing yellow silk fibre was very popular from the ancient time. However, according to European scholars Chinese bivoltine silk was introduced in India in 140 BC. Even during the British period, India was producing good quality of local silk fabric, which was exported to UK and France and mainly utilized in parachute industry.

India has a glorious sericulture tradition of its own, which no other country in the world can share. India is the only country where all the four types of commercially available silk, namely, Mulberry, Tasar, Eri and Muga are produced. Of these, Muga and tropical Tasar silk originated in India and Muga silk is exclusively confined to North Eastern region of India. Besides Muga and Tasar, Eri and Mulberry silks are also cultivated in large quantity.

In India, total silk production during the year 2003-2004 was 16,655 MT. Out of this 14,617 MT was mulberry (87.763 %), 352 MT Tasar (2.113 %), 1,533 MT Eri (9.204 %) and 153 MT Muga (0.920 %) silk (Anonymous, 2004) .

The bulk of mulberry silk comes from the states of Karnataka, Andhra Pradesh, Tamil Nadu, West Bengal and Jammu and Kashmir, whereas the Tasar silk is produced mainly in the states of Jharkhand, Chhattisgarh, Orissa, Madhya Pradesh, Maharashtra and Andhra Pradesh. The production of eri silk is restricted to Northern Bihar and North Eastern

States of India. Commercial production of Muga silk is confined to Assam.

The real strength of Indian sericulture is its strong domestic consumption. The commercial production of non-mulberry silks, are not only the rich heritage of Indian sericulture but they make the Indian sericulture distinct and hence the planners, administrators and scientists are giving greater emphasis and thrust for the overall development of non mulberry sericulture in India. This keeps the Indian sericulture unique in the world silk market.

Non-mulberry sericulture has certain advantages over the mulberry sericulture. It holds a good potential for forestry as a supplementary activity and can help in reducing the serious problem of forest destruction. In fact, forestation can be promoted by systematic plantation of non-mulberry food plants on specific lines through social forestry.

A large number of species (400-500) are used in the production of non-mulberry silks, but only about eighty are commercially exploited in Asia and Africa, chiefly by tribal communities (Nanavty, 1965). The most important species for Tasar industry belong to the genus *Antheraea* which is purely local Asiatic genus confined to Indo-Australian, Palaearctic regions extending eastward to China and Japan. Thirty-five sericigenous species have been recorded (Crotch, 1956), thirty-one in the Indo-Australian region (Seitz, 1933), three in the Palaearctic region (Seitz, 1913) and one in the USA (Collins and West, 1961). The list of commercially exploited non-mulberry sericigenous insects with their distribution in the world is presented (Table 1.3 & Fig.1.1).

Table 1.3. Important silk producing insects of the world.

Non-Mulberry silk varieties	Species	Distribution
Tasar silk	*Antheraea mylitta* Drury	Tropical India
	A. roylei Moore	Temperate zone of India
	A. pernyi Guénerin-Méneville	Amur - S.China, SWEU, Japan.
	A. yamamai Guénerin-Méneville	Japan, Ceylon, N.India, hu, au, Yugoslavia
	A. frithi Moore	India, Sulawesi, Sundaland
	A. polyphemus Cramer	North America
	A. helferi Moore	NE.Himalaya, Sundaland
Muga silk	*Antheraea assamensis* Helfer	India, Burma, Sundaland
Eri or Ailanthus silk	*Philosamia ricini* Wardle	India
	P. cynthia Schüsseler	China, India, New guinea
Anaphe silk	*Anaphe* moloney Druce	Madagascar
	A. panda Boisduval	Tanzania, Congo (Kinshasa)
	A. reticulata Walker	Nigeria
	Anaphe infracta Walsingham	Congo (Brazzaville), Zambia
	Anaphe venata Butler	Nigeria
Fagara silk	*Attacus atlas* Linnaeus	Southern and Central Africa, India, Australia, China, Sudan.
Coan silk	*Pachypasa otus* Drury	Southern Nali, Greece, Romania.

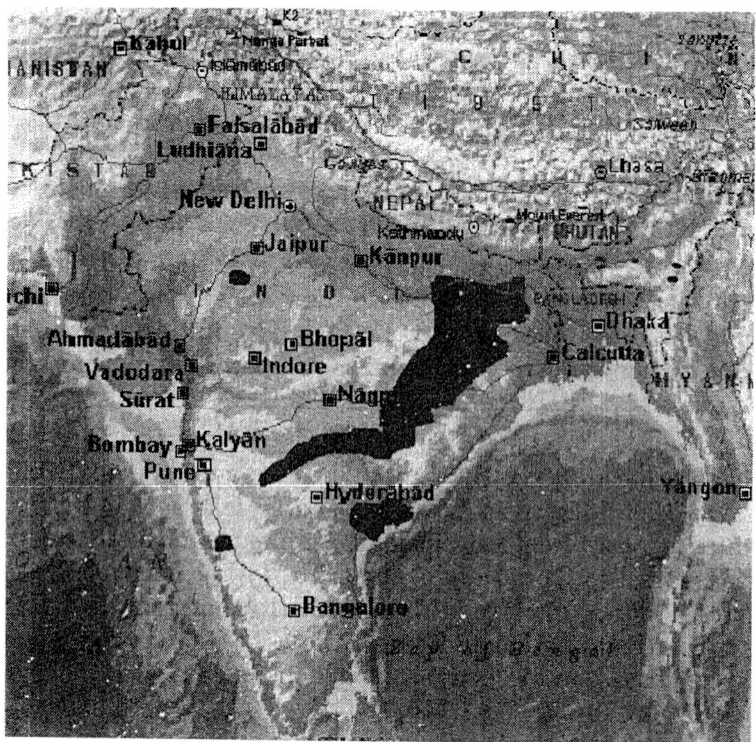

Fig 1.1. Distribution (red coloured area) of Antheraea mylitta in India.

Indian sericigenous insects, which produce silk, belong to families Bombycidae and Saturniidae of Order Lepidoptera. The natural silks are broadly classified as mulberry and non?mulberry. The mulberry sericulture involves the silkworm Bombyx mori, which primarily feeds upon Morus alba, M. indica and other Morus spp. Non-mulberry sericulture includes the silk obtained from insects which feed upon other than mulberry plants. In India, mainly three types of non-mulberry silks are produced namely, Eri, Muga and Tasar. The names of the sericigenous insect responsible for production of non-mulberry silk are Antheraea mylitta Drury (tropical Tasar), A. proylei Jolly (temperate Tasar), A. assama Westwood (Muga) and Samia cynthia ricinii Hutt (Eri silk).

Antheraea mylitta is highly polyphagous and it shows wide range of adaptation. Most harmoniously, it has adapted to feed and survive on Terminalia tomentosa (W & A) and Terminalia arjuna (Bedd.) in both semi-domesticated and wild conditions. A variety, which is reared outdoor by rearers, is grouped as semi-domesticated variety. Substantial shares of Tasar cocoon collections from trees of Shorea robusta (Roxburghii.), Anogeissus latifolia (Wall), Lagerstroemia spp., etc. in the forests of tropical India and are called "wild". These semi-domesticated and wild varieties available in different pockets of tropical India are also termed as ecoraces, eco-types or biotypes. The biotypes not only differ among themselves based on their host plants, but in their niche, behaviour, cocoon colours and com

Table 1.4. Biotypes of *A. mylitta* Drury found in India

SN	Forest Type		Biotype	Collection Site
1.	Tropical wet evergreen	1.	Nowgong	Nowgong (Assam)
		2.	NE1: 95	Hahim (Assam)
		3.	NE2, 95	Mendipathar (Meghalaya)
		4.	Jiribam	Jiribam (Manipur)
		5.	NG: 94	Dimapur (Nagaland)
		6.	Belgaum	Belgaum (Karnataka)
2.	Tropical moist deciduous	7.	Daba	Singhbum (Jharkhand)
		8.	Sarihan	Santhal Pargana (Jharkhand)
		9.	Munga	Santhal Pargana (Jharkhand)
		10.	Modia	Dhanbad (Jharkhand)
		11.	Laria	Peterbar Hazaribagh, (Jharkhand)
		12.	Barharwa	Simdega (Jharkhand)
		13.	Lodhma	Ranchi (Jharkhand)
		14.	Palma	Ranchi (Jharkhand)
		15.	Kowa	Palamau (Jharkhand)
		16.	Japla	Palamau (Jharkhand)
		17.	Modal	Keonjhar (Orissa)
		18.	Nalia	Sundergarh (Orissa)
		19.	Sukinda	Sukindagarh (Orissa)
		20.	Baodh	Phulbani (Orissa)
		21.	Simlipal	Simlipal (Orissa)
		22.	Omarkote	Kalahandi (Orissa)
		23.	Kurudh	Kurudh (Chhattisgarh)
		24.	Multai	Multai (Chhattisgarh)
		25.	Mandalla	Mandalla (Chhattisgarh)
		26.	Jhabua	Jhabua (Chhattisgarh)
		27.	Tira	Purulia (W. B.)
		28.	Bankura	Bankura (W. B.)
		29.	Dadar & Nagar Haveli	Dadar & Nagar Haveli (UT)
3.	Mountain sub-tropical	30.	Shiwalika	Batote (J & K) & Palampur (H.P.)
4.	Tropical Dry Deciduous	31.	Sukly	Khairpali (Orissa)Beramkela (Chhattisgarh)
		32.	Raily	Jagdalpur (Chhattisgarh)
		33.	Bhopalpatnam	Bhopalpatnam (Chhattisgarh)
		34.	Piprai	Piprai (Chhattisgarh)
		35.	Seoni	Seoni (Chhattisgarh)
		36.	Janghbhir	Bastar (Chhattisgarh)
		37.	Korbi	Korba (Chhattisgarh)
		38.	Bhandara	Bhandara (Maharashtra)
		39.	Andhra Local	Adilabad & Karimnagar (A. P.)
		40.	Monga	Dawaria (U. P.)
		41.	Mirzapur	Mirzapur (U. P.)
		42.	Sultanpur	Sultanpur (U. P.)
5.	Thorn Forest	43.	Tesera	Sahabad (Rajasthan)

mercial characteristics. The difference is so marked that they are popular with their local names. Examples of such semi-domesticated biotypes of *A. mylitta* are Daba, Sukinda, Laria, Raily, Sarihan, Modal, Tira, Bhandara, etc. Survey analysis reveals that major economically important ecoraces are distributed mainly in Central India within 24-16°N and 80-88°E, whereas occurrence has been reported within 12-30°N and 72-96°E. Though there is a notion that each biotype is specific to a particular region, overlapping of habitat at the edge zone can not be ruled out. Maximum production of Tasar cocoons comes from the regions of moist deciduous forests. Forty three ecotypes/ecoraces/ biotypes/variants have been reported from different places of India by Singh and Srivastava (1997) - (Table 1.4).

Table 1.5. Important host plants of A. mylitta Drury.

SN	Family	Scientific name	Local name
1.	Apocynaceae	1. *Carissa carandas* L.	Karaunda
2.	Anacardiaceae	2. *Semecarpus anacardium* L.	Bhelwa
3.	Caesalpinaceae	3. *Cassia lanceolata* L.	Kanchan
		4. *Bauhinia variegata* L.	-
4.	Celastraceae	5. *Celastrus paniculatus* Royle	Malkangni
5.	Combretiaceae	6. *Terminalia arjuna* Bedd.	Arjun, Sadar.
		7. *T. belerica* Roxb.	Bhaira, Behera
		8. *T. tomentosa* W&A	Asan, Ani, Saja
		9. *T. alata* Roth.	—
		10. *Anogeissus latifolia* Wall	Dhaura
		11. *T. crenulata* Kurz.	—
		12. *T. catappa* L.	Jungli badam
6.	Dipterocarpaceae	13. *S. talura* Roxb.	—
		14. *Shorea robusta* Gaertn.	Sal or Sakhua
7.	Euphorbiaceae	15. *Ricinus communis* L.	Castor plant
8.	Lythraceae	16. *Lagerstroemia indica* L.	Daiyeti, Telinga, China
		17. *L. parviflora* Roxb.	Banskli/Sidha
9.	Malvaceae	18. *Bombax heptaphyllum*	Semul
		19. *B. malabaricum* DC.	Silk cotton tree
10.	Meliaceae	20. *Cipadessa fruticosa* Bl.	Billu
11.	Moraceae	21. *Ficus benjamina* L.	Nandruck
		22. *F. religiosa* L.	Aswat, Peepal
		23. *F. retusa* L.	Kamrup
12.	Lecythidaceae	24. *Careya arborea* Roxb.	Kumbi
		25. *C. sphaerica* L.	—
13.	Myrtaceae	26. *Eugenia cuminii* Druce	Jamun
14.	Rhamnaceae	27. *Zizyphus jujuba* L.	Ber
		28. *Z. mauritiana* L.	—
15.	Rosaceae	29. *Prunus domestica* Pleem.	—
16.	Rhizophoraceae	30. *Rhizophora candelaria* DC	—
17.	Rubiaceae	31. *Canthium diecocum* Gaertn.	Merrill
18.	Sapindaceae	32. *Dodonaea visosa* Jacg. (L)	Sanatta
19.	Sapotaceae	33. *Bassia latifolia* Roxb.	Mohua
		34. *B. longifolia* L.	—
20.	Verbenaceae	35. *Tectona grandis* L.	Saguan

A. mylitta succeeds on nearly three dozen food plants (Arora and Gupta, 1979); however, the main food plants of commercial interest are *Terminalia arjuna*, *T. tomentosa* and *S. robusta*. Wild varieties are also able to thrive well on *S. robusta* in nature. The rearing of the semi-domesticated as well as wild varieties on Sal plants has not been found very successful. Only limited success in rearing of Laria ecotype on Sal has been found. Further, the wild races when reared on Asan, Arjun and other food plants tend to loose their richness of commercial traits. A few important food plants are listed in Table 1.5.

Though *T. arjuna*, *T. tomentosa* and *S. robusta* are present in abundance and are comparatively better exploited for rearing or cocoon collection of *A. mylitta*, its ecoraces have adapted to feed and survive on other food plant species of lesser importance. For example, Tira cocoons from Lagerstroemia and Anogeissus; Bhandara from *A. latifolia,* and *Lagerstroemia parviflora*, Dadar and Nagar Haveli from *Terminalia crenulata*, Sukly from *A. latifolia* and *Madhuca indica* are collected in nature. Besides above food plants, *Zizyphus* spp. (Ber) is also a source of cocoon production throughout the country.

It has been estimated that Tasar food plants cover 11.2 million hectares area in India. Of this, nearly 87% (9.7 million hectares) is covered by *S. robusta* (Sal) and rest area is occupied by *Terminalia spp.* and other food plants. Presently, only small fraction of the covered area by *T. tomentosa* and *T. arjuna* is used for the purpose of Tasar rearing by sericulturists. The use of Sal flora for Tasar rearing is only confined to the perpetuation of this insect species in nature. Nature-grown Tasar cocoons on Sal are collected from the forests of Bastar region of Madhya Pradesh and Simlipal in Orissa.

1.1. Dormancy in Insects

Insects as well as other organisms do not live in a constant environment, which is always perfect for optimum growth and survival. With the changes in environmental conditions, insects are able to adapt to and survive unfavourable seasons or conditions that do not support their continued development. It is known that insects have the capacity to stop development at their various life stages viz. egg, larvae, pupae and adult. Tauber et al. (1986) have defined this arrested period of development into following categories:

Dormancy - A general term referring to a seasonally recurring period in the life cycle during which growth, development and reproduction are suppressed. This term can include quiescence, diapause, hibernation or aestivation.

Quiescence or Non-diapause Dormancy - A state of suppressed metabolism or dormancy imposed by adverse recurring environmental conditions that are unfavourable for growth and reproduction. Recovery occurs soon after the end of adverse conditions.

Diapause or Diapause-mediated Dormancy: The term "diapause" was originated from a Greek word diapausis, derived from diapauein where dia- = between and pauein, = stop. Diapause is a period in life cycle of insects (and certain other animals) during which their physiological activity is very low and they are highly resistance to unfavourable external conditions. It takes place before the arrival of actual period of un-favourable conditions in response to token stimuli received by the specific stages of an insect. These token stimuli may be in the form of day-length, temperature, food etc.

Diapause lasting more than a year is known as "prolonged" or "extended" diapause and is reported in many insects. In fact, prolonged diapause seems to be common among insects.

Diapause may be obligatory or facultative. Obligatory diapause most often is seen in univoltine insects regardless of environmental conditions. "Obligatory" diapausing insects respond to photoperiod similar to those species which undergo facultative diapause (Gillott, 1995). Facultative diapause most often is seen in insects that are bivoltine or multivoltine. The most prevalent stimulus for undergoing facultative diapause is day length, temperature, food condition etc. In temperate species, diapause is governed by photoperiod/day length. The day length at which 50% of the population has entered diapause is called the critical day length. Insects entering diapause when the day length falls below this threshold are called long day insects. Those insects that develop normally when there are only a few hours of sunlight and which enter diapause when exposed to longer days are called short-day insects. The critical day length is a genetically determined property (Danilevskii, 1965; Tauber, et al., 1986).

Besides diapause that coincides with over wintering condition, aestivation can cause arrested development in the summer.

The physiological changes that take place during diapause are directed by a genetically programmed, neuro-hormonally mediated syndrome that prepares the insect for surviving the unfavourable conditions heralded by environmental cues. This syndrome of changes persists even when favourable conditions are transiently available.

Diapause can affect migratory and foraging behaviours of insects. Seasonal changes in colour or structure of insects also occur. These are the resultant of physiological, metabolic, endocrinological and molecular changes which take place during diapause in insects (Joplin et al., 1990).

In *Bombyx mori*, embryonic diapause appears to be regulated by diapause hormone, a 24-amino acid peptide that is produced in the sub-oesophageal ganglion (Fukuda, 1952, Hasegawa, 1952). This hormone acts on the maturing oocytes in the pupal stage and causes development to stop just prior to segmentation (Denlinger, 1985). Larval diapause appears to be controlled by the inhibition of Prothoracicotropic hormone (PTTH) production (Hasegawa, 1952, Fukuda, 1952). Similarly, in larval diapause, elevated titre of juvenile hormone plays a crucial role in induction and maintenance of diapause (Yin and Chippendale, 1976). Induction and maintenance of pupal diapause depends on the lack of PTTH and ecdysone secretion. Adult diapause occurs due to inhibitory action of brain on the corpora allata. It inhibits Juvenile Hormone (JH) secretion and this leads to imaginal diapause.

Thus factors responsible for induction, maintenance and termination of diapause vary e.g. season, temperature, or even tides are critical properties for many organisms to programme them to enter into diapause.

In case of *A. mylitta* diapause has been observed to be facultative. Based on nature of voltinism they have been placed under three categorics: univoltine, bivoltine and trivoltine. Normally, it behaves as bivoltine completing two generations in a year. However, in warmer climates, part of the population turns tri-voltine. The average larval period during first, second and third crops are around 30, 45 and 65 days, respectively. The winter crop harvested cocoons enter into diapause and emerge as moths giving rise to first crop during next year monsoon, which coincides with the sprouting of leaves in host trees. The life cycle passes through four stages egg, larva, pupa and adult. Out of four stages, the egg encompasses the embryonic development and larva feeds on different food plants, while the pupa and moths stages are non-feeding in nature. Moths emerge out of pupa, only for the purpose of reproduction. Diapausing pupae of *A. mylitta* remain in dormant stage to skip the unfavourable climatic conditions.

Voltinism in *A. mylitta* is not very strict and their behaviour changes based on the population raised at varying degrees of latitude, longitude, altitude and temperature (Sharan et al., 1994). During favourable season, humidity helps in the triggering of moth emergence in *A. mylitta* thereby humidity helps in avoiding pupal mortality due to desiccation of diapausing pupae. The voltinism has been found to be stable for a particular zone but on shifting the stock from one place having one regimen of day's length and temperature cycle to another, voltinism gets affected.

Therefore, knowledge on the physiological aspects of *A. mylitta* is necessary, as it has unstable voltinism. The best medium to understand the physiology of A. mylitta may be the study of its haemolymph. A complete study of the haemolymph of A. mylitta can enable us to understand its physiology in a better way in its non-diapausing and diapausing generations and between the sexes.

1.2. Present Study

Insect haemolymph is best described as the extra cellular circulating fluid that fills the body cavity or haemocoel. The haemolymph circulates freely within the body of insects bathing different tissues (Arnold, 1979a; Jones, 1979). It is physically isolated from direct contact with body tissues by a thin permeable membrane, which lines the haemocoel (Ashhurst, 1979). In insects, the haemolymph, like the blood of higher animals comprise two main components, the plasma and the corpuscles or haemocytes (Richards and Davies, 1977). The haemocytes or the blood cells of insects comprise a complex of several types of mesodermal cells, which are nucleated, comparable to leukocytes of other vertebrates (Andrew, 1965; Arnold, 1974). There is much confusion in the literature regarding haemocyte classification and the haemocyte classes are best distinguished for mature cells by bio-chemical tests and/ or by cytological parameters such as volume, relative size of nucleus and cytoplasm (Hrdy, 1958 ; Jones, 1962 & 1964 and Gupta, 1979 a & b). Most useful system for classifying insect haemocytes is the one developed by Jones (1962, 1964 and 1970). A variety of other elements occurs at times in haemolymph, such as plastids, fine granules (haemoconiae) and lipomicrons (Arnold, 1974). There are three well defined types of haemocytes in most of the insects (prohaemocytes, plasmatocytes and granulocytes) and one or more of four other types in some other insects (coagulocytes, spherulocytes, adipocytes and oenocytoids) (Arnold, 1979a and Jones, 1979). Quantitative methods have been applied by Arnold, in 1952 to record the population of these haemocytes such as total and differential haemocytes count, absolute number based on haemolymph volume, which depicts upon the stage of development, physiological state and sexual differences in insects. The population of haemocytes are reported to change during course of ontogenetic development either due to normal division of haemocytes or due to haemocytopoietic differentiation.

The haemocytes are reported to play a major role in synthesis, storage and transportation of nutritional material required for insect survival (Gupta, 1985). Their role in defence mechanism is also obvious by way of phagocytosis and nodule formation (Raichoudhury and Sen Gupta, 1959). The localisation of general proteins, bound lipids, nucleic acids, glycogen and phosphatases in the haemocytes' cytoplasm and nucleus provide a glimpse of the physiological state of an insect (Vaney and Maigon, 1905; Yeager and Munson, 1941; Nittono, 1960; Neuwirth, 1973; Ashhurst and Richards, 1964; Chippandale, 1970 a & b; Costin, 1975). Besides, Gupta (1985) has also reviewed haematology, histochemistry of haemocytes and biochemistry of blood plasma in different insect species.

Haemolymph acts as a storage reservoir for many materials essential for a variety of insects and its composition tends to vary in response to various conditions or activities such as ontogenetic effects (Jungreis et al., 1973) and distribution pattern (Buck, 1953; Pichon, 1970; Mullin, 1985). Physical properties, such as pH, specific gravity and osmotic pressure are always maintained at their optimum level during insects' development and survival state (Buck, 1953; Florkin and Jeuniaux, 1974) and substances available in dissolved form in haemolymph are proteins, amino acids, uric acids, carbohydrates, lipids and inorganic components etc.

Haemolymph cations plays a major role in maintaining the osmotic pressure of insect blood and in Endopterygotes (Lepidoptera, Hymenoptera and Coleoptera), the cations, such as $Na+$, $K+$, $Ca++$ and $Mg++$ are the major osmoeffectors (Buck, 1953; Mullin, 1985). They also play an important role in various conditions or activities of insects (Schin and Moore, 1977), diet (Jungreis et. al, 1973), etc.

Carbohydrates an important constituent in insects' haemolymph has high concentration of trehalose (Wyatt and Kalf, 1957), and trehalose levels are related to moulting, metamorphosis and diapause (Sakamoto and Horie, 1979).

Nitrogenous compounds, like proteins, amino acids, uric acid, etc. and their presence and role in different physiological state of insect haemolymph have been a subject of great interest (Buck, 1953; Fukuda and Hayashi, 1958; Florkin and Jeuniaux, 1974; Mullin, 1985). Variation in quantity of nitrogenous substances during developmental stages (Chen,1966; Wyatt and Pan, 1978; Wyatt, 1980); sex-specific proteins (Kawaguchi and Doira, 1965; Doira, 1968,; Engelmann, 1979; Kunkel and Nordin, 1985); precise analysis of haemolymph free amino acids in relation with season and dietary conditions (Mullin, 1985); moulting (Florkin and Jeuniaux, 1974); metamorphosis (Wirtz and Hopkins, 1974) and diapause (Boctor, 1981) have been reported in insects. Uric acid is reported to be a major nitrogenous end product (Mullin, 1985). Haemolymph concentrations of this nitrogenous end product may be influenced by feeding (Ramakrishna and Pawar, 1975) and development (Buckner and Caldwell, 1980).

Diapause associated proteins have been described in insects (Brown and Chippendale, 1978; Dortland and de Korte, 1978; Brown, 1980 and Koopmanschap et al., 1992). These proteins have been found in fat bodies and the haemolymph of diapausing insects. However, such proteins do not appear to be synthesised in haemolymph during diapause. These proteins disappear from haemolymph once diapause stage is over and are related to high molecular weight storage proteins (Levenbook, 1985), there is no evidence that they are involved in the initiation, regulation or control of diapause development (Nijhout, 1994).

Cholesterol is one of the main energy sources, transported through haemolymph plasma predominantly in a non-esterified form, and generally plays an important role in general histogenesis and development of estrogenic tissues (Downer and Chino, 1979; Chapman, 1980) and it helps in maturation of spermatocytes (Chino and Gilbert, 1971; Dumser, 1980; Hurkadli et al., 1989).

Such studies on the physiology of haemolymph in non-diapausing and diapausing generations of *A. mylitta* have not been carried out in its ontogenetic stages. Considering the importance of haemolymph in insects, it was felt essential to generate specific information on haematology of most widely exploited wild sericigenous insect A. mylitta because variation in haemocytes picture and haemolymph composition of the different insect species have shown a wide range of variation.

Therefore, the present study has been undertaken in non-diapausing and diapausing broods of *A. mylitta* in its various developmental stages viz. I, II, III, IV and V instar larvae, early, mid and late pupae and adults of both the sexes. Haematological parameters, such as types of haemocytes, their total and differential counts in relation to body weight and haemolymph volume of larvae, pupae and adults have been carried out in both of its non-diapausing and diapausing generations of *A. mylitta*.

Physical properties, such as pH and specific gravity of haemolymph and the presence of important osmolytes, such as Na+, K+, Ca++ and Mg++ of aforesaid life stages of A. mylitta have been studied.

Through histochemical tests, the presence of general proteins, DNA, RNA, PA/S substances, glycogen, bound lipids in cytoplasm and nucleus of different haemocytes has been confirmed in life stages of *A. mylitta*.

The studies on important organic constituents such as carbohydrates, especially trehalose; nitrogenous constituents of haemolymph, such as quantitative total proteins, total amino acids, qualitative proteins through disc gel electrophoresis, uric acid and lipids in the form of cholesterol have been undertaken in the present study in different life stages (larvae, pupae and adults) and the different sexes of non-diapausing and diapausing generations of *A. mylitta*.

2

Review of Literature

2.1. General

The blood or haemolymph of insects is contained in the general body-cavity and consists of liquid plasma and numerous blood cells or haemocytes (Jones, 1962; Richard and Davies, 1977; Gupta, 1979 a & b). The haemocytes originate from the mesodermal bands in the embryo (Jones, 1970). Post-embryonic multiplication in most insects is mainly by division of circulating cells (Tauber, 1936; Clark and Harvey, 1964; Lea and Gilbert, 1966 and Jones, 1970). The haemocytes of insects comprise a complex of several types of mesodermal cells which circulate within the haemolymph and sometimes attach loosely to other tissues or insinuate themselves within them (Arnold, 1974). A variety of other elements occur at times in haemolymph, such as plastids, fine granules (haemoconiae) and lipomicrons (Jones, 1962). In some ways insect haemocytes are closely comparable to the leukocytes of other invertebrates (Andrew, 1965). The features of haemocytes relate primarily to the main functions that the cells perform. They are nucleate, and some of them bear a superficial morphological resemblance and functional similarity to some of the wandering cells (amoebocytes) of invertebrates and certain leucocytes of vertebrates (Arnold, 1972).

According to Millara (1947), Cuénot (1896) was the first to classify the haemocytes into four categories, which was further followed by Hollande (1909 and 1911). Wigglesworth (1939), after summarising most of the earlier classifications, presented a classification that was widely accepted. Most useful system for classifying insect haemocytes is the one developed by Jones (1962, 1964 and 1970), though there is much confusion in the literature regarding haemocyte classification (Wigglesworth, 1959; Jones, 1962 & 1964; Gupta, 1969; Arnold, 1972).

The haemocyte classes are best distinguished for mature cells by bio-chemical tests and/or by cytological parameters such as volume, amount of DNA, relative size of nucleus and cell body, number and type of cytoplasmic inclusions (Hrdy, 1958, Wigglesworth, 1955; Jones, 1962; Ritter, 1965; Arnold, 1972).

In insects, several types of haemocytes are observed in haemolymph (Arnold, 1979 a and Jones, 1979). There is a disagreement among haematologists about the types of haemocytes in insects, i.e., from one or a few to as many as nine or more types (Ravindranath, 1978 and Gupta, 1979 a & b).

Based on the premise that there are three well defined cell types in most insects (prohaemocytes, plasmatocytes and granular haemocytes), one or more other types in many insects (cytocytes, spherule cells, adipohaemocytes and oenocytes) and two highly specialised cells viz., podocytes and vermiform cells (Arnold, 1974), a general description of various haemocyte types, based on both light and electron microscopy studies, their synonyms and interrelationships has been described after Gupta (1979 a & b). He portrayed that ultra-structurally only seven types of haemocytes have so far been identified in various insects and they are: prohaemocytes, plasmatocytes, granulocytes, spherulocytes, adipohaemocytes, oenocytoids and coagulocytes.

2.2. Types of haemocytes in different insects

Nittono (1960) classified the blood cells in the silkworm, *Bombyx mori* into six types, viz., prohaemocytes, plasmatocytes, granulocytes, spherulocytes, imaginal spherulocytes (observed only at the stage of adult, but occasionally in pupa on the day of emergence) and oenocytes. Raichoudhury and Sengupta (1959) named seven types of haemocytes in Bombyx mori. Yafaeva (1962) reported six categories in gypsy moth *Oenria dispar*. Lea and Gilbert (1966) reported five categories in *Hyalophora cecropia* (L.), *Antheraea polyphamous* (Cram.), and Samia cynthaea (Drury). Seven types of haemocytes were observed in the primary cultures and in fresh blood of *Chilo suppressalis* (Mitsuhashi, 1966). Recent electron microscopic studies have been proved helpful in distinguishing various haemocytes, viz., the larval plasmatocytes and granulocytes (Akai and Sato, 1973). Haemocytes of forest tent caterpillar *Malcosoma distria* Hubner (Lepidoptera: Lasiocampidae) have five categories of haemocytes. These were prohaemocytes, plasmatocytes, granular haemocytes, spherule cells and oenocytoids (Arnold, 1974). Prohaemocytes plasmatocytes, adipohaemocytes, spherule cells and oenocytoids have been observed in *Galleria mellonella* (Lea, 1986). Five types of haemocytes (prohaemocytes, plasmatocytes, adipohaemocytes, spherulocytes and oenocytes) were reported in *Agrotis segetum* (Bayram and Kilincer, 1987). A total of six cell types (prohaemocyte, granulocyte, adipohaemocyte, spherulocyte, plasmatocyte and Oenocytoid) were identified in *Diatraea saccharalis* (Barduco et al., 1988). Nine types of haemocytes were distinguished on the basis of their surface ultrastructure in *Spodoptera litura* (Saxena, et al., 1988). Four types of haemocytes (prohaemocytes, plasmatocytes, granulocytes and spherulocytes) were described in *Lymantria dispar* (Kim et al., 1990 a & b). A comparative study of haemocytes in 27 insect species belonging to 13 orders on the basis of haematological techniques indicated great variability in haemocyte types and haemograms (More and Sonawane, 1990). Through phase microscopic observations of the haemocytes, seven types were indicated in *Spodoptera litura*: namely, prohaemocytes, plasmatocytes, granulocytes, spherulocytes, oenocytes, podocytes and granular plasmatocytes (Kurihara, et al., 1992 a). Comparative study of haemocytes of three Lepidopterans viz., *Danaus chrysippus*, *Argina astrea* and *Earias* sp. by light and scanning electron microscopy was reported (Saxena, 1992). Five types of haemocytes (prohaemocytes, plasmatocytes, granular haemocytes, cystocytes and oenocytes) were reported in *Philosamia ricini* (*Samia cynthia ricini*), which change/degenerate in number after different humidity exposures (Begum et al., 1992 a & b). Based on ultrastructure, haemocytes were classified into six types in the final instar larvae of *Lucillia illustris* (Roe et al., 1993). Different types of haemocytes, viz. prohaemocytes, plasmatocytes, granulocytes, spherulocytes and oenocytes were reported in *Antheraea assama* (Bardoloi and Hazarika, 1992 and 1995). Eight types of haemocytes were morphologically differentiated in the haemolymph of larvae of *Agrotis ipsilon* (Ibrahim et al., 1993). Haemocytes in the fifth instar larvae and pupae of *Plusia orichalcea* were observed to be prohaemocytes, plasmatocytes, granulocytes, adipohaemocytes, spherulocytes, coagulocytes and oenocytes (Pathak and Saxena, 1994). Eight types of haemocytes were identified in *Spodoptera litura* (Kares, 1994). On the basis of the fine structure, seven distinct types of circulating cells: prohaemocytes, granulocytes, coagulocytes, adipohaemocytes, plasmatocytes, oenocytes and spherulocytes are reported in gypsy moth, Lymantria dispar (Butt and Shield, 1996). Larvae of *Bombyx mori*, once investigated by haemocytic differentiation, showed four types of haemocytes: prohaemocytes, plasmatocytes, granulocytes and oenocytes (Han et al., 1998).

2.3. Ultra-structure of Haemocytes:

There is a great deal of references available on the ultra-structure of haemocytes and their multiplication in insects, for example in Ephestia kühniella Zell. (Arnold, 1952); in *Euxoa campestris* Crote and *Euxoa decalarata* Walker (Arnold and Hinks, 1975); in *Hyalophora cecropia* (Lea, 1964 and Lea and Gilbert, 1966); in *Spodoptera littoralis* (Harpaz et al., 1969) and in *Galleria mellonella* (Shapiro, 1968); and on the basis of morphology of cells in forest tent caterpillar *Malcosoma disstria* Hübner (Arnold and Sohi, 1974); *Bombyx mori* L. (Akai and Sato, 1973) and *Antheraea pernyi* Guer. (Beaulaton and Monpeyssin, 1976).

Structure of the prohaemocytes described by various authors indicate that these are small, round, oval or elliptical cells with variable sizes, having low concentration of endoplasmic reticulum, mitochondria, Golgi bodies, and are generally found in groups (Gupta and Sutherland, 1966 and Gupta, 1969, 1979 a).

Plasmatocytes are large polymorphic cells of various sizes with micro-papillae in plasma membrane and with dense granules produced by Golgi bodies (Gupta, 1979 a). Plasmatocytes are regarded to be the primary types in insects (Gupta and Sutherland, 1966; Moran, 1971 and Gupta, 1979 a).

Granulocytes are small to large spherical or oval cells, laminar nature of plasma and nuclear membrane may not visible, membrane bound granules may be present (Goffinet and Gregoire, 1975; Gupta, 1979 a). Granulocytes are widely misinterpreted and confused with plasmatocytes and spherulocytes (Takada and Kitano, 1971; Arnold, 1974; Francois, 1975 a & b; Hinks and Arnold, 1977; Gupta, 1979 a).

Spherulocytes are ovoid and round cells, usually larger than granulocytes and the nucleus is generally small, rich in chromatin bodies with spherules having mucopolysaccharides in them (Nittono, 1960; Gupta and Sutherland, 1967 ; Costin, 1975).

Adipohaemocytes are observed to be small to large, spherical or oval cells with variable sizes, cytoplasm contain small to very large refractory fat droplets, have PAS-positive substance (Lea and Gilbert, 1966; Gupta, 1979 b).Though adipohaemocytes are not recognised by a few authors in insects (Witting, 1968; Akai and Sato, 1973; Raina, 1976).

Gupta and Sutherland (1966) have suggested that adipohaemocytes are transformed plasmatocytes; transforming after chilling, starvation and/or diapause.

Coagulocytes are spherical, hyaline, fragile and unstable cells combining the features of plasmatocytes and oenocytes (Arnold, 1974). They are involved in haemolymph coagulation (Gregoire, 1971; Francois, 1974), though they are reported to be transformed plasmatocytes (Gupta and Sutherland, 1966).

Oenocytoids found in Lepidoptera, are small to large thick, oval, spherical or elongate cells; plasma membrane with micro-papillae, cytoplasm thick, homogeneous with plate, rod or needle like inclusions, with tyrosinase in cytoplasm and they contain PAS-positive glycoproteins (Akai and Sato, 1973; Costin, 1975; Gupta, 1979 b). Insect haemocytes are involved in wound repair, coagulation, melanization and immobilisation of invading organisms, and/or phagocytosis (Ratcliffe and Rowley, 1979).

Further their structure is studied in Diatraea saccharalis (Barduco et al., 1988); *Spodoptera litura* (Saxena et al., 1988); *Lymantria dispar* (Kim et al., 1990 a & b); *Polistes lanio lanio* (Giannotti and Caetano, 1991); *Argina astrea*, *Danaus chrysippus* and *Earias* sp. (Saxena, 1992); *Simulium vittatum* (Luckhart et al., 1992 and Kurihara et al., 1992a & b); *Bacillus rossius* Ross (Scapigliati and Mazzini, 1992); *Phenacoccus manihoti* and *Phenacoccus citri* (Russo et al., 1993). Haemocytes are also studied on the basis of their ultra-structure, phase-contrast and light microscopy in pre-pupae of *Agrotis segetum* and *Armigeres subalbatus* (Zahedi, 1993); *Blattala germanica* (Pathak and Kulshrestha, 1993); *Plusia orchalacea* (Pathak and Saxena, 1994); *Lymantria dispar* (Butt and Shield, 1996); *Antheraea assama* (Hazarika et al., 1994); moths namely *Yponomeuta padellus*, *Y. cognatellus*, *Y. evony mellus* and *Y. malinellus* (Bartninkaite, 1995); *Mythimna unipuncta* (Ribeiro et al., 1996); *Dactylopius confusus* (Joshi and Lambdin, 1996); *Spilostethus hospes* (Sanjayan et al., 1996); *Dysdercus koenigii* (Sharma et al., 1998) and *Pseudoplusia includens* (Loret and Strand, 1998; Gardiner and Strand, 1999).

From the above published literature, it is apparent that the ultra-structure through various means in insects belonging to different orders and of Lepidoptera have been studied, however, such studies on the structure and type of haemocytes of Antheraea mylitta has not been carried out at length. Hence, the present study has been under taken in order to understand the structure and function of haemocytes in *Antheraea mylitta*.

2.4. Histochemistry of haemocytes

The haemolymph is both a controlled environment and a transport interface for all insect cells (Crossley, 1979). There are four well documented and generally accepted functions of haemocytes, viz., phagocytosis of small particles, encapsulation of large, foreign bodies, coagulation of blood by cellular agglutination or by contributing to plasma precipitation and storage and distribution of nutritive materials (Jones, 1979; Salt, 1970; Whitcomb et al., 1974; Crossley, 1975; Smith and Ratcliffe, 1978).

Insects appear to carry substantial levels of a number of metabolites dissolved in their plasma that other group of animals heap typically in their cells (Wyatt, 1963). Haemocytes are also capable of storing considerable bulk of material or nutritive substances (Poisson and Pesson, 1939; Gupta and Sutherland, 1967; Arnold and Salkeld, 1967). All types of haemocytes and especially the spherulocytes are considered as storage vehicle (Arnold, 1974). Glycogen has been reported in *Galleria* haemocytes (Ashhurst and Richards, 1964) and in granulocytes in the fifth instar of *Locusta* (Brehelin et al., 1975). Glycogen in its b form is reported in *Calliphora* (Crossley, 1968); in *Blaberus* (Moran, 1971) and in *Antheraea* (Beaulaton and Monpeyssin, 1976). Certain mucopolysaccharides are reported in Spherulocytes of several species of Dictyoptera (Gupta and Sutherland, 1967) and *Bombyx mori* (Akai and Sato, 1973); Granulocytes of *Galleria* (Asshurt and Richards, 1964; Neuwirth, 1973); in all haemocyte types except prohaemocytes in *Locusta* (Hoffmann, 1967 a).

Lipid droplets are formed in *Prodenia* haemocytes (Munson and Yeager, 1944) and role of haemocytes in lipid accumulation before pupation in *Ephestia kuhniella* (Arnold, 1952) and in *Galleria* (Ashhurst and Richards, 1964) have been reported. Presence of lipids in haemocytes has been also reported in *Ephestia* (Arnold, 1952; Grimstone et al., 1967); in *Calliphora* (Crossley, 1975) and in *Pectinophora gossypiella* (Raina, 1976).

The synthesis of haemolymph protein has been extensively studied in a variety of silkmoths, e.g., in *Bombyx mori* pupa (Sissakjan and Kuvajeva, 1957); in *Hyalophora* (Stevenson and Wyatt, 1962); in *Hyalophora* and *Samia*, the esterase, phosphatase, lipase, glucosidase, glactosidase and tyrosinase in cells have been reported (Laufer, 1960). Haemolymph also play a role in protein transit system (Price, 1961; Martin, 1969; Martin et al., 1969). Haemocytes are reported to synthesize cuticle proteins (Steinhauer and Stephen, 1959) and also in synthesis of injury proteins (Coles, 1965). Histochemically, the presence of glycoproteins in granulocytes in *Antheraea pernyi* (Beaulaton and Monpeyssin, 1976) is reported. The haemocytes are also involved in mucopolysaccharides and conjugated protein synthesis (Gupta and Sutherland, 1967; Crossley, 1979). Insect haemolymph contains abundant protein, both in cell fraction and in the plasma (Crossley, 1979). Haemocytes are considered to play a role in synthesis of glycol and lipoproteins (Brehelin, 1972), labelling amino acids and synthesizing vitellogenins and other proteins (Geiger et al., 1977) and are active sites for protein synthesis (Wigglesworth, 1979).

Ultrastructure of haemocytes during metamorphosis in *Lymantria dispar* has been studied and an increase in granules in pupal stage has been noticed and these granules were reported to be composed of polysaccharides, proteins and lipids (Kim et al., 1990 a & b). Granulocytes have also been reported to contain acid phosphatases in *Simulium vittatum* (Luckhart et. al., 1992). Granulocytes are reported to de-granulate and release nutrients for egg development in *Simulium vittatum* (Luckhart, et. al., 1992). Release of acid phosphatases from the haemocytes of *Melanoplus sanguinipes* during infection have been reported (Vincent et al., 1993). Haemocytes of fifth instar larvae of *Manduca sexta* collected under non-sterile conditions exhibited the presence of a novel high molecular weight protein (Beck et al., 1996). Some histochemical observations on the haemocytes of *Agrotis segetum* have been reported (Cebesoy and Ayvali, 1996).

2.5. Haemocyte population in Insects

Information on haemocyte population within an insect is essential for many types of physiological studies. The haemogram is a statement of the haemocyte population picture in an insect at a given time. Total haemocyte counts (THC) in a standard quantity of blood (usually mm3) or in a specific area

(Arvy et al., 1948; Lee, 1961) together with an estimate of the relative number of haemocytes in different categories (differential haemocyte count or DHC) in a random sample (Jones, 1962 and Arnold, 1972) give haemogram picture of the blood. Haemogram also includes blood volume, which has obvious bearings on THC and DHC and more meaningful way of expressing them is in absolute number (Wheeler, 1963; Jones, 1967a). This represents the total population as derived from total and differential counts in relation to blood volume. According to Shapiro (1979b), Tauber and Yeager (1934) were the first to study the THC in several insects and later (1935) in several orders. There are large differences in number of haemocytes in different insect species (Jones, 1962; Arnold, 1972).

The volume of the haemolymph varies widely according to age and developmental stages. A short time after ecdysis, the haemolymph volume decreases sharply (Loughton and Tobe, 1969). This absolute number may vary from one and one-half million in *Galleria sp.* (Jones, 1967d) to more than two millions in *Sarcophaga* (Jones, 1967a). It has been reported that most insects seem to have an average of 20,000 to 30,000 cells per mm3, but with a wide deviation (Laigo and Paschke, 1966; Gupta and Sutherland, 1968; Shapiro, 1968; Hoffmann, 1969). Studies on the total haemocyte count and haemolymph volume in *Periplaneta americana* have been reported (Wheeler, 1963). A Study of the haemocytes of the silkworm *Hyalophora cecropia* and their multiplication have been undertaken (Lea, 1964). Total haemocyte count in the *Halys dentata* has been observed (Bhadur and Pathak, 1971). Modifications of haemogram in *Locusta migratoria* after bacterial injection has been reported (Hoffmann et al., 1977). Arnold (1979a) has given a comparative study of the haemocytes of 16 species of cockroaches in relation with the size of haemocytes and their number. Total and differential haemocyte counts in the larvae of noctuid *Euxoa declarata* Walker have been reported by Arnold and Hinks (1976) and Hinks and Arnold (1977) and according to them, the haemocyte complex comprises plasmatocytes and granular haemocytes. Hazarika and Gupta (1987) have reported on haemocyte counts of *Blattella germanica*. Diets have been reported to influence the population of haemocytes and blood volumes in *Antheraea assama* (Hazarika et al., 1994). The total and differential haemocyte count and their morphometry were also studied in *Catamiarus brevipennis* (Ambrose and George, 1994). Effect of juvenile hormone on haemocyte counts have been reported in *Blattella germanica* (Hazarika and Gupta, 1997).

Many instances of reference are present in which marked changes in haemogram during the active growth period of insects are reported (Tauber and Yeager, 1934; Smith, 1938; Bhadur and Pathak, 1971; Hazarika and Gupta, 1987). Total and differential counts in virus infected Wax moth *Galleria mellonella* have been observed to be linked with life expectancy (Lea, 1986). Seasonal variation in blood volumes and haemocyte population of *Antheraea assama* Westwood has been observed (Bardoloi and Hazarika, 1992). Total and differential haemocyte counts (THC and DHC) were made during various larval instars of the saturniid moth *Antheraea assama* and gradual increase in THC was recorded with the increased number of instars (Bardoloi and Hazarika, 1995). Haemocyte count in the larvae of *Pericallia ricini* have been reported, relating it with nutrient quality of two caster varieties (Jeykumar et al., 1995). Total and differential haemocyte counts in three morphotypes of *Rhynoceris marginatus* were recorded and it was higher in *Niger* morphs (Ambrose and

George, 1996 b). Instar dependent haemocyte changes in early and late instars infested with *Cotesis glomerata* has been reported by Bauer et al. (1998). Total and differential haemocyte counts in *Spodoptera littoralis* parasitised by *Microplitis rufiventris* was reported by Hegazi et al., 1998. The THC decreases during detoxification of Deltamethrin in *Philosamia ricini* (Begum et al., 1998). An increasing haemogram pattern from early instar to adult has been reported in *Rhynocoris marginatus* (Ambrose et al., 1999).

A comparison of different counts on the same insect and an examination of the total haemocyte count throughout the life-cycle of the milkweed bug *Oncopeltus faciatus* was reported (Feir, 1964 a & b). Barduco et al., (1988) did not observe large variation in the total haemocyte count of different stages of larvae of *Diatraea saccharalis*. There was no difference in differential haemocyte count in between parasitised and non-parasitised *Pseudoplusia includens* (Strand and Noda, 1991). Variation in differential haemocyte count has been reported after Plumbagin treatment in *Dysdercus koenigii* (Saxena and Tikku, 1990). Changes in number of a few haemocytes during metamorphosis of Lymantria dispar have been reported (Kim et al., 1990 a). Effect of endocrine extracts on the blood volume and population of haemocytes in *Halys dentata* had no effect on blood volume, but it changed population of haemocytes (Pathak, 1991). A decrease in number of haemocytes in the noctuid *Mythimna separata* larvae treated with *Bacillus thuringiensis* Berliner was observed (Sha and Xie, 1992). Total and differential haemocyte count and their morphometry were studied in *Catamiarus brevipennis* (Ambrose and George, 1994). THC reduced in parasitised *Spodoptera littoralis* (Kares, 1994) and after B-ecdysone treatment in *Dysdercus cingulatus* (Ahmad, 1995). Effect of varying biochemical profiles on Ricinus communis Lin. and on the haemodynamics of *Pericallia ricini* Fabricius show haemocyte diversity (Jeyakumar et al., 1995). Haemocyte counts in healthy individual of *Apis mellifera* and *Apis cerana indica* collected in Jammu and Kashmir State of India were significantly higher than in those infested with microsporidian (Abrol, 1995). No impact on total haemocyte count in the *Aspergillus* infested *Atractomorpha crenulata* was observed, but there were quantitative changes in the haemogram (Mahalingam and Muralirangam, 1996). Changes in haemocyte profile of *Spilostethus hospes* Fabr. in relation to eclosion, sex and mating had varied range of effects on THC and DHC (Sanjayan et al., 1996). Effect of microsporidian infection in larvae of *Galleria mellonella* had bearings on DHC; the haemocytes showed phagocytised spores (Tonka, 1997). THC and DHC in braconid infested *Seasamia cretica* Led., both in non-diapausing and diapausing larvae have been reported (El Mandarawy, 1997). Instar dependent changes in THC and DHC in *Pieris brassica* have been reported after parasitisation (Bauer et al., 1998). A reduction in THC in parasitised *Spodoptera littoralis* and increased granular cells was recorded (Hegazi et al., 1998). In Deltamethrin treated larvae, pupae and adult *Philosamia ricini (Samia cynthia ricini)*, THC and DHC was observed to be decreasing (Begum et al., 1998). Total and differential haemocyte counts were estimated for the different life stages of the reduviid *Rhynocoris* marginatus showed a significant increase from the second nymphal instars. In different breeds of silkworm *Bombyx mori* L. and their changes during progressive infection has been reported (Balvenkatsubbaiah et. al., 2001) and in *Antheraea mylitta* changes in haemogram occur due to infection with *Nosema* sp. (Sharan et al., 2002).

Marked changes in haemogram during the developmental stages of *Prodenia eridania* has been reported (Yeager, 1945). Arnold (1952) observed that in *Ephestia kuhiniella*, the haemocyte count increased during larval life, but just prior to pupation, declined. According to him, the number of cells in haemolymph of holometabolous insects increased partly through growth and differentiation of prohaemocytes and partly through mitotic division of other cells among. Smaller fluctuations in number of haemocytes have been observed to occur with the growth of different life stages in milkweed bug *Oncopeltus fasciatus* (Feir and O'Connor, 1965). Nittono (1960) reported that in *Bombyx mori*, the peak in number of haemocytes occurred prior to moult in the fourth larval and pupal stages. Comparative variations in haemocyte population of various developmental stages of diapausing and non-diapausing generations have also been reported: such as in *Pectinophora gossypiella* (Clark and Chadbourne, 1960; Raina and Bell, 1974); in diapausing *Hyalophora cecropia* pupae following injury (Davis and Schneiderman, 1960; Lea and Gilbert, 1961; Clark and Harvey, 1964). Such variation has also been reported in *Schistocerca gregaria* (Lee, 1961).

Studies on the total haemocyte count and haemolymph volume in *Periplaneta americana* in relation to moulting have been done (Wheeler, 1961, 1963). Multiplication and increase in haemogram with instar has been reported in *Hyalophora cecropia* (Lea, 1964; Lea and Gilbert, 1966) and also in *Galleria mellonella* at various stages of development (Shapiro, 1968). Little changes in cell number occurred in six to ten days old *Heliothis zea* larvae (Shapiro et al., 1969). Transformation of prohaemocytes to other haemocyte types are also suggested (Yeager, 1945; Arnold, 1972; Shrivastava and Richards, 1965; Arnold, 1974; Gupta, 1979 a). Multiplication of circulating haemocytes occurred in *Euxoa decalarata* with the advancement of number of instars from I to VI (Arnold and Hinks, 1976). Depending on physiological state and external factors, great variations in the haemograms have been reported in 27 insect species belonging to 13 orders (More and Sonawane, 1990). Variation in haemocyte population from fifth moulting stage until sixth instar was observed in *Spodoptera litura* (Saxena, 1992). Such haemocytic population changes in fifth instar noctuid larvae *Mythimna separata* have also been reported (Sha and Xie, 1992). Rapid changes/degeneration at different instars due to variation in relative humidity has been reported in *Philosamia ricini* (Begum et al., 1992a & b). Haemocytes in the fifth instar larvae and pupae of *Plusia orichalcea* Fabr. have been studied (Pathak and Saxena, 1994). Total and differential haemocyte counts during various larval instars in *Antheraea assama* have been observed to fluctuate in stages (Bardoloi and Hazarika, 1995). Studies were carried out on the haemocyte structure and related aspects in relation to growth and moulting of larval to pupal stages in *Plusia orichalcea* Fabr. (Pathak and Saxena, 1994). A number of haemocytes have been reported in the life stages of *Catamiarus brevipennis* Senille (Ambrose and George, 1994) with highest number in adult females. Variation in population of haemocytes in different larval instars, pre-pupa and pupa was also reported in *Lymantria dispar* (Butt and Shield, 1996); fluctuation in the haemocyte population in life stages of *Rhynchoris marginalis* (Ambrose et al., 1999) and in different breeds of *Bombyx mori* (Balavenkatsubbaiah, et al. 2001) have been observed.

Minor differences in population, types, size, etc., of haemocytes have been reported between the sexes in some insect (Lea, 1964; Hoffmann, 1967 a & b). Differences in the two sexes of *Hyalophora cercropia* (Lea, 1964) and *Halys dentata* (Bhadur and Pathak, 1971) have also been reported. No quantitative difference was recorded in the haemocytes of *Diatraea saccharalis* (Barduco et al., 1988). Changes in the haemocyte profile of *Spilostethus hospes* Fab. in relation to eclosion, sex and mating have been reported (Sanjayan et al., 1996). An increase in haemocyte population after termination of diapause has been reported in *Sesmia cretica* (El Mandarawy, 1997).

Thus, it is seen that there are different types of haemocytes in various insect species, which may or may not specific to an insect order and changes in blood volume, haemogram, THC and DHC occur with the different developmental stages of insects in different sexes and diapausing and non-diapausing state. This may be true in case of *Antheraea mylitta* in its larval instars, pupae of different age groups and in adults of both the sexes in non-diapausing and diapausing generations. Therefore, this study has been taken up to collect first hand information on this physiological aspect of haemolymph of *Antheraea mylitta*.

2.6. Haemolymph composition.

2.6.1. Physical properties

Haemolymph acts as a storage reservoir for many materials essential for various insect life processes (Buck, 1953; Florkin and Jeuniaux, 1974; Mullin, 1985). Haemolymph acts as a reserve of water from tissues (Mellanby, 1939; Lee, 1961). The specific gravity of insects has been studied in view of its importance indicating the level of dissolved material in blood (Buck, 1953). Hydrogen ion concentration (pH) of blood indicates about buffering capacity of blood of insects and is reported to be slightly acidic near to neutral pH in insects (Buck, 1953; Wigglesworth, 1972; Chapman, 1980; Mullin, 1985). The osmotic pressure in the haemolymph are regulated as it remains constant even under circumstances where volume fluctuates (Dajajakusumah and Miles, 1966) or it also remains constant during metamorphosis in *Bombyx mori* (Jeuniaux, 1958); and in *Antheraea pernyi* (Fyhn and Saether, 1970). The osmotic pressure of haemolymph is influenced by the inorganic cations and anions like Na+ and Cl- pump which acts as osmotic effector in primitive pterygotes (Lockwood and Croghan, 1959). In case of Lepidopterans, the organic molecules or amino acids are the main osmolar effectors and they have very low value of Na+ index and very high value of the Mg+ and K+ indices (Duchateau et al., 1953). Definite relative proportion of the Na+, K+, Ca++ and Mg++ are essential in the cell medium of insects (Baldwin, 1962). Presence of these cations have been observed in the larvae of *Spilosoma lutae* and in the *Bombyx mori* third and fourth instar larvae and pupae and in *Antheraea mylitta* pupae (Duchateau, et al., 1953); in larvae of *Pieris rapae* (Ramsay, 1953); in larvae of *Actias selene* and in pupae of *Philosamia cynthia* (Duchateau, et al., 1953); in adult *Bombyx mori* (Bialaszewicz and Landan, 1938); in *Manduca sexta* and *Hyalophora cecropia* (Jungreis, et al.;1973; Jungreis,1974 & 1978) and in *Lymantria dispar* (Pannabecker et al., 1992). These workers have observed that cationic pattern in larvae is somewhat more specialised than adults. Similar studies on presence of cations in the haemolymph of *Morimus funereus* have been reported (Jankovic Hladni et al., 1992). The literature is rather scanty on the cationic pattern, pH and specific gravity of haemolymph in insects and it is essential to observe these parameters in blood of *Antheraca mylitta* in its different larval instars, early, middle and late age pupae and adults, both in non-diapausing and diapausing generations in order to have better understanding of haemolymph physiology. Hence, the present study has been undertaken.

2.6.2. Haemolymph biochemicals

Haemolymph is the only extracellular fluid in insects and it exits in an unbound, non-vascular state, in direct contact with tissue and organs. The haemolymph proteins of insects have been investigated from various points (Chen and Levenbook, 1966 a & b). These include (1) the mapping of protein parameters in various species for taxonomic purposes, (2) the identification of protein composition at successive developmental stages by both electrophoretic and immunological techniques, (3) proposal of possible functions of the protein components on the basis of enzymological and histological tests, and (4) the analysis of both site and mechanism of synthetic process by isotopic labelling. Buck (1953), Wyatt (1961) and Chen (1966) have reviewed considerably the blood haemolymph keeping in view the above objectives. The insect haemolymph does not markedly differ from that of vertebrates with respect to its protein nitrogen (Florkin and Jeuniaux, 1974). Characteristically, insect haemolymph is rich in amino acids, which play an important role as osmotic effectors.

The presence of free amino acids in insects belonging to different orders have been investigated, viz., Odonata (Raper and Shaw, 1948); Orthoptera (Benassi et al., 1959 & 1961); Hemiptera (Prat, 1950) and Lepidoptera (Auclair and Dubrenil, 1953a & b; Chen and Hadorn, 1954; Wyatt and Kalf, 1956; Pant and Agrawal, 1964, 1965; Mansingh, 1967). Amino acids present in the 15 species belonging to the genus *Citheroonia*, *Eacles*, *Saturnia*, *Antheraea*, *Actius*, *Hyalophora* and *Philosamia* have been reported(Florkin and Jeuniaux, 1974). Specific variation in the ontogenic protein pattern and free amino acids in haemolymph have been found during the larval development of a number of insect species (Chen, 1971). The chemical composition of haemolymph is highly variable among the diverse species examined and at different developmental stages of the same species (Florkin and Jeuniaux, 1974). The variations probably reflect the balance between the synthesis, storage, transport and degradation of structural and functional protein during ontogeny as well as a response to particular ecological and physiological conditions (Florkin and Jeuniaux, 1974). The concentration of free amino acid pool increases between the early and mid fourth larval stages followed by a decline during the late fourth larval period in *Chironomous tentans* (Firling, 1977). Variation in the concentration of amino acids in the haemolymph of *Bombyx mori* during the fifth larval instar, spinning and metamorphosis was reported (Jeuniaux, 1961). The concentration of tyrosine in haemolymph varies widely during the insects' whole life history, which is related to tanning and melanization of new cuticle (Duchateau-Bosson et al. 1962). Similar observation has been reported in pupation of *Sarcophaga* and *Ephestia kuhniella* (Blaich, 1969). Protein and amino acids are the primary source of the ten essential amino acids needed by most insects' viz., arginine, histidine, isoleucine, leucine, lysine, methionine, phenylalanine, threonine, tryptophan and valine. The quantitative balance of these essential amino acids is important. Larvae of flesh fly *Phormia regina* and the silkworm *Bombyx mori* require proline, glutamic acid or aspartic acid (Dadd, 1985). The essential free amino acids are required for egg production in adult stage. Some species acquire these amino acids during larval stages and do not have to ingest them as adults (Dadd, 1985). The role of most abundant amino acids during diapause (serine, alanine and praline) has been discussed by Morgan and Chippendale (1983). Large amount of serine in overwintering larvae or pupae of some insects have been detected during post diapause development (Osanai and Yonezawa, 1984; Story et al., 1986). The male specific proteins in *Galleria mellonella* have been reported to contain large amounts of asparagine, aspartic acid, glutamine, glutamic acid and lysine but small amounts of tyrosine, methionine and tryptophan (Vilcinskas et al., 1997). A diapause associated protein, pectinophorin, contains first 15 amino acids starting from amino terminus of the peptide chain: N-Ala-Lys-Thr-Ileu-val-Glu-Asn-Met-Pro-Pro-Thr-Pro-Len-Asn-Ala-C in Pectinophora gossypiella (Salama and Miller, 1992). Amino acid pools during embryonic development of the Japanese silkworm, Antheraea yamamai (Lepidoptera: Saturniidae) have been studied (Furusawa et al., 1993). Presence of amino acids and their ontogenic pattern in tasar silkworm *Antheraea mylitta* in relation with their age/stage is essential, hence the present study has been undertaken.

The protein concentration in the insect haemolymph is similar to that of the blood of main and other vertebrates, and generally higher than that of the internal fluids of other invertebrates. The average protein content is of 5 g/100 ml in Hymenoptera and 3-4 g/ 100 ml in Coleoptera, 2 g/100 ml in Lepidoptera and 1 g/100 ml in Orthoptera (Florkin, 1936a). The number of different proteins is highly variable according to species, caste (Lue and Dixon, 1967), sex (Stephen and Steinhauer, 1957; Kulkarni and Mehrotra, 1970), diet (Bodnaryk and Morison, 1966), starvation (Feir and Drazywda, 1969) or ontogenic stage (Florkin and Jeuniaux, 1974). Proteins (enzymes) play a central role in all metabolic process and in the structure and function of muscles and other tissues. Insect blood protein synthesis has been studied in several species during development and diapause (Telfer, 1954; Laufer, 1960; Wyatt, 1961 and Chen and Levenbook, 1966 a & b). Several possible role for the blood proteins have been postulated e.g., they may function as enzymes (Laufer, 1960); serve as amino acids reserves for adult tissues (Dinamarca and Levenbook, 1966) or be used intact in constructing adult structures (Loughten and West, 1965 and Fox and Mills, 1969). The synthesis of pupal blood proteins during diapause have been studied in *Cecropia* silk moth (Telfer and Williams, 1960); diluting the pupal proteins reactivates production of larval blood proteins (Patel, 1971) and blood protein synthesis in *Hyalophora cecropia* silk moth (Ruh and Willis, 1974). Diapause associated proteins have been characterised by Turunen and Chippendale (1980). During the pre-diapause, the total body content of lipids, proteins and carbohydrates increases (Grison and Lee Berre, 1953; El-Ibarashy, 1965) and diapausing proteins accumulate in haemolymph (De Loof and De Wilde, 1970). Diapause is also characterised by the cessation of the haemolymph protein synthesis and juvenile hormone (Dortland and De Korte, 1978). Changes in the concentrations of metabolites in haemolymph during and after diapause in female Colorado potato beetles, *Leptinotarsa decemlineata* has been reported (Lefevere et al., 1989).

Many references are available on protein titre of haemolymph. By using electrophoretic methods, identification of haemolymph proteins was done in many insects like *Ostrinia sp.* (Beck and Hanec, 1960; Chippendale and Beck, 1966), *Hyalophora cecropia* (Laufer, 1960, 1961) and *Periplaneta americana* (Siaktos, 1960). The blood proteins of pupae of *Hylophora cecropia* have also been studied by acrylamide gel electrophoresis in *Hylophora cecropia* (Ruh et al., 1972). The most striking features are haemolymph storage proteins in holometabolous insects (Thomson, 1975; Wyatt and Pan, 1978). These proteins comprise the bulk of polypeptides in larval haemolymph and later in pupal fat body. Developmental stage-specific alterations in tissue localisation have been documented for both Dipterans (Kinnear and Thomson, 1975; Thompson, 1975) and Lepidopterans (Chippendale and Kilby, 1969; Chippendale, 1970 a & b; Tojo et al., 1980; Kramer et al., 1978 a and Miller and Silhack, 1982). Lipoproteins have been reported in *Hyalophora gloveri*, *Hyalophora cecropia* and *Antheraea polyphamous* (Whitmore and Gilbert, 1974; Tojo et al., 1978) and in *Spodoptera litura* (Tojo et al., 1985); in *Triatoma infestans* (Rimoldi et al., 1990); in larvae of noctuid, *Mythimna separata* (Zhu et al., 1992); in larval weevil, *Diaprepes abbreviatus* (Shapiro et al., 1992); in larval and adult of *Leptinotarsa decemlineata* (Koopmanschap et al., 1992); in *Anopheles sinensis* (Ye et al., 1992). Diapause associated proteins have been studied in *Pectinophora gossypiella* (Salama et al., 1992). Heat shock proteins in haemolymph have been reported in insects (Peterson and Mithel, 1985; Linquist, 1986; Fittingoff and Riddifold, 1990; Sengupta et al., 1995). Many holometabolous insects have been shown to accumulate proteins during the last larval instar in preparation of non-feeding pupal stage (Telfer and Kunkel, 1991). These proteins are known as storage proteins, later these proteins are used for formation of adult structures (Koopmanschap et al., 1992). These haemolymph proteins are also reported in larval *Manduca sexta* (Beck et al., 1996); in last instar larvae of *Hyphantria cunea* (Jung Hee et al., 1996); in fall web worm, *Hyphantria cunea* (Song et al., 1995); antibacterial proteins in *Hyalophora cecropia* (Lockey and Ourth, 1996); protein profiles of larval haemolymph of *Bombyx mori* (Kawaguci et al., 1997); in *Lymantria dispar* (Masler and Kovaleva, 1997). Haemolymph proteins have been reported in relation with vitellogenin and egg production in *Heliothis virescens* (Zeng et al., 1997); in greater wax moth, *Galleria mellonella* (Vilcinakas et al., 1997) and in larvae of *Argyrogramma agenta* (Shu et al., 1997). Male specific proteins have been reported in *Galleria mellonella* (Lee et al., 1998) and lipoproteins have been identified during larval development of *Samia cynthia ricini* (Eri silkworm) (Saito, 1998). In the present study, attempts have been made to identify haemolymph proteins in non-diapausing and diapausing generations of *Antheraea mylitta* different life stages.

It has been known for a long time that insect haemolymph generally contains only small amount of fermentable sugars, almost no saccharose, and little, if any, glycogen (Florkin and Jeuniaux, 1974). Wyatt and Kalf (1956, 1957) reported existence of trehalose (a-glucose) in insects. The presence of trehalose in different insect species of diverse orders has been compiled after Florkin and Jeuniaux, 1974. In most insects, the trehalose of the haemolymph is absorbed and used by the cells of most tissues due to an intra-cellular trehalose (Howden and Kilby, 1960; Clegg and Evans, 1961). In the fat body, an inverse relation exists in between glycogen and trehalose, the former disappearing at each moult, while trehalose remains at constant level (Saito, 1963).

Amino acids of *Tenebrio molitor* (Marcuzzi, 1956), *Hyalophora cercropia* (Carey and Wyatt, 1960) and of *Bombyx mori* (Wyatt et al., 1956) larvae has been reported in their haemolymph. Haemolymph glucose is known to be incorporated into the chitin of growing cuticle in several insect orders (Bade and Wyatt, 1962). High haemolymph sugar levels are reported in a number of species of Hymenoptera and Diptera (Florkin and Jeuniaux, 1974). In insects' diapausing embryonic stage, sugars play the major role in diapause metabolism and there is direct relationship between the cold hardiness and levels of polyols (Salt, 1957 & 1959; Somme, 1967; Baust and Miller, 1970), carbohydrates (Tanno, 1964; Somme, 1967) and unsaturated fatty acids have been reported for several hibernating insects. Many diapausing species utilise stores of glycogen to generate cryoprotectants: such as glycerol, sorbitol, or trehalose (Chino, 1957, 1958; Wyatt, 1967). Conversion of glycerol to glycogen during end of diapause has been reported (Hayakawa and Chino, 1982 a & b). Changes in carbohydrate related to cryo-protection in Pieris brassicae (Pullin, 1992); during pupal moult of Manduca sexta (Siegert, 1995); on metal effects on carbohydrate pH of *Lymantria dispar* (Ortel, 1995). The haemolymph and total body composition viz. carbohydrates, lipids, proteins and free amino acids have been reported in gypsy moth larvae *Lymantria dispar* (Bischof and Ortel, 1996); in nutrient content of Agrotis ipsilon (EL Ghar et al., 1996); in larval instars of non-diapausing and diapausing generations of *Antheraea mylitta* (Sinha et al., 1998).

Haemolymph generally contains excretory end products in the form of uric acid (Florkin and Jeuniaux, 1974). Its concentration depends upon feeding (Ramakrishna and Pawar, 1975) and development (Buckner and Caldwill, 1980). Therefore, it was also observed in different ontogenic stages of tasar silkworm, *Antheraea mylitta* in non-diapausing and diapausing generations as well as in the two sexes of fifth instar, pupae and adults of both the generations.

Lipids are transported by the haemolymph from the mid-gut to the fat body, where they are stored in more or less modified form and from fat bodies to tissues/organs. Role of lipids in post embryonic diapause have been described (Lees, 1955). Wide variations of the lipid's concentration are observed during muscular activity (flight), development or metamorphosis (Nowosielsky and Patton, 1965; Nelson et al., 1967; Mayer and Candy, 1969). Proportional quantification of phospholipids, total lipids, sterols, un-sterified fatty acids and neutral glycerides have been reported in *Galleria mellonella* (Wlodawer, and Wisniewska, 1965) and in *Acheta domesticus* (Wang and Patton, 1969). Metabolic relation between lipid transportation by haemolymph and place of storage has been reported in wax moth (Wlodaner and Wisniewska, 1965) and in *Hyalophora cecropia* (Martin, 1969). The effect of chilling and integumentory injury on carbohydrate and lipid metabolism in diapause and non-diapause pupae of *Manduca sexta* has been observed (Siegert, 1986). Similar studies have also been reported in *Diatraea grandiosella* (Kuthiala and Chippendale, 1989) and whole body lipid content of *Plathypena scabra* have been reported on the basis of daily and seasonal changes (Mason et al., 1990). Comparison of fatty acid composition in lipids of diapause and non-diapause eggs of *Bombyx mori* has been observed (Shimizu, 1992). Carbohydrate and lipid metabolism during the last larval moult of tobacco hornworm, *Manduca sexta* have been reported after Siegert et al., 1993. Reduction in lipid contents in *Lymantria dispar* has been observed due to metal intoxication (Ortel,

1995). Effect of parasitism on lipid content in *Lymantria dispar* was also reported (Bischof and Ortel, 1996).

Thus, it is observed that a complete account of difference in the haemolymph picture on aspects related to haematology of haemocyte, histochemical studies on different nutritional aspects and enzymes, ontogenic proteins through electrophoretic studies and related biochemical studies on different nutritional reserves in a single insect species is not available. Under the present study, an attempt has been made to generate information on theses aspects and different life forms in different generations of tropical Tasar silkworm *Antheraea mylitta*, which may be helpful to have better understanding of haemolymph of this insect.

3

Materials and Methods

3.1. THE EXPERIMENTAL ANIMAL

3.1.1. Life Cycle:

Antheraea mylitta Drury [Lepidoptera: Saturniidae] stocks used in this experiment were maintained at Central Tasar Research and Training Institute, Field Laboratory, Piska-Nagri (23.21°N, 85.20°E, 654.16 meters AMSL), Ranchi, India, where it completes two generations in a year. Generations are commonly known as first and second crops in sericulture parlance.

The life cycle (Fig. 3.1.) passes through four stages: egg, larva, pupa and adult. Out of four stages, the egg encompasses the embryonic development, which lasts for about 7-8 days. The larval stage has five instars and the total larval duration is of 28 ± 2 days. A. mylitta shows a facultative pupal diapause. Its first generation's "non-diapause destined larvae (NDD larva)" form pupae, which do not diapause (NDD pupae). Pupal stage remains for about a period of 17-25 days, followed by emergence of adult moths (NDD adults). The male and female moths couple and reproduce, since they are only reproductive stages.

The pre-winter or second crop larvae are "diapause-destined (DD)", feed for about 45 days and formed pupae remain in diapausing condition for about 190-220 days. Moth emergence takes place next monsoon, which coincides with the sprouting of leaves of host plants in nature.

The experiments were carried out during the first and second crop rearing seasons of July-August and September-October, respectively for larval stages. After fourth moult, the fifth instar larvae were sexed based on Ishiwata and Harold's glands for further experimental studies (Fig. 3.2). Similarly, pupae and adults were sexed and used in the present experimental studies.

3.2. GENERAL HAEMATOLOGY

3.2.1. Obtaining haemolymph samples from larvae, pupae and adults

From I, II and III instar larvae, haemolymph was obtained by severing one of the prolegs with sterilised razor blade, whereas, from IV and V instars haemolymph was obtained by puncturing the prolegs with a fine needle. The haemolymph from pupae was obtained by pricking near the cephalic window with a fine needle and from adults by cutting the middle or hind leg with a safety-razor blade.

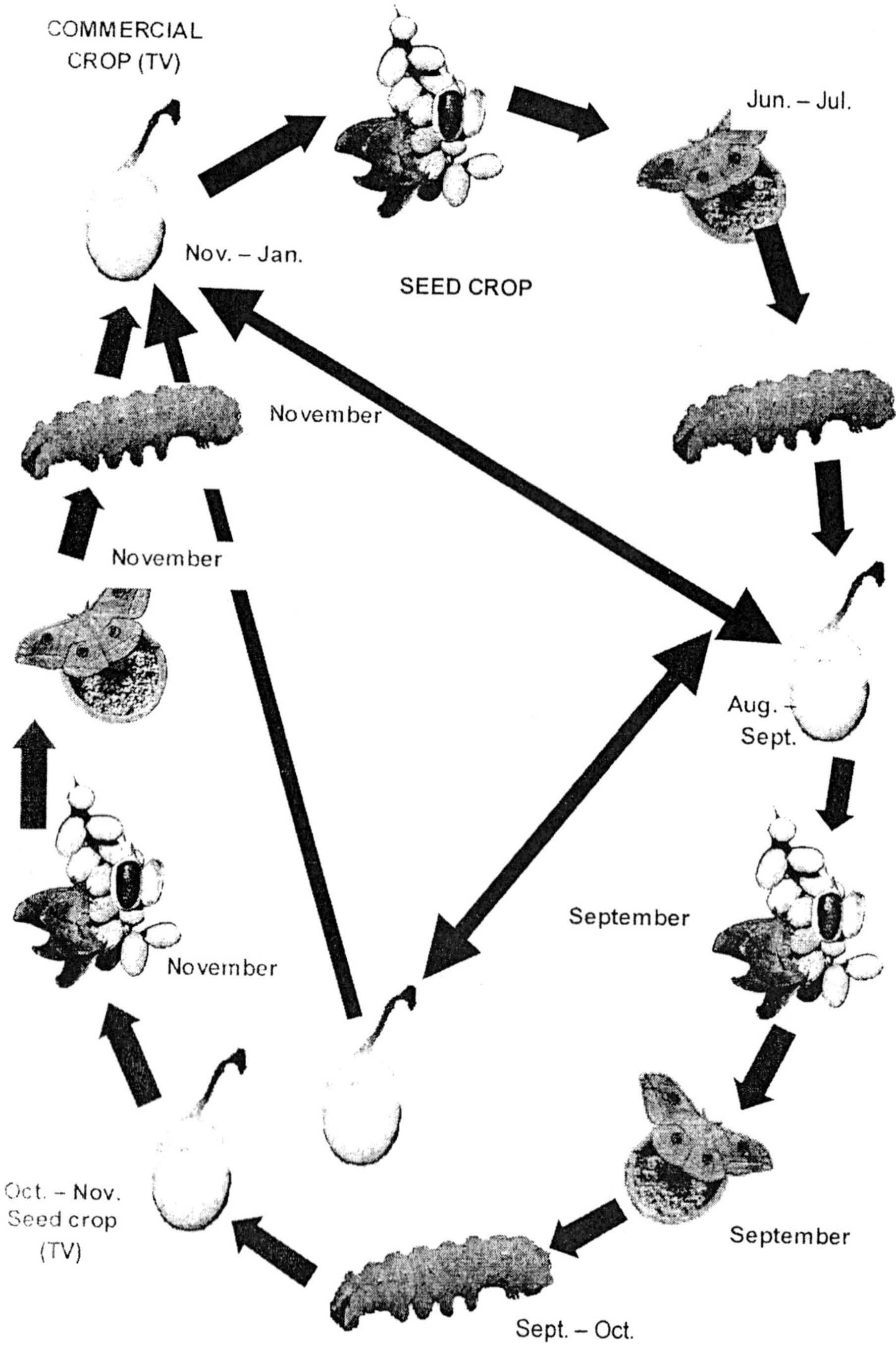

Fig 3.1. Life cycle of *Antheraea mylitta* Drury

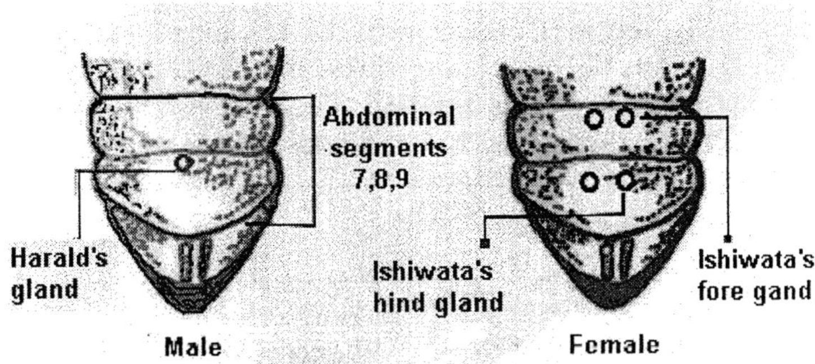

Fig. 3.2. Sex markings of V instar larvae

3.1.2. Stages of life-cycle selected for observations

The stages of life cycle selected for experiments and the abbreviations used in text and tables are detailed in Table 3.1.

Table 3.1. Stages of life cycle used in experiments and abbreviations used.

Sl. No.	Milestone in life-cycle	Abbreviation used	
		Diapause destined	Non-diapause destined
1.	Fully grown stage of the first instar	I-DD	I-NDD
2.	Fully grown stage of the second instar	II-DD	II-NDD
3.	Fully grown stage of the third instar	III-DD	III-NDD
4.	Fully grown stage of the fourth instar	IV-DD	IV-NDD
5.	Freshly moulted fifth instar: Female	VEF-DD	VEF-NDD
6.	Freshly moulted fifth instar: Male	VEM-DD	VEM-NDD
7.	Mid stage of the fifth instar: Female	VMF-DD	VMF-NDD
8.	Mid stage of the fifth instar: Male	VMM-DD	VMM-NDD
9.	Fully grown fifth instar: female	VLF-DD	VLF-NDD
10.	Fully grown fifth instar: Male	VLM-DD	VLM-NDD
11.	Spinning fifth instar larva: Female	SLF-DD	SLF-NDD
12.	Spinning fifth instar larva: Male	SLM-DD	SLM-NDD
13.	Immediately after pupation: Female	PPF-DD	PPF-NDD
14.	Immediately after pupation: Male	PPM-DD	PPM-NDD
15.	Mid pupa: Female	MPF-DD	MPF-NDD
16.	Mid pupa: Male	MPM-DD	MPM-NDD
17.	Late pupa: Female	LPF-DD	LPF-NDD
18.	Late pupa: Male	LPM-DD	LPM-NDD
19.	Adult on the day of emergence: Female	AF-DD	AF-NDD
20.	Adult on the day of emergence: Male	AM-DD	AM-NDD

3.2.2. Fixation of haemocytes

Since unfixed haemolymph and haemocytes did not give consistent results during study of the morphology and morphometry of the haemocytes, heat fixation method suggested by Jones, 1962 was followed for fixation of haemocytes by submerging the insect into hot water (55 - 60°C). The durations used were as follows:

Early instars (I-III):	1-2 minutes
Late instars (IV-V), pupae and adults:	3-5 minutes

This method was found most satisfactory as it was (a) friendly to the insect and was effective, (b) simple to use and (c) did not caused excessive regurgitation. Haemolymph was (d) easily obtained, it did not agglutinate or precipitate, (e) cells did not agglutinate and (f) cells did not changed in size, shape, or form. For fixation, absolute methanol (BDHTM) was used and the films/smears were air-dried for staining.

3.2.3. Staining of haemocytes for general study

For general histological study, smears were stained with Giemsa stain. The Giemsa stain was prepared by adding a pinch of Giemsa stain concentrate to 70% ethanol and allowing it to mature for one week before use. Before use, the stain was filtered. Unstained smears were used for examination under phase contrast microscope. The smears were prepared by oozing haemolymph from the proleg of larvae, cephalic region of pupa or severed legs of adults. The haemolymph was directly taken on a clean slide, and with the help of the edge of another slide, it was spread into a thin film or smear. The smear prepared was fixed with 70% methanol for about 5 minutes. The methanol was removed by keeping the slide in inclined position for a few seconds and then the smear was fully covered with Giemsa stain for 10-12 minutes. The slide was immersed in 70% ethanol for a few seconds and then dehydrated in 100% ethanol for 30-60 minutes. The smear was then immersed in xylene (E. MerckTM) for 5 minutes and mounted in DPX (BDHTM). In Giemsa stain, the nucleus stained deep purple while cytoplasm was from purple to light purple.

3.2.4. Scanning Electron Microscopy for general study

Haemolymph was collected on a clean glass coverslip and the cells were fixed by immersing the coverslip in chilled 3 % (v/v) glutaraldehyde in 0.1 M sodium cacodylate, 2 mM Ca2+ buffer, pH 7.2 for 15 minutes. After washing in buffer for 5 minutes, they were post fixed in 1 % (w/v) buffered osmium tetraoxide, washed as before, then dehydrated through a graded ethanol series, and critical point dried in a Sorvall critical point drying system. Cells were coated with gold and examined in Philips scanning electron microscope at 10 kV.

3.3. TOTAL COUNT

Diluting fluid for Total Count

A. Routine fluid for Total count

Sodium acetate (E. MerckTM)	3.0 g
Distilled Water	100.0 ml
Gentian Violet	A pinch

B. Toisson's fluid (Gilliam and Shimanuki, 1967)

Sodium Chloride (BDHTM, AR)	1.000 g
Sodium Sulfate (BDHTM, AR)	8.000 g
Glycerol (BDHTM, AR)	30.000 ml
Crystal Violet (BDHTM)	0.015 g
Distilled Water	100.000 ml

Procedure

With the help of clean graduated WBC pipette sucked the solution up to mark 0.5 (larger pipette with markings at 0.5, 1 and 101 and a white bead inside it). Excess solution sticking to the tip of the pipette was wiped out, keeping the level of solution exactly constant at 0.5. It was followed by sucking the diluting fluid up to mark 101. The tip of the pipette was cleaned of excess diluting fluid. Then the end of the pipette was held firmly between the forefinger and thumbs and shaken for a minute for a thorough mixing of the solution with the diluting fluid. After that, the finger was removed and 2-3 drops of the fluid from the capillary tube was blown out from the pipette chamber. Then a drop was transferred to the platform of the counting chamber (Neubaure improved haemocytometer, consisting of a thick glass slide with a transverse bar), the surface of which was sunk 1/10 mm below the surface of the rest of the slide (Fig. 3.2). A thick cover slip glass was applied over the bar, before transferring the fluid mixture from the pipette. Care was taken that no air bubble passes inside the space between the counting chamber and cover slip. The fluid usually runs into counting space under the cover slip by capillary action, if the pipette tip held at 45° to the surface of the chamber. The solution was diluted to 1 in 200 (as the fluid in the capillary part up to mark 101 was used). So the total number of cells is 1 mm3 will be 4000a X 200 = 8,00,000x (x = number of cells in each squire).

Area of squares abcd = 1 mm X 1 mm = 1 mm2.

Square abcd contains 400 (16 X 5 X 5) small squares.

Therefore, area of small square is 1/400 square mm.

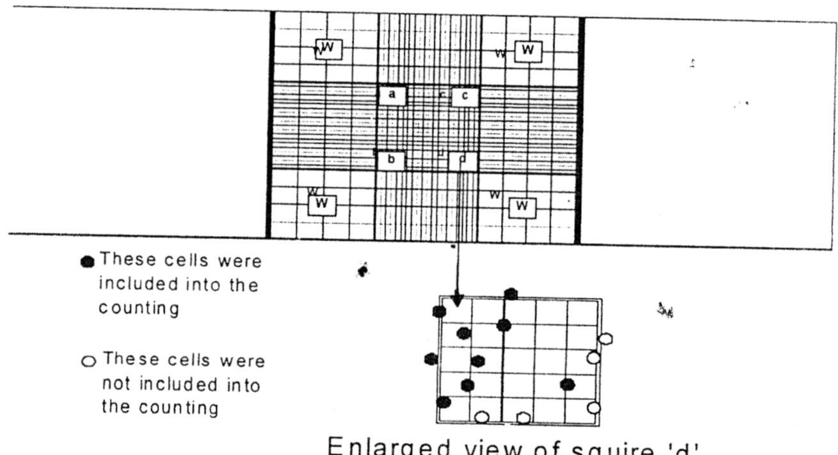

Fig. 3.2. Counting chambers of Neubaure improved haemocytometer

Under microscope, the squares were marked off into sets of sixteen by double lines for counting, at least five sets of sixteen (one at each corner and the central one) were counted under high power (1/6 objective) of the microscope. Cells, which lie on the boundary lines, were not counted except those on the upper and left hand lines.

Calculations

Total Number of cells counted in five sets of sixteen squares was divided by 80 to get the count in each square. The Total Volume of each small square is 1/400 X 1/10 = 1/4000 mm3. Therefore, if 'a' is the number of cells in each square, the total number will be 4000 X a in one mm3.

3.4. BODY WEIGHT

All the experimental animals were weighed by SimadzuTM 3-digit electronic balance after selecting according to stage and age as mentioned in section 3.1.2. Only those individuals were used whose weight did not vary more than 5 % of mean weight of the stage. The mean weight was used for calculation of absolute number of haemocytes.

3.5. BLOOD VOLUME

Blood volume was determined by the method of Richardson et al., 1931. In earlier instars (I-III), ten larvae were heat fixed and weighed together, accurately. In late instars (IV-V), pupae and adults, individual insects were weighed. The insects were introduced in a weighing bottle with stopper to prevent desiccation, if any. The insects were taken out one by one and a mid-dorsal incision was given along the entire length exposing the internal organs. The haemolymph was then blotted quickly and as completely as possible with the weighed strips of filter paper. The blood soaked filter papers were collected in a pre-weighed weighing bottle with stopper until no haemolymph from the previously weighed insect could be blotted out. Weighing bottle containing the blood soaked filter paper strips were then accurately weighed to find out the weight of the blood taken. The volume of haemolymph per gram of larval body weight was then calculated as v = m/d, where v = the volume of haemolymph, m = mass and d = density of the haemolymph which is numerically equal to the specific gravity (Section 3.8). The experiment was repeated 5 times with each age group and analysed statistically.

3.6. DIFFERENTIAL COUNT

For differential count of haemocytes, Leishmann's stain was used in the proportion mentioned below and it was dissolved in methanol before 24 hours of use:

A. Preparation of Stain

Leishmann's stain (BDHTM/GurrTM)	1.50 g
Absolute Methanol (E. MerckTM)	100.00 ml

B. Buffer for Staining

Di-Sodium-hydrogen phosphate (Anhydrous) (BDHTM)	5.447 g
Potassium-di-hydrogen phosphate (Anhydrous) (BDHTM)	4.752 g

To prepare the buffer, the salt mixture was pulverized thoroughly in a mortar and pestle and 1 g of this pulverized mixture was dissolved in 1 litre of distilled water to get the required pH 7.00 (Anonymous, 1958).

Staining

For differential count, thin and uniform film was prepared on clean and grease free slides with haemocytes collected from experimental stages of A. mylitta. The films were partially air-dried then 8 to 10 drops of the filtered staining solution were poured over the film so that whole film could be covered with the stain. The stain was allowed to act for 2 minutes. After that, 16 to 20 drops of above-mentioned buffer solution were put over the slide. Both staining and buffer solution were mixed thoroughly by gentle agitation and the mixture was allowed to act for another 10 minutes. The dilute stain was drained off and the stained smear was washed with the same buffer solution until the smear appeared pink to naked eye. The smear was then blotted dry with a strip of filter paper and finally sealed with DPX (Anonymous, 1958).

Counting

In general, the method given by Shapiro (1968) was used. The smear was examined under oil immersion, and 200 cells per slide were differentiated using the Yeager's classification as modified after Jones (1959). The following types of haemocytes were counted: Prohaemocytes (PRs), Plasmatocytes (PLs), Granulocytes (GRs), Spherulocytes (SPs), and Oenocytes (OEs). Cells in mitosis (MC) and degenerating cells (Deg) were also counted. In addition, certain cells found in the microscopic field, which could not be identified, were not counted. Fifteen slides were prepared for each stage and 200 cells were counted in each slide, using mechanical counter used by clinical laboratories. This meets the recommendations of Jones (1962) which states that a minimum of five insects of a given stage and physiological status should be counted and whenever possible, minimum 200 cells should be classified (Fig. 3.3).

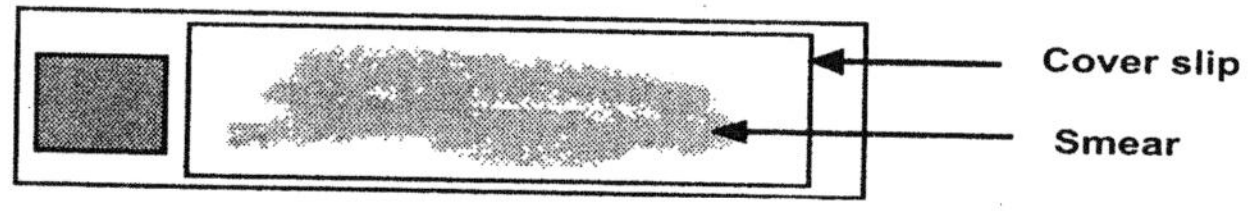

Fig. 3.3. A slide showing smear for differential haemocyte count

3.7. HISTOCHEMISTRY

3.7.1. General Protein

Smears of haemolymph from non-diapausing and diapausing stages as per section 3.1.2. were fixed in Lillie's buffered formalin chilled at 0°C for 24 hours (Lillie,1954).

Preparation of buffered formalin (Lillie, 1954)

37 to 40 % Formaldehyde (BDHTM)	10.00 ml
Distilled water	90.00 ml
Sodium-di-hydrogen phosphate (E. MerckTM)	0.40 g
Di-sodium hydrogen phosphate (E. MerckTM)	0.65 g

The salt mixture was accurately weighed and dissolved completely in diluted formalin. The buffered formalin thus prepared, was first chilled to 0°C and thereafter the haemolymph films were kept in the fixative at 0°C for 24 hours. After the stipulated time, the temperature of the fixative was gradually adjusted to room temperature and then the smears were washed in tap water for 1 hour. This was followed by washing in distilled water with two changes, each of 20 minutes duration. The smears were then stained for 2 hours at room temperature by Bonhog's (1955) process using mercury bromophenol blue reagent.

Preparation of mercury bromophenol blue reagent:

Solution A: 1 % aqueous solution of mercuric chloride (M&BTM)

Solution B: 0.05 % solution of bromophenol blue (E. MerckTM) in 2 % aqueous acetic acid (BDHTM) was mixed for preparation of solution B.

Staining

Before use, the solutions A and B were mixed in equal proportions. After staining for general protein, the smears were rinsed in buffered water (pH 7.00) and finally dehydrated in tertiary butyl alcohol with two changes until the smears acquired an intense blue colour. The dehydrated smears were then cleared in xylene and finally sealed with DPX.

Preparation of buffered water

Sodium di-hydrogen phosphate	0.40 g
Di-sodium hydrogen phosphate	0.65 g
Distilled water	90 ml

Result: Proteins are stained in deep clear blue colour.

3.7.2. Nucleic Acids

3.7.2.1. DNA - Feulgen-Schiff's Method (Feulgen and Rossenbeck, 1924):

Smears of haemolymph from control and infected larvae were fixed in buffered formalin chilled at 0°C for 24 hours as suggested by Lillie's (1954).

Preparation of buffered formalin (Lillie, 1954)

37 to 40 % Formaldehyde (BDH)	10.00 ml
Distilled water	90.00 ml
Sodium-di-hydrogen phosphate (EM)	0.40 g
Sodium hydrogen phosphate (EM)	0.65 g

The above salt mixtures were dissolved completely in diluted formalin. The buffered formalin thus prepared, was first chilled to 0°C and thereafter the haemolymph films were kept in the fixative at 0°C for 24 hours. Thereafter, the temperature of the fixative was gradually adjusted to room temperature. Fixed smears were washed in tap water for 1 hour. This was followed by washing in distilled water with two changes, each of 20 minutes duration.

Staining

Preparation of Schiff's Reagent (A)

Basic Fuchsin (BDHTM)	1.000 g
Distilled Water	200 ml
N-HCl	20 ml
Potassium metabisulphite (E. MerckTM)	1.000 g
Activated Charcoal	2.000 g

Basic Fuchsin was dissolved in measured quantity of boiling distilled water by gentle shaking. The solution was allowed to cool down to 50°C when the desired amount of N-HCl acid was added. The acidified dye solution was further cooled to 25°C and then potassium meta-bisulphite was mixed with it. The reagent prepared was kept in a tightly stoppered amber coloured bottle in dark for 24 hours. After the stipulated time, the reagent was decolourised by shaking with 2 gm. of activated charcoal. The reagent was stored in dark at 3-4°C and before use, filtered in dark.

Preparation of bisulphide reagent (B)

10% aqueous potassium meta-sulphite	5 ml.
N-HCl	5 ml.
Distilled water	100 ml.

Preparation of N-HCl (C)

Conc. HCl (sp. gr. 1.16, BDHTM, AR)	31.465 ml.
Distilled water	1 litre

The properly fixed and washed smears were rinsed in chilled (4°C-5°C) N-HCl and then carefully hydrolysed for 10 minutes in N-HCl at 60°C. The smears were then again rinsed in cold N-HCl. This was followed by rinsing with distilled water. The smears were then treated with Schiff's reagent for 1 hour at room temperature in dark. Thereafter, the smears were rinsed in freshly prepared bisulphite reagent (B) and were finally washed in distilled water with three changes of 30 seconds duration each. The smears were then dehydrated in alcohol, cleared in xylene and sealed in DPX.

3.7.2.2. DNA and RNA - methyl-pyronin Y method (Kurnick,1955)

Smears of haemolymph from control and infected larvae were fixed in Lillie's (1954) buffered formalin chilled at 0°C for 24 hours.

Preparation of buffered formalin (Lillie, 1954)

37 to 40 % Formaldehyde (BDHTM)	10.00 ml
Distilled water	90.00 ml
Sodium-di-hydrogen phosphate (E. MerckTM)	0.40 g
Sodium hydrogen phosphate (E. MerckTM)	0.65 g

The salt mixture was accurately weighed and dissolved completely in diluted formalin. The buffered formalin thus prepared, was first chilled to 0°C and thereafter the haemolymph films were kept in the fixative at 0°C for 24 hours. After the stipulated time, the temperature of the fixative was gradually adjusted to room temperature and then the smears were washed in tap water for 1 hour. This was followed by washing in distilled water with two changes, each of 20 minutes duration.

Staining

Preparation of methyl green-Pyronin Y Solution

Solution A

Methyl Green (BDHTM)	1.000 g
Distilled Water	50.000 ml

Solution B

Pyronin Y (GurrTM)	2.500 g
Distilled Water	50.000 ml

Both the solutions A and B were taken in two separating funnel, washed repeatedly with chloroform until chloroform layer became colourless. The washed solution of methyl green and Pyronin Y thus obtained were mixed in the following proportions before use:

Working Solution

Methyl-green Solution	10.0 ml
Pyronin Y Solution	17.5 ml

Procedure

The previously fixed and washed smears were immersed in the above-mentioned staining solution for 10 minutes. The smears then taken out and the excess stain was drained off and blotted with a filter paper. The stained films were dehydrated in n-butyl alcohol with two changes, each of 5 minutes duration. These were cleared in xylene and finally sealed with DPX.

3.7.3. Localisation of PA/S substances

3.7.3.1. PA/S reactive substances through Schiff's reagent

PA/S reactive substance was demonstrated according to methods given by Hotkiss (1948).

Fixation

Films of haemolymph were fixed in 10 % neutral aqueous formalin for 20 minutes at room temperature. The smears were then washed in distilled water with two changes of 15 minutes each. This was followed by rinsing in 50 % ethanol.

Staining

Prior to staining with Schiff's reagent, the smears were oxidised in freshly prepared 0.4% aqueous periodic acid to break the C-C bonds of the polysaccharides, mucopoly-saccharides, muco-proteins and glycol-proteins etc., having 1:2 glycol groups into the respective di-aldehydes. After oxidation, the smears were treated with acidified sodium thiosulphate solution to remove traces of periodic acid or periodate, which might have been present in the smears because of periodic oxidation.

Preparation of Oxidising Reagent

Periodic acid crystals (ReidelTM)	0.4 g
Double distilled ethanol (E. MerckTM)	35.0 ml
N/5 aqueous sodium acetate	5.0 ml

The ingredients were mixed together before use and shaken to dissolve the periodic acid crystals.

Preparation of reducing solution

Potassium iodide (E. MerckTM)	1.0 g
Sodium thiosulphate (Sarabhai MerckTM)	1.0 g
Double distilled ethanol (E. MerckTM)	30.0 ml
Distilled water	20.0 ml
2N-Hydrochloric acid	0.5 ml

The ingredients were mixed together and shaken until the crystals of potassium iodide and sodium thiosulphate was dissolved. Turbidity due to precipitation of sulphur, if any, was removed by filtration. Fresh solution was always prepared before use.

Procedure

The fixed and washed smears were first oxidised in periodic acid solution at room temperature for 2-3 minutes. This was followed by rinsing of the smears in 70% ethanol and reduction in sodium thiosulphate solutions for 1 minute at the same temperature. The smears were then again rinsed in 70% ethanol and treated with Schiff's reagent for 1 hour at room temperature in dark. After this the smears were rinsed in tap water with 2 changes, each of 5 minutes duration and then dehydrated in grades of alcohol, cleared in xylene and sealed with DPX.

Preparation of Schiff's reagent.

Schiff's reagent [A]:

Basic Fuchsin (BDHTM)	1.000 g
Distilled Water	200 ml
N-HCl	20 ml
Potassium metabisulphite (E. MerckTM)	1.000 g
Activated Charcoal	2.000 g

Requisite Basic Fuchsin was dissolved in 200 ml boiling distilled water by gentle shaking. The solution was allowed to cool down to 50°C when the desired amount of N-HCl acid was added. The acidified dye reagent thus prepared was kept in a tightly stoppered amber coloured bottle in dark to 24 hours. Thereafter, the reagent was decolourised by shaking with 2 gm. of activated charcoal. The reagent was stored in dark at 3-4°C and before use filtered in dark.

Bisulphide reagent [B]:

10% aqueous potassium metasulphite	5 ml.
N-HCl	5 ml.
Distilled water	100 ml.

N-HCl [C]:

Conc. HCl (sp. gr. 1.16, BDHTM, AR)	31.465 ml.
Distilled water	1 litre

The properly fixed and washed smears were rinsed in chilled (4-5°C) N-HCl and then carefully hydrolysed for 10 minutes in N-HCl at 60°C. The smears were then again rinsed in cold N-HCl. This was followed by rinsing with distilled water. The smears were then treated with Schiff's reagent for 1 hour at room temperature in dark. After that, the smears were rinsed in freshly prepared bisulphite reagent (B) and finally washed in distilled water with three changes of 30 seconds duration each. The smears were then dehydrated in alcohol, cleared in xylene and sealed in DPX.

Double distillation methods used

(a) Ethanol

Briskly boiling water baths were used for boiling of the solvent. Re-distillation was done over sodium hydroxide pellets (50 g/dm3). The procedure was repeated if the specific gravity of the distilled ethanol was found above 0.81, which is equivalent to 95% (v/v).

Absolute alcohol (not more than 0.8% water) was super dried by mixing 7 g of Na/litre of absolute alcohol. To it added 30 g (26.9 cm3) ethyl phthalate per litre and refluxed for 2 hours. Distillation was done using long Vortex column. Storage was done under dry conditions.

(b) Methyl alcohol

Distillation was done over sodium hydroxide pellets (25 g/dm3). It was dried and refluxed over barium oxide (BaO). Absolute methyl alcohol was dried by the same procedure as for absolute ethanol.

(c) Acetone

Acetone was put to distillation and only the vapours were collected which came out within the temperature range of 70 to 85°C. It was washed with an equal volume of 5% sodium carbonate and then with saturated calcium chloride solution. The solvent was then dried over anhydrous potassium carbonate or magnesium sulphate. Distillation was done using a fractioning column collecting at 77°C.

(d) Chloroform

After distillation, chloroform was washed twice with concentrated HCl, once with water and twice with 2M NaOH. It was washed again with water until washings were neutral. The chloroform was dried over CaCl2, collecting from 75 to 78°C.

3.7.3.2. Through Acridine Orange Staining

Procedure

A small drop of fresh haemolymph from fixed insect was taken onto a quartz microscopic slide. The same was mixed with an equal volume of acridine orange solution, 0.1 mg/ml in 0.9% NaCl as suggested after Arnold and Hinks (1979).

A cover glass was placed over haemolymph preparation and immediately examined under fluorescence microscope (Carl ZiessTM Model FLUOVAL® 2). Filters were used as per need.

The filter set consisted of two KP-490 multi-layer interference filters of 2 mm thickness each (short wave pass filters) and three B-226, B-228 and B-229 coloured glasses (1 mm, 2 mm and 3 mm thick).

Results

The inclusions of Spherulocytes (Carbohydrate rich substances) gave intense red to orange fluorescence. Nuclei of all haemocytes gave yellowish fluorescence.

3.7.4. Bound Lipids

Acetone-Sudan Black B Method

Fixation

Haemolymph smears were fixed in chilled 10% neutral formalin for 24 hours at 0°C (Pearse, 1961).

Preparation of 10 % neutral formalin

40 % Formaldehyde (BDHTM)	25 ml
Distilled water	75 ml
Magnesium Carbonate	In excess

The mixture was shaken for five minutes and then kept inside a freezing chamber until the mixture attained the ambient temperature (0°C).

Staining

The bound lipids were demonstrated by staining the smears with Acetone-Sudan Black B solution as suggested by Berenbaum (1958).

Preparation of Acetone-Sudan Black B solution

Sudan Black 'B' (BDH)	2.0 g
Absolute acetone	100.0 ml

Sudan Black 'B' was completely dissolved in the measured quantity of acetone and the solution was kept in an incubator at 30°C ± 1°C until it attained the ambient temperature.

Procedure

The fixed and washed smears were immersed in the acetone Sudan Black B solution at 35°C and were incubated for 12 hours at same temperature. After incubation, the smears were washed in xylene with 5-6 changes of two minutes each. The cleared smears were sealed with DPX.

Results: Bound lipids stained black.

3.7.5. Glycogen

Best's Ammonium Carmine Method (Best, 1906) was used for demonstration of glycogen.

Fixation

The haemolymph smears were prepared as usual and fixed in absolute ethanol for one minute at room temperature.

Staining

Preparation of stain

(A) Stock Solution

Carmine powder (BDHTM)	2.0 g
Potassium carbonate (E. MerckTM)	1.0 g
Potassium chloride (E. MerckTM)	5.0 g
Distilled Water	600.0 ml

The ingredients were mixed in distilled water and then solution was gently boiled for five minutes and then cooled. After that, 20 ml of strong ammonium hydroxide solution (sp. gr. 0.91) was stored in a stoppered bottle as stock solution.

(B) Staining solution

Stock solution (A)	15.0 ml
Ammonium hydroxide (sp. gr. 0.91)	12.5 ml
Absolute methanol	12.5 ml

(C) Differentiator

Absolute Ethanol	8.0 ml
Absolute Methanol	4.0 ml
Distilled water	10.0 ml

Procedure

The fixed smears were rinsed with water and kept in staining solution (solution B) for 30 minutes at room temperature. After that, the smears were differentiated in solution (C) for 2-3 seconds. This was followed by quick dehydration in 90% and absolute ethanol. Finally, the smears were cleared in xylene and sealed with DPX. For confirmation of the result, a similar pair of fixed smears was digested for three minutes at room temperature with saliva and similarly stained, differentiated, dehydrated, cleared and sealed.

Results: Nuclei stained dark blue. Cytoplasm stained in red colour.

3.7.6. Alkaline phosphatases

The activity of this enzyme was demonstrated by the method of Denielli (1958). This method incorporates the deposition of calcium phosphate by appropriate incubation in a proper substrate and subsequently calcium ion is exchanged for cobalt, which is rendered visible by incorporation of sulphide ion. By this process, the phosphate salt is converted into respective sulphide salt, which is black. The cobalt sulphide thus formed is deposited in a site positive of the enzyme activity.

Fixation

The haemolymph smears from specimens were fixed for 24 hours in chilled absolute acetone (0°C ± 2°C). The smears were subsequently passed through absolute acetone and then 40 % acetone at room temperature.

Incubation

The smears after subsequent washing through down grades of acetone at room temperature and were incubated at 37°C ± 1°C in following substrate:

Preparation of substrate

2 percent aqueous sodium barbitone (M&BTM)	20 ml
2 percent b-glycerophosphate (BDHTM)	20 ml
2 percent aqueous calcium nitrate (BDHTM)	20 ml
0.8 percent aqueous magnesium chloride (BDHTM)	2 ml
Distilled water	38 ml

The substrate was mixed and prepared as above and the pH was adjusted to 9.3 (optimum pH for alkaline phosphatase activity) with a few drops of N-sodium hydroxide solution. The substrate was then taken in a closed "couplin jar" and kept inside the incubator until the solution attained the ambient temperature (37°C). The smears were then immersed in the substrate and kept again in the incubator for 1 hour. After that, the smears were rinsed for 5 minutes in 2 % calcium nitrate solution prepared as mentioned below:

Preparation of calcium nitrate solution

2 percent aqueous calcium nitrate	100 ml
2 percent aqueous sodium barbitone	0.5 ml

After treatment in the calcium nitrate solution for five minutes, the smears were dipped in cobalt nitrate solution for 2 minutes.

Preparation of calcium nitrate solution

2 percent aqueous cobalt nitrate	100 ml
2 percent aqueous sodium barbitone	0.5 ml

The treatment with cobalt nitrate was followed by rinsing in distilled water with two changes, each of two minutes duration. The smears were then treated for one minute with 1 % aqueous solution of yellow ammonium sulphide. It was followed by gradual dehydration through higher grades of acetone and the cleared in xylene with a brief rinse. The smears were sealed with DPX.

Result

Clean black deposits of cobalt sulphide indicate sites of alkaline phosphatase activity.

3.8. SPECIFIC GRAVITY

Principle

The specific gravity of the haemolymph was determined following Pilmer (1944), who adopted Van Slyke's copper sulphate method for determination of specific gravity of whole blood, plasma and serum. The principle involved in the method is that when blood or serum is dropped into a solution of copper sulphate, it was enclosed in a sac of copper proteinate and remains as a discrete drop without changing for 10 to 15 seconds. During this time, the position of the drop in the solution indicates the gravity relative to the solution.

Procedure

A stock of saturated solution of copper sulphate at room temperature (27-30°C) was prepared with distilled water. From this, solutions of different specific gravities, viz., 1.01. 1.02, 1.03, 1.04, 1.05, 1.06, 1.07, 1.08, 1.09, and 1.10 were prepared with the help of a hygrometer (ZealTM, England). The solutions were taken in small beakers. Droplets of haemolymph taken with a micropipette from the insects were gently put into the beakers containing the copper sulphate solution of known specific gravity as prepared. The specific gravity of the haemolymph in which the droplets of haemolymph remained suspended, i.e., neither sank to bottom nor floated on the surface, was considered at par with the specific gravity of the solution. The experiment was repeated 5 times for each specimen and minimum of five specimens in each category was examined.

3.9. pH VALUE

The larvae were anesthetised with chloroform vapour and the proleg was snipped. Pupae were pierced with a fine needle and one of the legs of the adults was severed off to obtain haemolymph. The haemolymph, which oozed out was tested immediately with E. Merk'sTM narrow range indicator papers ("Tarbskala Zum Special indikaterpapier MerkTM").

Prior to use, the indicator papers were tested for their accuracy with standard buffer solutions of different pH and the experiment was replicated 15 times. Following solutions were used in different ratios to obtain desired pH:

ml1.0 M-HCl	pH	ml1.0 M-NaOH	pH
10	0.83	0	1.64
9	0.90	1	1.84
8	0.92	2	2.16
7	0.94	3	2.90
6	1.02	4	3.42
5	1.10	5	3.76
4	1.14	6	3.98
3	1.18	7	4.07
2	1.29	8	4.18
1	1.44	9	4.27
0	1.64	10	4.35

3.10. BIOCHEMISTRY

3.10.1. Haemolymph cations

For estimation of different cations SYSTRONICSTM Flame Photometer Model MK-II, Type 125 with digital display unit was used. Cations were measured following the method of Dean (1960) for Na+, Wootton (1964) for K+, Stadt (1964) for Ca++ and Neill and Neely (1956) for Mg++ .

3.10.1.1. Estimation of Sodium

Stock Solution of Sodium for calibration of Flame Photometer

For estimation of Na+, analytical grade NaCl (BDH) is used for the preparation of standards. One mole of a compound weighs the formula weight in grams. Using formula weight in calculation, stock solution containing 50 mEq/litre solutions was prepared by dissolving 2.925 g of NaCl in 1 litre of distilled water. By taking various volumes of this stock solution, solutions of following concentrations were prepared.

Working Solutions for calibration of Flame Photo-meter

Concentration in mEq/litre	Volume of Stock Solution in ml	Distilled water added in ml	Net volume in ml
50	100	-	100
45	90	10	100
40	80	20	100
35	70	30	100
30	60	40	100
25	50	50	100
20	40	60	100
15	30	70	100
10	20	80	100
05	10	90	100
04	08	92	100
03	06	94	100
02	04	96	100
01	02	98	100

3.10.1.2. Estimation of Potassium

Stock Solution of Potassium for calibration of Flame Photometer

For estimation of K+, analytical grade KCl (BDHTM) is used for the preparation of standards. One mole of a compound weighs the formula weight in grams. Stock solution containing 50 mEq/litre solutions was prepared by dissolving 3.728 g of KCl in 1 litre of distilled water. By taking various volumes of this stock solution, solutions of following concentrations were prepared:

Working Solutions for calibration of Flame Photo-meter

Concentration in mEq/litre	Volume of Stock Solution in ml	Distilled water added in ml	Net volume in ml
50	100	-	100
45	90	10	100
40	80	20	100
35	70	30	100
30	60	40	100
25	50	50	100
20	40	60	100
15	30	70	100
10	20	80	100
05	10	90	100

3.10.1.3. Estimation of Calcium

Stock Solution of Calcium for calibration of Flame Photometer:

2.497 g of CaCO3 dissolved in 300 ml distilled water and 10 ml of concentrated HCl gave (final volume = 1 litre) 50 mEq/litre stock solution. By taking various volumes of this stock solution, solutions of following concentrations were prepared:

Working Solutions for calibration of Flame Photo-meter

Concentration in mEq/litre	Volume of Stock Solution in ml	Distilled water added in ml	Net volume in ml
50	100	-	100
45	90	10	100
40	80	20	100
35	70	30	100
30	60	40	100
25	50	50	100
20	40	60	100
15	30	70	100
10	20	80	100
05	10	90	100

3.10.1.4. Estimation of Magnesium

Titan Yellow method of Neill and Neely (1956) was followed for the estimation of Magnesium.

Principle

The magnesium in a protein-free filtrate complexed with titan yellow (a dye) in an alkaline medium and the resulting 'red lake' is measured photometrically.

Reagents Required.

2/3 V sulphuric acid

10 per cent sodium tungstate

0.05 per cent polyvinyl alcohol.

0.125 g of polyvinyl alcohol (low viscosity type) was dissolved in distilled water by stirring in 65°C water bath. The solution was diluted to 250 ml with water and filter.

0.05 percent Titan yellow: 0.1 g of titan yellow powder was dissolved in 200 ml distilled water.

4-N Sodium hydroxide: 160 g of sodium hydroxide was dissolved in enough water to make the final volume to 1 litre.

Calcium chloride solution

16.13 mg of calcium chloride (CaCl2.H2O) was dissolved in enough water to make the final volume to 100 ml.

Stock Magnesium standard

40.4 g of magnesium ammonium phosphate ($MgNH_4PO_4.6H_2O$) was dissolved in 0.1 1 litre N HCl.

Working Magnesium Standard

1 ml of the stock standard was diluted to 200 ml with distilled water. This solution contains 0.02 mg magnesium per ml.

Procedure

The serum was separated from the clot as soon as possible to avoid passage of magnesium from red cells into serum. A protein-free filtrate was prepared by adding 1 ml of non-haemolysed serum to 5 ml water, 2 ml of 10 per cent sodium tungstate, and 2 ml of 2/3 N sulphuric acid which was mixed well and centrifuged for 5 minutes at high speed. 5 ml of the filtrate was transferred to a test tube containing 1 ml of water and 1 ml of polyvinyl alcohol. Again the solution was mixed well and 1 ml of titan yellow and 2 ml of 4N sodium hydroxide were added to it. The solution was shaken to mix the reagents well. Simultaneously, a reagent blank using 1 ml of calcium chloride solution in place of serum was prepared and a standard was made using 1 ml of the working standard in place of serum. Both were carried through the entire procedure. Approximately, one minute after adding the sodium hydroxide, reading was taken at 540 mμ, after setting the spectrometer to zero density with water.

Calculation

$$\frac{\text{Density of unknown} - \text{Density of Blank}}{\text{Density of Standard} - \text{Density of Blank}} \times 2 = \text{mg magnesium per 100 ml serum}$$

$$\frac{\text{Mg magnesium per 100 ml} \times 10}{12.2} = \text{mEq. magnesium per litre}$$

3.10.2. Estimation of Trehalose

The estimation of trehalose was done by the method of Wyatt and Kalf (1957) which takes advantage of the exceptional stability of trehalose to both acid and alkali. Five percent homogenate of haemolymph was prepared in double distilled water. The homogenate was centrifuged at 2,800 rpm for 10 minutes and the supernatant was de-proteinised by keeping in boiling water for 10 minutes. The supernatant was then used for the estimation of trehalose.

Samples containing 20 to 75 μl of haemolymph samples were evaporated to dryness in Pyrex tubes and re-dissolved in 0.2 ml of 0.1 N H_2SO_4. The tubes were evaporated with foil and boiled for 10 minutes resulting in the complete hydrolysis of any colorimetrically interfering sucrose or glucose-1-phosphate. Such samples were cooled to room temperature and were made alkaline by adding 0.15 ml of 6N sodium hydroxide. These samples were boiled for 10 minutes at 100°C to destroy all reducing sugars. Later, these samples were chilled at 0°C for two minutes and a 0.2ml aliquot was added to 1.5 ml anthrone reagent (0.2% anthrone in 95% H_2SO_4). Following 30 minutes of incubation at 25°C, the optical density was determined at 590 nm in a spectrophotometer. Peak amplitudes of standard curves were plotted against trehalose concentration. Unknown levels were extrapolated from this curve.

3.10.3. Haemolymph Protein

3.10.3.1. Quantitative estimation of Protein

The estimation of quantitative protein was done following the method of Lowry et. al. (1951).

Preparation of reagents

a) Protein reagent (alkaline copper solution): It was prepared by adding 1.0 ml of a mixture of 0.5 % copper sulphate (w/v) in 1 % sodium tartarate to 50 ml of a mixture of 2 % $Na2CO3$ (w/v) in 0.1 N NaOH.
b) Folin-Ciocalteu reagent: A mixture consisting of 100 g of sodium tungstate, 25 g of sodium molybdate, 700 ml of distilled water, 50 ml of 85 % phosphoric acid and 100 ml of concentrated hydrochloric acid (sp. gr. 1.16) were refluxed gently for 10 hour in 1 to 1.5 litre conical flask. After this, 150 g of lithium sulphate, 50 ml of distilled water and a few drops of bromine water were added. This whole mixture was boiled for 15 minutes without condenser to remove excess bromine. It was then cooled and diluted to one litre with distilled water.
c) Protein standards: The protein standards were prepared from Bovine serum by dissolving 100 µg of it in 1.0 ml distilled water. 1.0 ml, 0.9 ml, 0.7 ml and 0.3 ml of standard solution were mixed with 0.1 ml, 0.3 ml, 0.5 ml and 0.7 ml of distilled water. After reaction, optical density (OD) vs. concentration curve was plotted.

Procedure for colour development:

a) 0.2 to 0.5 ml of haemolymph was taken for the analysis of haemolymph protein and the protein was first precipitated with 10 percent TCA. For collection of the precipitate, the sample was centrifuged at 5,000 rpm for 10 minutes.
b) The supernatant fluid was discarded and the protein precipitate was washed with 5 % TCA.
c) The content was centrifuged again and the supernatant was discarded.
d) The precipitate so obtained was dissolved in 2.0 to 4.0 ml of 0.1 NaOH depending upon the amount and the type of tissue taken.
e) An aliquot of 0.1 ml of the sample was taken and the volume was made up to 0.5 ml with distilled water in each case.
f) 5.0 ml of protein reagent (alkaline copper solution) was added in each sample and after stirring, they were allowed to remain at room temperature for 10 minutes.
g) In each tube, 0.05 ml of Folin's reagent (diluted 1:1) was added and after 30 minutes, the optical density was measured at 750 nm calorimetrically.

Calculation

From the standard curve, the value of one optical density with respect to the protein concentration was calculated and these subtracted with experimental values to gain the protein value. Values were noted in mg/ml of wet tissue.

3.10.3.2. Qualitative Proteins of haemolymph.

Polyacrylamide disc gel electrophoresis was used for detection of qualitative pro-teins.

Collection of Samples for Gel Electrophoresis

I to IV instars

In the early stages of life cycle and in adult stages, two problems, first, less protein content in the haemolymph and second, very little volume of available haemolymph, led to adoption of 2.7 mm diameter gel rods for separation. For collection of sample from I, I and III larval stages of life cycle, one of the 1st pro - legs was punctured with needle and the (larva was washed previously with sterile distilled water and dried by blotting on to filter papers) bleeding proleg was kept directly above the gel-tube. The haemolymph thus accumulated on the top of gel rod by capillary action. The depths of the accumulated haemolymph were measured before adding of tracking dye and tank buffer to calculate volume. The depths were so adjusted that 5 μl of the haemolymph remained for gel electrophoresis.

V instar and pupa

Collections of haemolymph for gel electrophoresis were done in small polyethylene sterile tubes provided with lids for V instar to pupal stages. One hour before collection, tubes were kept in ice to retard melanization of haemolymph during collection and application of the samples to the gel rods. 20 μl of the haemolymph were taken in micropipettes with disposable tips from each of the stage and applied at the origin 7.5% acrylamide gel rods of 5 mm diameter, prepared previously.

Adults

With adults, same 2.7 mm gel - rods were used but haemolymph were collected by cutting off one of the middle legs and pushing the stump into the gel - tube's orifice. The haemolymph were collected by capillary action. The legs were washed with 70 % ethanol and sterile distilled water and dried before cutting of the legs to avoid contamination of gels.

Preparation of Acrylamide gel rods

Separation gel rods were prepared according to method described by Davis (1964) as modified by Datta et al., 1980. The major work was done with 5 % gel. The following solutions were prepared for gel casting:

Solution A

TRIS (2-Amino-2-(Hydroxymethyl)-1,3propaediol) (BDHTM)	36.67 g
1 N HCl (BDHTM, AnalarTM)	48 ml
TEMED (N,N,N',N'-Tertramethylenediamine) SigmaTM (USA)	0.23 ml
Distilled water to	100 ml (pH 8.9)

Solution B

Acrylamide (E. MerckTM, Germany)	30 g
BIS (N,N'-Methylene bissacrylamide)	1 g
Distilled water to	100 ml

Solution C

Ammonium persulphate (E. MerckTM, Germany)	400 mg
Distilled water to	100 ml

Solutions A, B and C were mixed in the ratio of 1: 2: 2 and to it 1 part distilled water was added. The mixture was administered in 8 cm long of 5 mm diameter gel-tubes by the use of a hypodermic syringe. Needle (21 no.; long stem) was used for separation of haemolymph proteins of V to pupal stages. 8 cm long 2.7 mm diameter gel-tubes were used for separation of haemolymph proteins of I, II, III and IV instar larvae and also of adult stages. These tubes were procured from PharmaciaTM Biotechnology International AB, 575182, Uppsala, Sweden and were kept in gel-rod casting stand GRC-16 of same make. Before use, these tubes were cleaned in detergent solution (TeepolTM) and ringed in ample quantity of distilled water. To facilitate better flow of gel-mixture solution, the tubes were finally ringed in 1 % KODAKTM Photoflo® solution and then drained and dried. These cleaning procedures were always repeated before re-use of gel tubes. About 1 cm space was left empty on the top of each gel tube for water layering and sample application. This was accomplished by means of a syringe with thin needle (25 no.). The needle with the attached syringe containing 1 to 2 ml of distilled water was introduced into the top of the gel tube so that the needle rested against the wall of the tube. The tip was kept just above the gel solution and water was allowed to flow very slowly and evenly down the inner wall of the gel tube to form a smooth layer of water on the top of denser gel solution. Such water layer of 3 mm to 4 mm height was found adequate.

Photo-polymerisation was affected by placing the stand containing gel tubes directly under a daylight fluorescent tube (6300°K) of 40 watts; the fluorescent tube was so positioned that it was about 5 to 8 cm above the tips of the gel tubes. A sample of same gel solution in 50 ml beaker was also kept and examined for photo-polymerisation after about 30 minutes. When the sample was found photo-polymerised (usually 30 to 45 minutes were found sufficient), then only the stand containing gel tubes were disturbed.

The photo-polymerised gel tubes were kept in small beakers containing diluted tank buffer so that their lower ends after removal from stoppers were always dipped in this buffer (Fig. 3.7). The water layer on the top was removed just prior to application of samples of haemolymph. For this, the tubes were inverted and the tip kept on a filter paper so that the water and any inhibited gel solution could flow down the walls of the tubes to the open ends and absorbed in the filter paper leaving tip clean and dry. The tubes were inserted into the grommets of upper buffer reservoir of PharmaciaTM Fine Chemical's Gel Electrophoresis apparatus Model GE/2 after wetting the lower end of the tubes with diluted tank buffer solution.

A. Tank buffer stock solution

TRIS	6.00 g
Glycine	28.60 g
Distilled water to	1 l

B. Working tank buffer solution

Stock solution (A)	100.00 ml
Distilled water	900.00 ml
pH	8.3

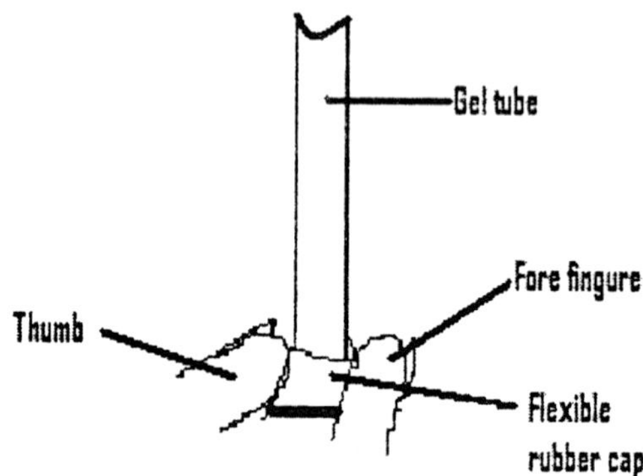

Fig. 3.7. A diagrammatic representation of "removal" of gel tube from flexible cap

The following tank buffer solution was used:

The lower reservoir was kept ready filled with above diluted (1:10 = Stock tank buffer: water) tank buffer and its cooling coil was kept connected through a pump to a cryostat containing water at about 0°C - 4°C (in contact with ice). The temperature of the tank was thus maintained at about 5°C ± 2°C throughout the experiment.

Electrophoresis

The samples (20 µl for 5 mm diameter gel rod and 5 µl for 2.7 mm diameter gel rods) were applied to the origin of gel rods with the help of micropipettes having disposable tips of directly inserting punctured prolegs/adult middle thoracic legs. After application of the samples (5 replications for each stage of the insect), 1 µl of 0.001 % Bromophenol blue solution in saturated sucrose solution were added as marker dye. The remaining space was then filled fully and carefully with buffer solution so as not to disturb the denser sample on the top of the gel rods. The upper buffer reservoir was then filled with cooled tank buffer with the help of pump provided in the apparatus. The buffer was delivered at the floor level of the upper buffer reservoir by this pump facilitating minimum disturbance to the already filled up gel tubes with buffer after application of sample and marker dye. The upper buffer reservoir was filled until the buffer's level becomes 1 cm above the tips of the gel tube.

The upper reservoir's platinum wire was then connected to negative of power supply (cathode) and lower platinum wire was connected to positive of the power supply (anode). The electrophoresis was then towards the anode. The temperature of the tank buffer was maintained at 5°C ± 2°C by constantly circulating ice-cold water from the cryostat.

The power supply unit used was Pharmacia'sTM model EPS-500/400. At the beginning, the current was applied @ one mA/tube. After migration of the marker dye to about 1 cm, the current was increased to three mA/tube. The time taken by the marker dye to reach about 0.5 cm behind the lower end of the gel rod were then about one hour.

When marker dye reached at the destined level, power was switched off and the gel rods were rimmed out with the help of blunt needle of thin gauge (25 no.) fitted in a syringe containing distilled water. The distance travelled by the marker dye was recorded and then the gel rods were stained in the following protein stain for 5 minutes at room temperature:

Protein staining solution

Amido Black Powder (BDH, Poole)	700.00 mg
7 % acetic acid to	100.00 ml

At the end of 5 minutes, the stain was decanted from the test tubes, in which staining were carried out separately for each rod, and excess stain was removed by several changes of 7 % acetic acid through next 48 hours. When no stain was found coming out from the gel rods, these were preserved in same 7 % acetic acid for data recording.

Analysis of the electrophorogram: Relative mobility

The term relative mobility refers to the movement of a type of polypeptide through a gel relative to other protein bands in the gel. Relative mobility is the distance migrated by a band divided by the distance migrated by the dye front. Absolute mobility would be the distance travelled in a particular time. Relative mobility is useful because it can be used to compare the migration of a protein from gel to gel, regardless of the physical length of the gel or duration of electrophoresis. A common abbreviation for relative mobility is Rf, for "retention factor."

Each stained gel, kept within a test tube of narrow diameter, was placed against bright but diffused light and the distance of each band from the origin was recorded. From the mean values obtained from the analysis of the five gels, the relative mobility values were calculated and plotted on a normal graph.

$$\text{Relative mobility (Rf)} = \frac{\text{Distance traversed by a band}}{\text{Distance traversed by the marker dye}}$$

For broad bands, measurements were made for both anterior and posterior ends and the average was taken as Rf value. The intensities of staining were recorded through visual estimates (Datta et al., 1980).

3.10.4. Quantitative estimation of total free amino acids.

The quantitative estimation of total free amino acids was done by the modified procedure based on Moore and Stein (1948).

Reagents

a) Glycine solution: 200 µg/ml solution was prepared in distilled water.
b) Ninhydrin reagent: 1.0 g of ninhydrin in 20 ml of absolute ethyl alcohol and 0.04 g of stannous chloride in 25 ml of citrate buffer of pH 5.0 was dissolved.
c) Citrate buffer: It was prepared by mixing 20 ml of 0.05 g citric acid in 50 ml distilled water and 29.5 ml of 2.94 g sodium citrate in 100 ml of distilled water.

Procedure

a) 50 µl of haemolymph from each age group of insects (pooled from 10 larvae of first instar) was added separately in 2.0 ml of 96% ethyl alcohol. The mixture was centrifuged at 5000 rpm for 5 minute and the supernatants were collected in test tubes.
b) Supernatants collected from each insect were divided in five test tubes @ of 0.1 ml/tube. To it, 0.1 ml distilled water and 2.0 ml of ninhydrin reagent were added separately. Thus, total volume came to 2.2 ml. The test tubes were kept in boiling water bath for exactly 15 minutes. Cooled and added 2.0 ml of 50 % ethyl alcohol.
c) Violet blue colour developed in the solution, the intensity of which depended on the concentration of amino acids present in the mixture. The concentration was measured in the terms of optical density by spectrophotometer at 575 nm against tube containing standard solution (control).
d) Standard solution was prepared by adding 2.00 ml Ninhydrin reagent in 0.2 ml distilled water.

Standardisation of amino acid calibration curve

50 µl, 100 µl, 150 µl and 200 µl Glycine standard was mixed with 50 µ, 100 µl, 150 µl and 200 µl Ninhydrin reagent, respectively. After colour development, optical density (amino acid Gly. ine concentration) was noted for standard calibration curves.

Calculation

From standard curve, the amino acid concentration was calculated after calculating the value of one optical density. The tissue amino acid concentrations are written as µg/ml of wet tissue.

3.10.5. Estimation of Uric acid

For the estimation of Uric acid, the procedure as described in Hawk's Physiological Chemistry (1979) was followed.

[1] 0.5 ml of haemolymph was diluted to 5 ml with water. It was mixed with 1 ml each of 10% w/v Sodium Tungstate and 0.67 N H2SO4. The mixture was shaken well, diluted to 10 ml with water and centrifuged. To 5 ml of this protein-free supernatant were added 2 ml of 12 % w/v NACN and 2 ml of 50 % w/v urea, followed by 1 ml of Uric acid reagents (see below). After incubation at 38°C for 1 hour, the volume was made up to 25 ml and the

colour density was measured in SimadzuTM Spectrophotometer at 608 (red) mμ.

[2] Following samples and reagents were processed for reading in the spectrophotometer:

1:10 ml protein free blood filtrate	1.0 ml
Urea-cyanide solution (see below)	2.0 ml
Uric acid reagent	0.8 ml

The mixture of above solutions were mixed by lateral shaking and allowed to stand for 20 min, diluted to the 10 ml mark with distilled water, mixed well by inversion and read in the spectrophotometer within 20 minutes or so against the blank at 0°C.

Uric acid Reagent

To 100 gm Na-Tungstate, 150 ml of water and 32 ml of 85% H3PO4 was added to make "Uric acid Reagent". The solution was boiled gently for 1 hour with a funnel in neck after adding quartz. Standard solution of uric acid was prepared by adding 2.5 mg of 100 ml of above solution.

Urea-cyanide Solution

75 gm of NaCN was taken in 2-litre beaker and dissolved in 700 ml distilled water. To this solution, 300 gm urea was added. It was stirred and then 4-5 gm CaO was added, stirring was continued for at least 10 minutes. The mixture was allowed to stand for overnight. Next day, 2 gm Lithium oxalate was added. Shaken well for 15 minutes and filtered.

Standard curve of uric acid in ShimadzuTM spectrophotometer

(608 nm filter):

Sl. No.	Amount in mg	Reading	Sl. No.	Amount in mg	Reading
1.	0.005	66	12.	0.060	340
2.	0.010	96	13.	0.065	-----
3.	0.015	129	14.	0.070	400
4.	0.020	170	15.	0.080	425
5.	0.025	216	16.	0.090	435
6.	0.0.30	234	17.	0.100	400
7.	0.035	220	18.	0.150	556
8.	0.040	264	19.	0.200	-----
9.	0.045	298	20.	0.300	750
10.	0.0.50	305	21.	0.400	725
11.	0.055	350	22.	0.500	725

3.10.6. Estimation of Total Cholesterol

For the estimation of Cholesterol, the procedure as described in Hawk's Physiological Chemistry (1979) was followed. Chilled haemolymph were homogenized on CHCl3/MeOH (2:1 v/v) until complete extraction. Extract was centrifuged and supernatant made up to

25 ml. This lipid extract (1 ml) was pipetted into a series of glass stoppered tubes evaporated to dryness in vacuum and re-dissolved in ethyl-acetate-absolute alcohol (1 ml 1:1), FeCl3 (2.5 mg, 0.1 %, w/v in ethyl acetate) was added to each tube and mixed. Concentrated H2SO4 (2 ml sp. gr. 1.84) was then added. The contents mixed, cooled and optical density was measured at 550 mμ against a reagent blank (SimadzuTM Spectrophotometer). A standard cholesterol solution was also run simultaneously.

3.11. STATISTICAL ANALYSIS

For statistical analysis, MS EXCEL 10Ò software were used on a Pentium IVÒ 2.4 MHz computer. Means and standard errors were determined for each parameter. Student's 't' test was applied to test for the difference between means, of each parameter. Variability was illustrated through use of co-efficient of variance percent. Comparison between non-diapausing and diapausing generations was done for all the observations in this work. ANOVA was used in some cases where differences of mean in more than two populations were needed to be determined.

4

Observations

4.1. General Histology

In the present study, five types of haemocytes namely Prohaemocytes (PRs), Oenocytoids (OEs), Plasmatocytes (PLs), Spherulocytes (SPs) and Granulocytes (GRs) were identified in tropical tasar silkworm, *Antheraea mylitta* Drury.

4.1.1. Prohaemocytes

Prohaemocytes were usually round in shape and relatively smaller than other haemocyte types. They were characterised by light purple staining of cytoplasm by Giemsa stain and a low concentration of intracellular granules in their cytoplasm (Fig. 4.1). PRs measured 8.500 to 18.063 µm (mean 11.794 ± 1.138 µm) in diameter and their nuclei, which were deeply stained in deep purple colour in Giemsa stain, were relatively large in comparison to the cell size and occupied most of the interior volume of the cell. The cell area ranged in between 56.768 to 256.342 µm2 (mean 117.527 ± 23.582 µm2) and their volume ranged in between 321.685 to 3086.362 µm3 (mean 1068.113 ± 325.688 µm3). The nucleus was normally spherical as all measurements of nucleus axes were almost equal;

Fig. 4.1 : Prohaemocyte under light microscope (Giemsa stain)

longer axis ranging from 6.375 to 14.875 µm (mean 10.519 ± 0.931 µm) and smaller axis ranging from 5.525 to 14.875 µm (mean 9.648 ± 1.099 µm) in length. The calculated area

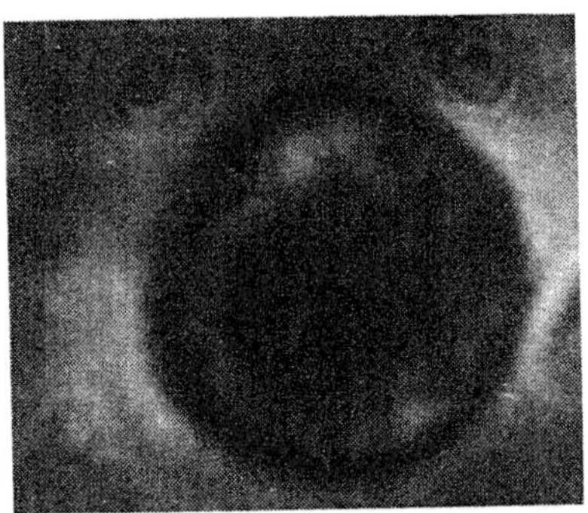

Fig. 4.2 : Prohaemocyte under phase contrast optics

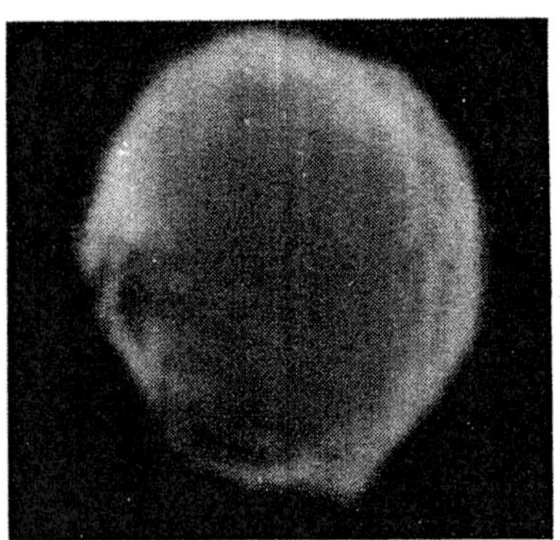

Fig. 4.3 : SEM photomicrograph of Prohaemocyte.

of nucleus ranged from 31.932 ± 173.852 µm2 (mean 86.053 ± 17.290 µm2). The mean ratio of nucleus vs. cell area was 0.732 ± 0.027, ranging in between 0.563 to 0.810 (Table 4.1). Fixed unstained cells examined under phase contrast optics (Fig. 4.2) showed bluish white cytoplasm with deep brown coloured nucleus. Under scanning electron microscope, spherical surface was evident; however, it had several depressions (Fig 4.3).

4.1.2. Oenocytoids

Oenocytoid cells were characterised by their spherical or ellipsoidal shape, large amount of cytoplasm and small nuclei. The mean length of their large axis was 21.548 ± 1.322 µm (ranged between 14.875 to 25.500 µm) and small axis was 19.975 ± 1.344 (ranged between 14.875 to 25.500 µm). Their cytoplasm stained little yellowish red in cells stained in Giemsa stain (Fig. 4.4, Table 1). The cell area ranged in between 173.852 to 510.911 µm2 (mean 349.051 ± 42.121 µm2) and cell volume ranged in between 1724.028 to 8685.482 µm3 (mean 4951.227 ± 858.025 µm3).

Fig. 4.4 : Oenocytoid under light microscope (Giemsa stain)

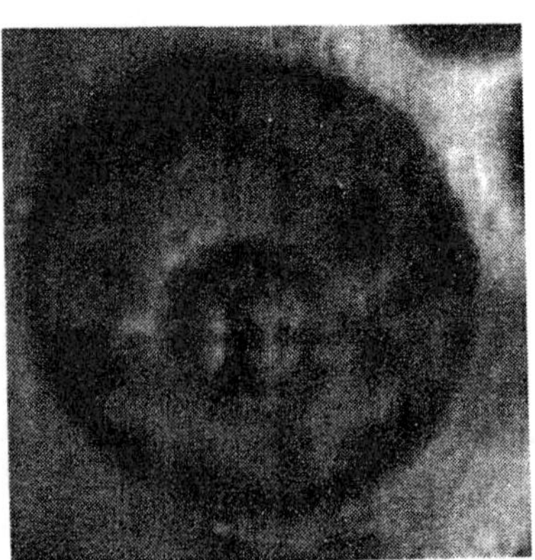

Fig. 4.5 : Oenocytoid under phase contrast optics.

The nucleus was comparatively smaller (area ranging from 14.192 to 31.932; mean area 20.224 ± 2.790 μm2), spherical and its diameter in different cells ranged from 4.250 to 6.375 μm (mean 5.100 ± 0.366 μm). The ratio of nuclear area vs. cellular area ranged in between 0.028 to 0.100 (mean 0.058 ± 0.007). The cytoplasm stained variedly with Giemsa stain in crimson colour. Under phase contrast microscope, their spherical shape, relatively large amount of cytoplasm and small nuclei were evident (Fig. 4.5). Under scanning electron microscope, Oenocytoids were either spherical or ellipsoidal in shape. A clean smooth surface was very much evident under gold coating as shown in Fig. 4.6.

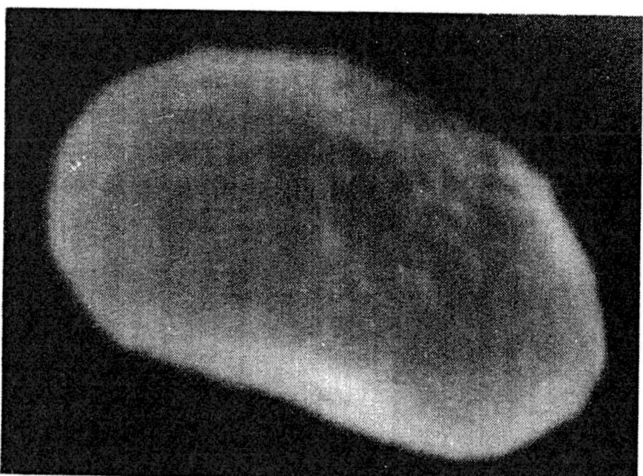

Fig. 4.6 : SEM photomicrograph of Oenocytoid.

4.1.3. Plasmatocytes

Plasmatocytes were round to elliptical in shape (Fig 4.7). However, several fusiform plasmatocytes (PLFs) were also observed. The longer axis of elliptical forms ranged from 12.750 to 23.375 μm (mean 18.913 ± 1.130 μm) and in fusiform cells it varied in between, 17.000 to 34.000 μm (mean 23.375 ± 2.138 μm). The smaller axis measured at cell's broadest

Fig. 4.7. Plasmatocyte under light microscope (Giemsa stain)

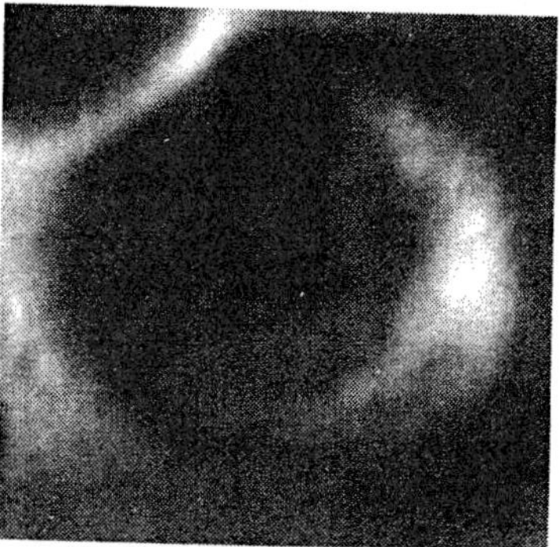

Fig. 4.8 : Plasmatocyte under phase contrast optics.

point ranged from 10.625 to 19.125 µm (mean 15.406 ± 0.903 µm) in elliptical forms, while in fusiform cells this axis was 8.500 to 17.000 µm (mean 12.538 ± 1.026 µm). The calculated area of elliptical cells ranged from 106.440 to 319.319 µm2 (mean 234.877 ± 25.167 µm2) (Fig 4.8) and of fusiform from 113.536 to 397.375 µm2 (mean 237.715 ± 34.354 µm2). Similarly, the volume of elliptical cells ranged from 753.948 to 4071.320 µm3 (mean 2519.192 ± 381.475 µm3) and of fusiform from 643.369 to 3940.635 µm3 (mean 2150.260 ± 441.872 µm3). The area of nucleus in elliptical cells ranged from 56.768 to 106.440 µm2 (mean 78.765 ± 5.513 µm2) and fusiform cells from 56.768 to 283.839 µm2 (mean 150.009 ± 24.722 µm2). The mean ratio of nucleus vs. cell area was double in fusiform cells (mean 0.631 ± 0.076, range 0.333 - 1.125) than that in elliptical cells (mean 0.335 ± 0.049; range 0.200 - 0.625). The size of normal Plasmatocytes was less variable as compared with its fusiform form, which has large variation as evident from calculated co-efficient of variance % (Table 4.1). A Plasmatocyte, as seen under phase contrast optics, is shown in Fig. 4.8. The Plasmatocyte having fusiform shape with tapering ends were very clear under SEM (Fig. 4.9).

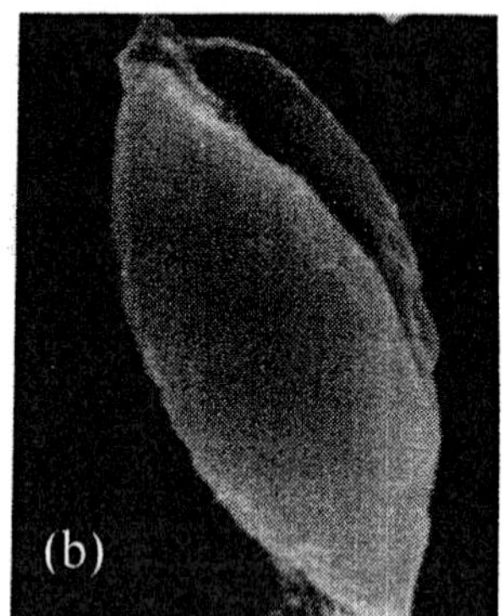

Fig. 4.9 : Photomicrograph of Fusiform Plasmatocyte (a) under optical microscope (b) under scanning electron microscope.

4.1.4. Spherulocytes

Spherulocytes were round to oval in shape. The long axis of SPs ranged from 12.750 to 29.75 µm (mean 20.740 ± 1.610 µm) and small axis ranged from 4.250 to 12.750 µm (mean 9.350 ± 1.117 µm). The cell area ranged from 42.576 to 212.879 µm2 (mean 117.793 ± 19.827 µm2) and volume ranged from 120.632 to 1809.475 µm3 (mean 848.443 ± 207.978 µm3). The shape of the nucleus was irregular and so measurement of the same was not done (Table 4.1.). The ratio of nucleus to cell was not calculated as the nuclei of SPs were irregular. However, nuclei occupied a considerable

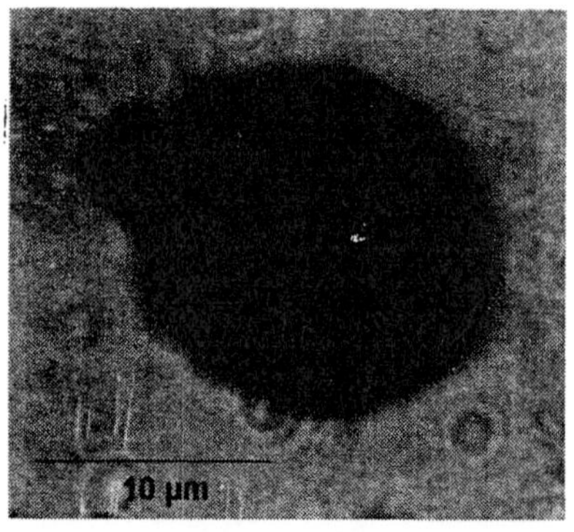

Fig. 4.10. Spherulocyte under light microscope (Giemsa stain)

olume of cell. The cells stained deep purple; nucleus was far deeper than cytoplasm. The intensity of the stain of nucleus sometimes reached very near to blackish tinge due to its high affinity towards Giemsa stain (Fig. 4.10). Under phase contrast optics, unstained spherulocytes had light orange irregular nucleus and cytoplasm had bluish white colour. The cell boundaries were not clear (Fig. 4.11.). Under scanning electron microscope, the gold-coated surface of the Spherulocytes showed undulation. A number of blister-like elevations were also seen (Fig. 4.12).

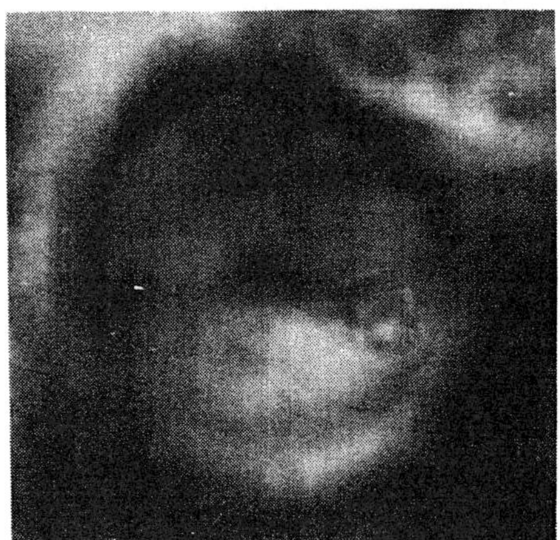

Fig. 4.11 : Plasmatocyte under phase contrast optics

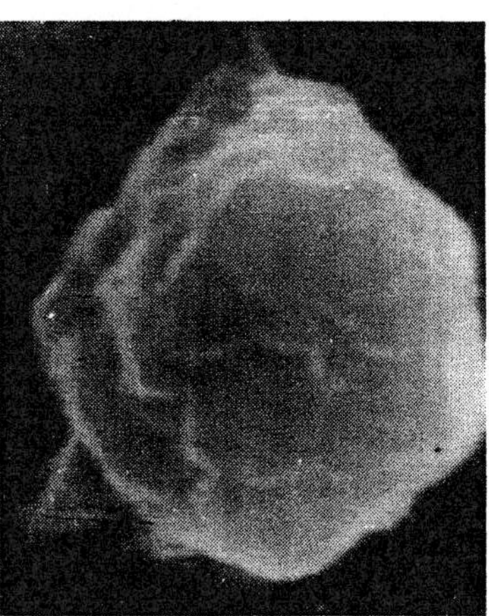

Fig. 4.12 : SEM photomicrograph of Spherulocyte.

4.1.5. Granulocytes

Granulocytes were recognised as spherical or oval shaped haemocytes, which varied considerably in size. The cytoplasm had numerous vacuolar bodies distributed around the nucleus. The nucleus stained deep purple and was irregular in shape like Spherulocytes so ratio of cell to nucleus was not calculated. The long axis of GRs ranged in between 12.750 to 25.500 μm (mean 19.975 ± 1.344 μm). The short axis length ranged between 8.500 to 23.375 μm (mean 15.088 ± 1.473 μm). The cell area ranged from 85.152 to 468.335 μm2 (mean 247.650 ± 39.271 μm2). The cell volume ranged from 482.527 to 7298.218 μm3 (mean 2797.650 ± 725.581 μm3) - (Table 4.1).

Fig. 4.13 : Granulocyte under light microscope (Giemsa stain)

Densely dispersed cytoplasmic inclusions clearly distinguish these haemocytes from other types. Under phase contrast optics, numerous inclusions were clearly visible and had irregular outer boundary and shape. The cytoplasmic inclusions almost obscured the visibility of nucleus, which was also irregular in shape (Fig. 4.14). Under scanning electron microscope, highly irregular surface was very much evident. The surface had many elevations and depressions and the cell itself was irregular in shape (Fig. 4.15).

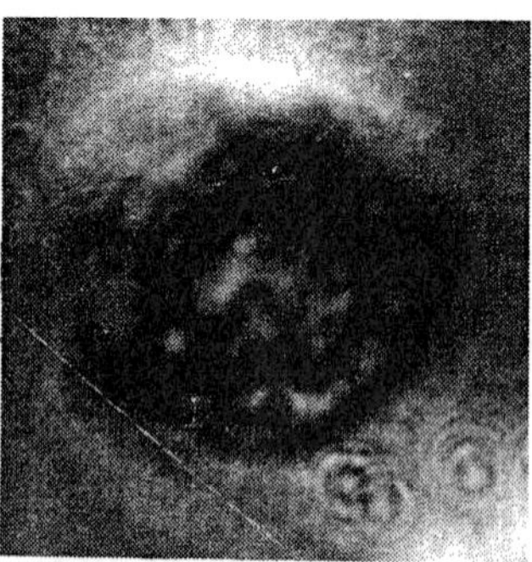

Fig. 4.14 : Granulocyte under phase contrast optics.

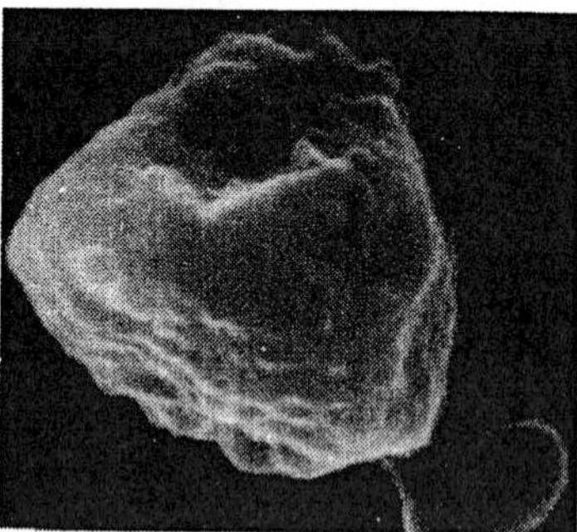

Fig. 4.15 : SEM Photomicrograph of Granulocyte.

4.1.6. Mitotic cells and degenerated cells

Mitotic cells and degenerated cells were counted for recording of their population in the haemolymph of various stages of life-cycle in non-diapausing and diapausing generations of the insect under study. A few examples of mitotic cells and degenerated cells, photographed during the course of investigation are shown in Figs. 4.16 and 4.17.

ANOVA and critical difference (CD) at $p < 0.05$ and $p < 0.01$ was calculated (except where the nucleus was irregular in shape) for testing whether means for each haemocyte differs from each other in cell area, cell volume and nucleus area (Table 4.2). Each type of the haemocyte was found significantly different from all other types in their means of area and volume of the cells as well as in the volume of nucleus. Co-efficient of Variance percentage was also calculated for the above mentioned parameters.

Table 4.1. Major dimension, area of cells and nucleus and volume of haemocytes

Haemocyte type	Parameters	Calculated from micrometer reading								Ratio of nucleu to cell
		Cell					Nucleus			
		a* (µm)	b* (µm)	c* (µm)	Area (μm^2)	Volume (μm^3)	a* (µm)	b* (µm)	Area (μm^2)	
1. Prohaemocyte (PRs)										
	MINIMUM	8.500	8.500	8.500	56.768	321.685	6.375	5.525	31.932	0.56
	MAXIMUM	18.063	18.063	18.063	256.342	3086.362	14.875	14.875	173.852	0.81
	MEAN	11.794	11.794	11.794	117.527	1068.113	10.519	9.648	86.053	0.73
	STD. ERROR	1.138	1.138	1.138	23.582	325.688	0.931	1.099	17.290	0.02
	CV%	28.944	28.944	28.944	60.196	91.476	26.550	34.189	60.277	11.15
	n = 10									
2. Oenocytoids (OEs)										
a	MINIMUM	14.875	14.875	14.875	173.852	1724.028	4.250	4.250	14.192	0.02
	MAXIMUM	25.500	25.500	25.500	510.911	8685.482	6.375	6 375	31.932	0.10
	MEAN	21.548	19.975	19.975	349.051	4951.227	5.100	4.888	20.224	0.05
	STD. ERROR	1.322	1.344	1.344	42.121	858.025	0.342	0.366	2.790	0.00
b	CV%	18.410	20.185	20.185	36.202	51.989	21.002	21.517	41.393	38.67
	n = 10									
3. Plasmatocytes (PL)										
	MINIMUM	12.750	10.625	10.625	106.440	753.948	8.500	8.500	56.768	0.20
	MAXIMUM	23.375	19.125	19.125	319.319	4071.320	12.750	10.625	106.440	0.62
	MEAN	18.913	15.406	15.406	234.877	2519.192	11.475	8.713	78.765	0.33
	STD. ERROR	1.130	0.903	0.903	25.167	381.475	0.684	0.224	5.513	0.04
	CV%	17.923	17.583	17.583	32.145	45.428	17.891	7.713	21.000	44.04
	n = 10									

Table 4.1 (Continued)

Haemocyte type	Parameters	Calculated from micrometer reading								Ratio of nucleu to cell
		Cell					Nucleus			
		a* (μm)	b* (μm)	c* (μm)	Area (μm^2)	Volume (μm^3)	a* (μm)	b* (μm)	Area (μm^2)	
4. Fusiform Plasmatocytes (PLFs)										
	MINIMUM	17.000	8.500	8.500	113.536	643.369	8.500	7.650	56.768	0.33
	MAXIMUM	34.000	17.000	17.000	397.375	3940.635	25.500	17.000	283.839	1.12
	MEAN	23.375	12.538	12.538	237.715	2150.260	15.640	11.815	150.009	0.63
	STD. ERROR	2.138	1.026	1.026	34.354	441.872	1.768	1.200	24.722	0.07
	CV%	27.441	24.562	24.562	43.355	61.649	33.911	30.466	49.442	36.34
	n = 10									
5. Spherulocyte (SPs)										
	MINIMUM	12.750	4.250	4.250	42.576	120.632	Irregular	Irregular	Irregular	Irregular
	MAXIMUM	29.750	12.750	12.750	212.879	1809.475	Irregular	Irregular	Irregular	Irregular
	MEAN	20.740	9.350	9.350	117.793	848.443	Irregular	Irregular	Irregular	Irregular
	STD. ERROR	1.610	1.117	1.117	19.827	207.978	Irregular	Irregular	Irregular	Irregular
	CV%	19.422	35.855	35.855	50.497	73.539	Irregular	Irregular	Irregular	Irregular
	n = 10									
6. Granulocyte (GRs)										
	MINIMUM	12.750	8.500	8.500	85.152	482.527	Irregular	Irregular	Irregular	Irregular
	MAXIMUM	25.500	23.375	23.375	468.335	7298.218	Irregular	Irregular	Irregular	Irregular
	MEAN	19.975	15.088	15.088	247.650	2797.650	Irregular	Irregular	Irregular	Irregular
	STD. ERROR	1.344	1.473	1.473	39.271	725.581	Irregular	Irregular	Irregular	Irregular
	CV%	20.185	29.282	29.282	47.573	77.806	Irregular	Irregular	Irregular	Irregular
	n = 10									

a = greater radius; b = smaller radius; b = c. = 22/7 or = 3 1428571. 11 ocular micrometer readings = 4.25 stage micrometer scale graduated in 1 μm
Area of an ellipse = x a x b. Volume of an ellipsoid = (4/3) x () x a x b x c

Table 4.2. Analysis of variance (ANOVA) of cell area & volume and of nuclear volume of haemocytes of *Antheraea mylitta*

Source of Variation	SS	df	MSS	F	P-value	F crit	CD p < 0.01	CD p < 0.05
Cell Area								
Between Groups	388613.5307	5	77722.7061	8.5233	0.000005	2.3861	99.3334	70.2508
Within Groups	492419.2728	54	9118.8754					
Total	881032.8035	59						
Cell Volume								
Between Groups	7022171563.8637	5	1404434312.7727	8.3880	0.000006	2.3861	13460.0744	9519.2701
Within Groups	9041474510.2712	54	167434713.1532					
Total	16063646074.1349	59						
Nuclear Volume								
Between Groups	84560.2259	3	28186.7420	13.2099	0.000006	2.8663	48.0504	33.9823
Within Groups	76815.1670	36	2133.7546					
Total	161375.3929	39						

4.2. Total Haemocyte Counts (THCs)

Details of mean THC of different stages of both the generations are presented in Tables 4.3 and 4.4.

4.2.1. THC of NDD and DD generations

In non-diapausing generation, an increasing trend in THC was recorded from I instar to middle aged V instar larvae of both the sexes (Table 4.3). The mean values increased from 43,620.36 ± 210.46/mm3 (I-NDD) to VMF-NDD (75,234.66 ± 427.94/mm3) and VMM-NDD (77,526.36 ± 221.10/mm3). This trend was then reversed and started decreasing up to SLF-NDD (56,627.20 ± 117.58/mm3) and SLM-NDD (57,054.12±80.82/mm3). However, an increase was noticed in PPF-NDD (73,290.82 ± 136.33/mm3) and PPM-NDD (78,265.44 ± 492.80/mm3). Thereafter, the level of THC went down abruptly from MPF-NDD (30,510.34 ± 39.19/mm3) and MPM-NDD (32,477.38 ± 37.82/ mm3) to AF-NDD (8,756.52 ±20.38/ mm3) and AM-NDD (8,533.80 ± 46.18/mm3)

Table 4.3. Total haemocyte counts (mean ± SE) of NDD generation (numbers/mm3)

SN	Age	Mean	±	SE
1	I-NDD	43,620.36	±	210.46
2	II-NDD	46,498.08	±	39.63
3	III-NDD	57,996.10	±	91.43
4	IV-NDD	55,955.10	±	84.14
5	VEF-NDD	69,167.96	±	405.15
6	VEM-NDD	60,824.26	±	1,617.94
7	VMF-NDD	75,234.66	±	427.94
8	VMM-NDD	77,526.36	±	221.10
9	VLF-NDD	60,904.80	±	84.80
10	VLM-NDD	62,259.10	±	21.08
11	SLF-NDD	56,627.20	±	117.58
12	SLM-NDD	57,054.12	±	80.82
13	PPF-NDD	73,290.82	±	136.33
14	PPM-NDD	78,265.44	±	492.80
15	MPF-NDD	30,510.34	±	39.19
16	MPM-NDD	32,477.38	±	37.82
17	LPF-NDD	11,543.86	±	139.86
18	LPM-NDD	11,528.32	±	44.43
19	AF-NDD	8,756.52	±	20.38
20	AM-NDD	8,533.80	±	46.18

n = 50 (Abbreviations used are as per section 3.1.2.)

In Diapause destined (DD) life forms also, it was observed that the level of THC gradually increased from I instar (I-DD) larvae (43,370.30 ± 136.68/mm3) to 75,097.96 ± 79.77/mm3 (VMM-DD) and 70,967.98 ± 78.19/mm3 (VMF-DD). These values went down in spinning larvae of both the sexes i.e., 55,456.72 ± 45.57/mm3 (SLF-DD) and 56,073.16 ± 82.83/mm3 (SLM-DD). A rise in THC level was again recorded in both the prepupal forms DD females (75,846.08 ± 94.14/mm3) and males (75,678 ± 97.06/mm3). Thereafter, THC went down abruptly in mid pupal form (MPF-DD: 16,646.54 ± 119.06/mm3 & MPM-DD: 21,283.86 ± 227.79/mm3) and this trend continued until emergence of adults (AF-DD: 8,740.84 ± 21.37/mm3 & AM-DD: 8,536.56 ± 41.42/mm3) - (Table 4.4.).

Table 4.4. Total haemocyte counts (mean ± SE) of DD generation (numbers/mm3)

SN	Age	Mean	±	SE
1	I-DD	43,370.30	±	136.68
2	II-DD	45,542.40	±	39.19
3	III-DD	55,536.56	±	41.42
4	IV-DD	54,990.32	±	87.03
5	VEF-DD	67,363.22	±	222.99
6	VEM-DD	62,876.82	±	1,214.80
7	VMF-DD	70,967.98	±	78.19
8	VMM-DD	75,097.96	±	79.77
9	VLF-DD	69,894.90	±	80.13
10	VLM-DD	60,085.24	±	73.35
11	SLF-DD	55,456.72	±	45.57
12	SLM-DD	56,073.16	±	82.83
13	PPF-DD	75,846.08	±	94.14
14	PPM-DD	75,678.84	±	97.06
15	MPF-DD	16,646.54	±	119.06
16	MPM-DD	21,283.86	±	227.79
17	LPF-DD	11,656.94	±	411.31
18	LPM-DD	11,084.82	±	78.38
19	AF-DD	8,740.84	±	21.37
20	AM-DD	8,536.56	±	41.42

n = 50 (Abbreviations used are as per section 3.1.2.)

The THC values of NDD and DD larval instars, pupal forms and adults were put for one way ANOVA. THC values were significantly different ($p < 0.05$) in I-DD & I-NDD larvae. Similarly, the mean difference between AM-DD & AF-DD and AF-NDD & AM-NDD were also significant ($p < 0.05$) - (Table 4.5).

From ANOVA, the differences in the mean THC values were observed to be significant at $p < 0.01$ in most of the life stages studied, except in between VMM-DD & VMF-NDD, VLF-NDD & VEM-NDD, SLF-DD & III-NDD, SLM-DD & IV-DD, PPM-DD & PPF-DD, LPF-NDD & LPF-DD, LPM-NDD & LPF-DD, LPM-NDD & LPFNDD, AF-NDD & AF-DD, AM-DD & AF-DD, AM-NDD & AF-DD and lastly AM-NDD & AM-DD, where difference was observed to be non-significant (Table 4.3, 4.4 and 4.5).

Table 4.5 : Matrix showing significant mean among stages/age of both NDD and DD generations of *A. mylitta*.

Stage/Age	I-DD	I-NDD	II-DD	II-NDD	III-DD	III-NDD	IV-DD	IV-NDD	VEF-DD	VEF-ND	VEM-D	VEM-N	VMF-D	VMF-N	VMM-D	VMM-N	VLF-DD	VLF-ND	VLM-D	VLM-N	SLF-DD	SLF-ND	SLM-D	SLM-N	PPF-DD	PPF-ND	PPM-D	PPM-N	MPF-D	MPF-N	MPM-D	MPM-N	LPF-DD	LPF-ND	LPM-D	LPM-N	AF-DD	AF-NDD	
I-DD																																							
I-NDD	*																																						
II-DD	**	**																																					
II-NDD	**	**	**																																				
III-DD	**	**	**	**																																			
III-NDD	**	**	**	**	**																																		
IV-DD	**	**	**	**	**	**																																	
IV-NDD	**	**	**	**	**	**	**																																
VEF-DD	**	**	**	**	**	**	**	**																															
VEF-NDD	**	**	**	**	**	**	**	**	**																														
VEM-DD	**	**	**	**	**	**	**	**	**	**																													
VEM-NDD	**	**	**	**	**	**	**	**	**	**	**																												
VMF-DD	**	**	**	**	**	**	**	**	**	**	**	**																											
VMF-NDD	**	**	**	**	**	**	**	**	**	**	**	**	**																										
VMM-DD	**	**	**	**	**	**	**	**	**	**	**	**	**	NS																									
VMM-NDD	**	**	**	**	**	**	**	**	**	**	**	**	**	**	**																								
VLF-DD	**	**	**	**	**	**	**	**	**	**	**	**	**	**	**	**																							
VLF-NDD	**	**	**	**	**	**	**	**	**	**	**	NS	**	**	**	**	**																						
VLM-DD	**	**	**	**	**	**	**	**	**	**	**	**	**	**	**	**	**	**																					
VLM-NDD	**	**	**	**	**	**	**	**	**	**	**	**	**	**	**	**	**	**	**																				
SLF-DD	**	**	**	**	**	NS	**	**	**	**	**	**	**	**	**	**	**	**	**	**																			
SLF-NDD	**	**	**	**	**	**	**	**	**	**	**	**	**	**	**	**	**	**	**	**	**																		
SLM-DD	**	**	**	**	**	**	NS	**	**	**	**	**	**	**	**	**	**	**	**	**	**	**																	
SLM-NDD	**	**	**	**	**	**	**	**	**	**	**	**	**	**	**	**	**	**	**	**	**	**	**																
PPF-DD	**	**	**	**	**	**	**	**	**	**	**	**	**	**	**	**	**	**	**	**	**	**	**	**															
PPF-NDD	**	**	**	**	**	**	**	**	**	**	**	**	**	**	**	**	**	**	**	**	**	**	**	**	**														
PPM-DD	**	**	**	**	**	**	**	**	**	**	**	**	**	**	**	**	**	**	**	**	**	**	**	**	NS	**													
PPM-NDD	**	**	**	**	**	**	**	**	**	**	**	**	**	**	**	**	**	**	**	**	**	**	**	**	**	**	**												
MPF-DD	**	**	**	**	**	**	**	**	**	**	**	**	**	**	**	**	**	**	**	**	**	**	**	**	**	**	**	**											
MPF-NDD	**	**	**	**	**	**	**	**	**	**	**	**	**	**	**	**	**	**	**	**	**	**	**	**	**	**	**	**	**										
MPM-DD	**	**	**	**	**	**	**	**	**	**	**	**	**	**	**	**	**	**	**	**	**	**	**	**	**	**	**	**	**	**									
MPM-NDD	**	**	**	**	**	**	**	**	**	**	**	**	**	**	**	**	**	**	**	**	**	**	**	**	**	**	**	**	**	**	**								
LPF-DD	**	**	**	**	**	**	**	**	**	**	**	**	**	**	**	**	**	**	**	**	**	**	**	**	**	**	**	**	**	**	**	**							
LPF-NDD	**	**	**	**	**	**	**	**	**	**	**	**	**	**	**	**	**	**	**	**	**	**	**	**	**	**	**	**	**	**	**	**	NS						
LPM-DD	**	**	**	**	**	**	**	**	**	**	**	**	**	**	**	**	**	**	**	**	**	**	**	**	**	**	**	**	**	**	**	**	**	**					
LPM-NDD	**	**	**	**	**	**	**	**	**	**	**	**	**	**	**	**	**	**	**	**	**	**	**	**	**	**	**	**	**	**	**	**	NS	NS	**				
AF-DD	**	**	**	**	**	**	**	**	**	**	**	**	**	**	**	**	**	**	**	**	**	**	**	**	**	**	**	**	**	**	**	**	**	**	**	**			
AF-NDD	**	**	**	**	**	**	**	**	**	**	**	**	**	**	**	**	**	**	**	**	**	**	**	**	**	**	**	**	**	**	**	**	**	**	**	**	NS		
AM-DD	**	**	**	**	**	**	**	**	**	**	**	**	**	**	**	**	**	**	**	**	**	**	**	**	**	**	**	**	**	**	**	**	**	**	**	**	NS	*	
AM-NDD	**	**	**	**	**	**	**	**	**	**	**	**	**	**	**	**	**	**	**	**	**	**	**	**	**	**	**	**	**	**	**	**	**	**	**	**	NS	*	[illegible]

CD 1% 302.25884 n = 50

CD 5% 213.76431

4.2.2. THC between sexes of NDD and DD generations

In Table 4.6., the mean THC values of females and males of NDD generation of all the stages were compared using Student's t Test (n = 15). All the means among different stages were observed to be significant ($p < 0.01$) except between LPF-NDD & LPM-NDD, which were non-significant.

Most of the means between the sexes of DD generation were significantly different from each other at $p < 0.01$ except between PPF-DD & PPM-DD and LPF-DD and LPM-DD, which were non-significant (Table 4.7).

Table 4.6. Total haemocyte counts (numbers/mm3) of different sexes of NDD generation

SN	Stage	Mean	±	SE	t Stat	
1	VEF-NDD	69,167.960	±	405.150	4.935	**
2	VEM-NDD	60,824.260	±	1,617.941		
3	VMF-NDD	75,234.660	±	427.944	-4.923	**
4	VMM-NDD	77,526.360	±	221.097		
5	VLF-NDD	60,904.800	±	84.796	-15.665	**
6	VLM-NDD	62,259.100	±	21.077		
7	SLF-NDD	56,627.200	±	117.577	-2.875	**
8	SLM-NDD	57,054.120	±	80.822		
9	PPF-NDD	73,290.820	±	136.332	-9.728	**
10	PPM-NDD	78,265.440	±	492.799		
11	MPF-NDD	30,510.340	±	39.195	-33.380	**
12	MPM-NDD	32,477.380	±	37.822		
13	LPF-NDD	11,543.860	±	139.858	0.108	NS
14	LPM-NDD	11,528.320	±	44.427		
15	AF-NDD	8,756.520	±	20.379	4.580	**
16	AM-NDD	8,533.800	±	46.178		

n = 15 (Abbreviations used are as per section 3.1.2.)

4.3.1. Body weight

In non-diapausing generation the mean weight increased with increase in larval instars viz., 81.200 ± 3.966 mg in I-NDD, 470.769 ± 42.401 mg in II-NDD, 2,182.974 ± 143.857 mg in III-NDD and 11,098.00 ± 516.569 mg in IV-NDD. The body weight of VEF-NDD and VEM-NDD, VMF-NDD and VMM-NDD, VLF-NDD & VMF-NDD also showed an increasing trend and this was reversed in early, mid and late age male and female pupae as it decreased gradually until emergence of adult females and males (Table 4.8).

Table 4.7. Total haemocyte counts (numbers/mm3) of different sexes of DD generation

SN	Stage	Mean	±	SE	t Stat	
1	VEF-DD	67,363.220	±	222.988	3.766	**
2	VEM-DD	62,876.820	±	1,214.797		
3	VMF-DD	70,967.980	±	78.192	-39.593	**
4	VMM-DD	75,097.960	±	79.773		
5	VLF-DD	69,894.900	±	80.130	90.961	**
6	VLM-DD	60,085.240	±	73.350		
7	SLF-DD	55,456.720	±	45.567	-6.787	**
8	SLM-DD	56,073.160	±	82.831		
9	PPF-DD	75,846.080	±	94.136	1.264	NS
10	PPM-DD	75,678.840	±	97.060		
11	MPF-DD	16,646.540	±	119.057	-17.611	**
12	MPM-DD	21,283.860	±	227.790		
13	LPF-DD	11,656.940	±	411.308	1.425	NS
14	LPM-DD	11,084.820	±	78.377		
15	AF-DD	8,740.840	±	21.374	4.194	**
16	AM-DD	8,536.560	±	41.420		

n = 15 (Abbreviations used are as per section 3.1.2.)

4.3. Absolute haemocytes counts in relation to body weight and haemolymph volume.

In diapause destined generation also, similar trend of increase in weight was recorded. I-DD larvae weighted 71.800 ± 2.080 mg, which went up to 36,592.000 ± 2,435.415 mg in VLM-DD and 43,376.000 ± 1,184.062 mg in VLF-DD. After gut purging, the weight of spinning larvae decreased in both the sexes and this trend was noticed till emergence of adults. Adult female weighed 5,378.000 ± 266.362 mg and was heavier than adult male, which weighed 4,218.000 ± 266.362 mg (Table 4.9).

4.3.2. Haemolymph volume

In NDD life stages, the total volume of haemolymph increased with increase in weight and larval instars and it was recorded to be highest in late larval fifth instar i.e., 21,897.833 ± 679.304 µl in female and 19,979.538 ± 612.558 µl in male. Drastic fall in haemolymph volume was observed in both the sexes of NDD in early pupal age i.e., in female (6,590.519 ± 204.261 µl) and male (4,592.233 ± 152.098 µl) and lowest volume was recorded in adult forms AF-NDD (1,330.186 ± 100.081 µl) and AM-NDD (927.400 ± 71.024 µl) - Table 4.8.

In diapause destined generation highest volume was recorded in VLF-DD (25,552.189 ± 1,298.742 µl) and VLM-DD (21,457.085 ± 1,504.495 µl) and thereafter the volume gradually decreased until emergence of adults. The volume of adult female was higher (1499.582 ± 106.168 µl) than adult male (843.600 ± 41.904 µl) in diapausing generation also (Table 4.9).

4.3.3. Haemolymph volume as percent of body weight

The percent of haemolymph volume of body weight in I-NDD was 80.490 ± 2.435 % of body weight (range 70.290 to 89.169 %), 77.904 ± 0.374 % in II-NDD (range 76.494 to 80.133 %) and 78.045 ± 0.222 in III-NDD (range 79.050 to 77.292). The ratio further went down with increase of larval instars, pupae and adults viz., VEF-NDD (69.849 ± 0.249) & VEM-NDD (70.766 ± 0.276 %), VLF-NDD (59.678 ± 0.251 %) & VLM-NDD (60.195± 0.301), PPF-NDD (49.801± 0.264) & PPM-NDD (50.842 ± 0.262) & LPF-NDD (29.808 ± 0.322%), LPM-NDD (29.658 ± 0.248 %), AF-NDD (21.247 ± 0.688 %), AM-NDD (20.063 ± 0.311 %). Marked difference was not observed in percent volume values of late pupal and adults (Table 4.8.).

The percent volume of body weight was comparatively higher in NDD than DD larvae in II (DD - 75.342 ± 1.363 % & NDD - 77.904 ± 0.374 %), III (DD - 74.788 ± 1.200 % & III NDD - 78.045 ± 0.222 %); IV (DD - 73.140 ± 1.273 % & NDD - 74.462 ± 0.491 %). In both NDD and DD generations, the percent haemolymph volume against body weight attained its peak in I instar. Thereafter, this ratio gradually declined during subsequent stages of development, such as middle and late V instar, pupae and adults (Tables 4.8. and 4.9). When compared between the sexes, the females had comparatively more percent of haemolymph than males.

4.3.4. Absolute number of haemocytes

Highest absolute number was recorded in mid-aged male (VMM-NDD - 1,357,562,600) and females (VMF-NDD - 1,343,513,585). Absolute number of haemocytes gradually decreased until emergence of adults. Adult females had comparatively higher (AF-NDD - 11,647,800) number of cells than adult males (AM-NDD -7,914,246). The trend of decrease or increase in absolute number of haemocytes in DD generation larval instars, pupal stages and adults was like NDD generation. Highest absolute number was recorded in VLF-DD (1,785,967,121) and VLM-DD (1,289,254,125). However, AF-DD (13,107,603) had higher number of haemocytes in comparison with AF-NDD (11,647,800) and AM-DD (7,201,442) had less absolute number of haemocytes than AM-NDD (7,914,246) - (Tables 4.8 and 4.9).

4.4. Differential Haemocyte Counts (DHC)

Differential haemocyte counts were carried out to record the population of identified cells as percentage of 200 cells counted in each of the 15 slides prepared for every stage of life cycle studied in both non-diapausing and diapausing generations. The clearly identified types of haemocytes were prohaemocytes (PRs), Oenocytoids (OEs), Spherulocytes (SPs), Granulocytes (GRs) and Plasmatocytes (PLs). In addition to these mitotic cells (MCs) and degenerated cells (Deg.) were also counted and their percentage was calculated. The cells, which could not be classified under above types, were not taken into account.

Table 4.8. Body weight, volume, volume as percent of body weight, total haemocyte counts and absolute number of haemocytes in different stages of non-diapause destined generation

Table 4.8. Body weight, volume, volume as percent of body weight, total haemocyte counts and absolute number of haemocytes in different stages of non-diapause destined generation

Stage	Mean weight (mg ± SE)	Mean volume (µl ± SE)	Volume as percent of body weight			THC (Mean ± SE) (numbers/mm^3)	Absolute number of haemocytes
			Max	Min	Mean ± SE		
I-NDD	81.200 ± 3.996	65.403 ± 4.051	89.169	70.290	80.490 ± 2.435	43,620 ± 210.46	2,852,888
II-NDD	470.769 ± 42.401	366.876 ± 29.719	80.133	76.494	77.904 ± 0.374	46,498 ± 39.63	17,059,046
III-NDD	2,182.974 ± 143.857	1,704.664 ± 96.291	79.050	77.292	78.045 ± 0.222	57,996 ± 91.43	98,863,848
IV-NDD	11,098.000 ± 516.569	8,257.327 ± 367.045	77.065	73.034	74.462 ± 0.491	55,955 ± 84.14	462,039,548
VEF-NDD	10,464.000 ± 433.039	7,307.355 ± 294.248	71.131	68.757	69.849 ± 0.249	69,168 ± 405.150	505,434,818
VEM-NDD	9,804.700 ± 452.124	6,936.208 ± 293.361	72.551	69.139	70.766 ± 0.276	60,824 ± 1617.941	421,889,726
VMF-NDD	27,573.300 ± 426.122	17,857.641 ± 234.524	65.663	63.528	64.781 ± 0.227	75,235 ± 427.944	1,343,513,585
VMM-NDD	26,710.300 ± 686.753	17,510.981 ± 390.491	66.993	64.202	65.545 ± 0.347	77,526 ± 221.097	1,357,562,600
VLF-NDD	36,682.100 ± 1,353.965	21,897.833 ± 679.304	61.329	58.342	59.678 ± 0.251	60,905 ± 84.796	1,333,683,139
VLM-NDD	33,184.800 ± 983.790	19,979.538 ± 612.558	61.342	58.971	60.195 ± 0.301	62,259 ± 21.077	1,243,908,078
SLF-NDD	26,834.900 ± 659.848	16,133.900 ± 374.550	61.981	58.954	60.134 ± 0.323	56,627 ± 117.577	913,617,558
SLM-NDD	23,932.900 ± 403.702	14,444.848 ± 220.867	62.346	58.921	60.374 ± 0.342	57,054 ± 80.822	824,138,093
PPF-NDD	13,232.900 ± 434.157	6,590.519 ± 204.261	50.665	48.476	49.801 ± 0.264	73,291 ± 136.332	483,024,568
PPM-NDD	9,038.600 ± 367.538	4,592.233 ± 152.098	51.911	49.081	50.842 ± 0.262	78,265 ± 492.799	359,413,173
MPF-NDD	10,066.600 ± 138.507	4,533.984 ± 62.391	45.696	43.574	45.044 ± 0.234	30,510 ± 39.195	138,333,390
MPM-NDD	7,519.200 ± 160.864	3,392.832 ± 75.020	46.892	43.488	45.116 ± 0.289	32,477 ± 37.822	110,190,304
LPF-NDD	9,652.800 ± 234.792	2,880.305 ± 88.718	31.686	28.946	29.808 ± 0.322	11,544 ± 139.858	33,249,835
LPM-NDD	7,061.800 ± 166.761	2,096.133 ± 60.998	30.829	28.646	29.658 ± 0.248	11,528 ± 44.427	24,164,891
AF-NDD	6,046.300 ± 456.941	1,330.186 ± 100.081	24.892	19.248	21.247 ± 0.688	8,757 ± 20.379	11,647,800
AM-NDD	4,637.000 ± 356.822	927.400 ± 71.024	21.470	19.008	20.063 ± 0.311	8,534 ± 46.178	7,914,246

Table 4.9. Body weight, volume, volume as percent of body weight, total haemocyte counts and absolute number of haemocytes in different stages of diapause destined generation

Stage	Mean weight (mg ± SE)	Mean volume (µl ± SE)	Volume as percent of body weight			THC (Mean ± SE) (numbers/mm^3)	Absolute number of haemocytes
			Max	Min	Mean ± SE		
I-DD	71.800 ± 2.080	56.306 ± 1.461	83.381	71.538	78.517 ± 2.199	43,370 ± 136.682	2,441,995
II-DD	418.900 ± 18.243	313.982 ± 8.656	83.728	70.587	75.342 ± 1.363	45,542 ± 39.192	14,299,485
II-DD	2,171.300 ± 67.283	1,621.399 ± 45.010	79.285	70.879	74.788 ± 1.200	55,537 ± 41.420	90,046,945
V-DD	8,937.000 ± 59.616	6,537.316 ± 123.041	79.072	69.140	73.140 ± 1.237	54,990 ± 87.029	359,489,109
F-DD	10,657.000 ± 692.171	7,374.129 ± 563.726	75.275	58.536	68.747 ± 1.900	67,363 ± 222.99	496,745,041
-DD	10,278.000 ± 727.598	6,952.953 ± 448.658	74.688	62.505	67.768 ± 1.484	62,877 ± 1214.80	437,179,550
F-DD	34,215.000 ± 770.994	21,838.661 ± 644.843	69.457	56.583	63.811 ± 1.220	70,968 ± 78.19	1,549,845,655
-DD	26,352.000 ± 1,030.454	16,560.320 ± 760.558	69.738	56.634	62.792 ± 1.614	75,098 ± 79.77	1,243,646,211
F-DD	43,376.000 ± 1,184.062	25,552.189 ± 1,298.742	66.747	50.312	58.732 ± 1.996	69,895 ± 80.13	1,785,967,721
-DD	36,592.000 ± 2,435.415	21,457.085 ± 1,504.495	68.972	50.498	58.655 ± 2.377	60,085 ± 73.35	1,289,254,125
F-DD	29,998.911 ± 2,741.599	17,151.389 ± 1,496.487	65.726	51.126	57.460 ± 1.965	55,457 ± 45.57	951,159,757
-DD	28,945.033 ± 1,783.929	16,971.664 ± 1,182.516	69.787	50.526	58.906 ± 2.429	56,073 ± 82.83	951,654,858
F-DD	12,579.300 ± 456.941	6,069.873 ± 315.775	53.620	41.076	48.126 ± 1.379	75,846 ± 94.14	460,376,090
-DD	8,915.400 ± 356.761	4,359.181 ± 212.476	54.103	41.370	48.911 ± 1.465	75,679 ± 97.06	329,897,785
F-DD	10,104.200 ± 397.197	4,305.022 ± 182.161	50.712	34.693	42.841 ± 1.804	16,647 ± 119.06	71,663,716
-DD	7,735.000 ± 235.452	3,273.069 ± 156.397	48.981	34.423	42.288 ± 1.550	21,284 ± 227.79	69,663,546
F-DD	9,549.100 ± 263.773	2,697.827 ± 209.201	35.872	20.789	28.161 ± 1.875	11,657 ± 411.31	31,448,410
-DD	7,105.500 ± 367.763	1,915.966 ± 124.247	34.213	22.716	27.053 ± 1.269	11,085 ± 78.38	21,238,136
F-DD	5,378.000 ± 266.362	1,499.582 ± 106.168	38.761	19.463	28.296 ± 2.270	8,741 ± 21.37	13,107,603
-DD	4,218.000 ± 266.362	843.600 ± 41.904	25.819	15.726	19.963 ± 1.172	8,537 ± 41.42	7,201,442

4.4.1. DHC of NDD and DD generations

4.4.1.1. Prohaemocytes

In I instar larvae of non-diapausing generation (I-NDD), on an average only 8.1667 ± 0.2073 % of haemocytes were comprised of Prohaemocytes. Their mean percentage dipped in II instar (5.8333 ± 0.2639 %), but rose in III instar (8.5328 ± 0.2786 %). In IV instar, the mean percent of Prohaemocytes was 4.5325 ± 0.2866 %, which gradually declined until spinning larval stages of both the sexes i.e., SLF-NDD (0.7333 ± 0.1311 %) and SLM-NDD (0.7667 ± 0.1615 %). A rise in their mean percent population was observed during PPF-NDD (2.9333 ± 0.1985 %) and MPM-NDD (1.0333 ± 0.2051 %). The mean percent population of Prohaemocytes was generally high in adult females i.e., in AF-NDD it was 1.000 ± 0.1633 % as compared to only 0.1000 ± 0.0989 % in AM-NDD (Table 4.10.).

In larvae of diapausing generation, mean percent Prohaemocytes were 7.5667 ± 0.2971 % in I instar with a little lower percent (6.1667 ± 0.1721 %) in II instar, and abruptly higher percent (10.4333 ± 0.1870 %) in III instar. Their percentage went down in IV instar (4.2000 ± 0.1116 %) and this lowering trend continued till VMM-DD (1.3333 ± 0.2639 %) and VLF-DD (1.6000 ± 0.1789%). A rise was noticed during pupal stages, it was 3.1667 ± 0.2073 % in PPM-DD and 2.2000 ± 0.3451 %) in PPF-DD. Thereafter, the mean percentage of Prohaemocytes went down until emergence of adults: 1.1000 ± 0.2129 % in AF-DD and 0.7330 ± 0.0911 % in AM-DD (Table 4.10.).

The mean percent values of the population of Prohaemocytes of non-diapausing generation were put for one way ANOVA to determine the difference among all the stages of life cycle studied. The results of such analysis is tabulated in the lower left diagonal of Table 4.11 (PR) for non-diapausing generation (CD 0.3453 at 5 % and 0.4883 at 1 %) and for diapausing generation (CD 0.3461 at 5 % and 0.4894 at 1 %) of Table 4.12. Statistical analysis showed that the mean percent of Prohaemocyte population was significantly different in most of the life stages at $p < 0.01$.

The mean percent values of the population of Prohaemocytes of diapausing generation were put for one way ANOVA to determine the difference among all the stages of life cycle studied. The results of such analysis are tabulated in the lower left diagonal of Table 4.12 for diapausing generation (CD 0.3461 at 5 % and 0.4894 at 1 %). Statistical analysis showed that the mean percent of Prohaemocyte population was significantly different ($p < 0.01$) among most of the life stages.

Each of the life stages of non-diapausing generation under study were compared with same stage of diapausing generation using Student's t-test and the results of the test were tabulated in Table 4.10. The mean percent population of Prohaemocytes in between I-NDD & I-DD and VMM-NDD & VMM-DD differed significantly at 5 %. Difference of Prohaemocyte percent population was observed to be significant in between III-NDD & III-DD, VMF-NDD & VMF-DD, VLM-NDD & VLM-DD, SLM-NDD & SLM-DD, PPF-NND & PPF-DD, MPF- NDD & MPF-DD, MPM-NDD & MPM-DD, LPF-NDD & LPF-DD, LPM-NDD & LPM-DD, AM-NDD & AM-DD. However, no significant difference was recorded in between second & fourth instars, early fifth instar female & early fifth instar male larval stages and adult females of both generations.

Table 4.10. Percent Prohaemocytes in different stages of NDD & DD generations of *A. mylitta*

SN	Age	Mean	±	SE	t Stat	
1	I-NDD	8.1667	±	0.2073	1.9394	*
2	I-DD	7.5667	±	0.2971		
3	II-NDD	5.8333	±	0.2639	-1.5404	NS
4	II-DD	6.1667	±	0.1721		
5	III-NDD	8.5328	±	0.2786	-8.0889	**
6	III-DD	10.4333	±	0.1870		
7	IV-NDD	4.5325	±	0.2866	1.4311	NS
8	IV-DD	4.2000	±	0.1116		
9	VEF-NDD	3.3667	±	0.2622	-0.8367	NS
10	VEF-DD	3.5333	±	0.1940		
11	VEM-NDD	2.2667	±	0.1471	-0.2112	NS
12	VEM-DD	2.3000	±	0.1978		
13	VMF-NDD	1.2667	±	0.2738	-2.8084	**
14	VMF-DD	1.9333	±	0.1615		
15	VMM-NDD	1.8667	±	0.1559	2.1246	*
16	VMM-DD	1.3333	±	0.2639		
17	VLF-NDD	1.1333	±	0.1695	-3.2868	**
18	VLF-DD	1.6000	±	0.1789		
19	VLM-NDD	1.0000	±	0.2108	-12.9087	**
20	VLM-DD	2.7667	±	0.1471		
21	SLF-NDD	0.7333	±	0.1311	-1.0000	NS
22	SLF-DD	0.9000	±	0.1366		
23	SLM-NDD	0.7667	±	0.1615	-6.8590	**
24	SLM-DD	1.9000	±	0.1366		
25	PPF-NDD	2.9333	±	0.1985	2.8470	**
26	PPF-DD	2.2000	±	0.3451		
27	PPM-NDD	0.8333	±	0.2073	-11.0680	**
28	PPM-DD	3.1667	±	0.2073		
29	MPF-NDD	1.2000	±	0.1738	-5.0020	**
30	MPF-DD	1.9667	±	0.1241		
31	MPM-NDD	1.0333	±	0.2051	-9.7228	**
32	MPM-DD	2.8000	±	0.1862		
33	LPF-NDD	1.0333	±	0.1241	-4.0359	**
34	LPF-DD	1.4000	±	0.1193		
35	LPM-NDD	0.4667	±	0.1940	-4.7042	**
36	LPM-DD	1.8667	±	0.3430		
37	AF-NDD	1.0000	±	0.1633	-0.5092	NS
38	AF-DD	1.1000	±	0.2129		
39	AM-NDD	0.1000	±	0.0989	-6.9711	**
40	AM-DD	0.7333	±	0.0911		

n = 15 (Abbreviations used are as per section 3.1.2.)

4.4.1.2. Oenocytoids

In NDD life stages and instars, scanty percentage of population of Oenocytoids was recorded. Among the larval instars, percent population of OE existed to the tune of 0.5667 ± 0.1311 % in I instar NDD, slightly higher 1.2333 ± 0.1311 % in II instar and highest 4.5325 ± 0.2866 % in IV instar. Another peak of percentage was recorded in VLF-NDD (3.8333 ± 0.2073 %) and VLM-NDD (4.1333 ± 0.1695 %). Thereafter, the mean percentage of Oenocytoids came down and this decreasing trend continued until emergence of adults (Table 4.13).

Overall data of Oenocytoids were subjected to one-way ANOVA analysis. The CD worked out to be was 0.6574 at 5% and 0.9295 at 1%. The difference in mean value of Oenocytoid percentage in haemolymph was observed to be significant ($p < 0.05$ or $p < 0.01$) in all the stages except in between II-NDD & III-NDD, VEF-NDD & VEM-NDD, VEM-NDD & VMF-NDD, VLF-NDD & VLM-NDD, VLM-NDD & SLF-NDD, SLF-NDD & SLM-NDD, SLM-NDD & PPF-NDD, MPF-NDD & MPM-NDD, MPM-NDD & LPF-NDD, where the differences were observed to non-significant (Table 4.11).

Table 4.11 : Test of significance of DHC mean using CD at 5% and 1% in each age group in NDD generation.

Age	I-NDD	II-NDD	III-NDD	IV-NDD	VEF-NDD	VEM-NDD	VMF-NDD	VMM-NDD	VLF-NDD	VLM-NDD	SLF-NDD	SLM-NDD	PPF-NDD	PPM-NDD	MPF-NDD	MPM-NDD	LPF-NDD	LPM-NDD	AF-NDD	AM-NDD
I-NDD	PR	*	NS	**	**	**	**	**	**	**	**	**	**	**	**	**	**	*	**	NS
II-NDD	**		NS	**	NS	NS	*	**	**	**	**	**	**	**	NS	NS	**	NS	**	*
III-NDD	*	**		**	*	*	**	**	**	**	**	**	**	**	*	*	**	NS	**	NS
IV-NDD	**	**	**		**	**	**	**	*	NS	*	**	**	NS	**	**	**	**	**	**
VEF-NDD	**	**	**	**		NS	NS	**	**	**	**	**	**	**	NS	NS	*	NS	**	**
VEM-NDD	**	**	**	**	**		NS	**	**	**	**	**	**	**	NS	NS	*	NS	**	**
VMF-NDD	**	**	**	**	**	**		*	**	**	**	**	*	**	NS	NS	NS	NS	NS	**
VMM-NDD	**	**	**	**	**	*	**		**	**	**	NS	NS	**	**	**	NS	**	NS	**
VLF-NDD	**	**	**	**	**	**	NS	**		NS	NS	NS	**	**	**	**	**	**	**	**
VLM-NDD	**	**	**	**	**	**	NS	**	NS		NS	*	**	*	**	**	**	**	**	**
SLF-NDD	**	**	**	**	**	**	**	**	*	NS		NS	**	**	**	**	**	**	**	**
SLM-NDD	**	**	**	**	**	**	**	**	*	NS	NS		NS	**	**	**	*	**	NS	**
PPF-NDD	**	**	**	**	*	**	**	**	**	**	**	**		**	**	**	NS	**	NS	**
PPM-NDD	**	**	**	**	**	**	*	**	NS	NS	NS	NS	**		**	**	**	**	**	**
MPF-NDD	**	**	**	**	**	**	NS	**	NS	NS	*	*	**	*		NS	NS	NS	*	**
MPM-NDD	**	**	**	**	**	**	NS	**	NS	NS	NS	NS	**	NS	NS		NS	NS	*	**
LPF-NDD	**	**	**	**	**	**	NS	**	NS	NS	NS	NS	**	NS	NS	NS		**	NS	**
LPM-NDD	**	**	**	**	**	**	**	**	**	**	NS	NS	**	*	**	**	**		**	**
AF-NDD	**	**	**	**	**	**	NS	**	NS	NS	NS	NS	**	NS	NS	NS	NS	**		**
AM-NDD	**	**	**	**	**	**	**	**	**	**	**	**	**	**	**	**	**	*	**	OE
	PR	Lower diagonal														Upper diagonal			OE	
CD 5%	0.3453																		0.6574	
CD 1%	0.4883																		0.9295	

Similar to NDD generation, in DD generation also, Oenocytoids population was recorded to be scanty. As illustrated in Table 4.13, there were 0.6667 ± 0.0861 % of Oenocytoids in I instar, 1.9333 ± 0.1311 % in II instar, 2.2333 ± 0.1311 % in III instar and on slightly higher side in IV instar (3.3333 ± 0.2802 %). A fluctuating trend in the population of Oenocytoids was observed during different age fifth instars larvae and pupae of both the generations. Most decreased level of population in pupae was recorded in MPF-DD (1.2667 ± 0.2198 %), though an elevation in OE population was noticed in LPF-DD (4.2667 ± 0.4314 %). Adult females of DD generation had comparatively less percentage of OE (2.2333 ± 0.3993) than males (2.4667 ± 0.1241). Overall data of Oenocytoids of DD generation of different stages were subjected to one-way ANOVA analysis. The CD worked out was 0.4234 ($p < 0.05$) and 0.5986 ($p < 0.01$). The Oenocytoids ratio in haemolymph was observed to be significant ($p < 0.05$ or $p < 0.01$) in between all the stages of DD except in between II-DD & III-DD, VMF-DD & VMM-DD, VLM-DD & SLF-DD, SLF-DD & SLM-DD, SLM-DD & PPF-DD, PPF-DD & PPM-DD, AF-DD & AM-DD, where no significant difference was observed (Table 4.12).

Table 1.2 : Test of significance of DHC mean using CD at 5% and 1% in each age group in DD generation.

Age	I-DD	II-DD	III-DD	IV-DD	VEF-DD	VEM-DD	VMF-DD	VMM-DD	VLF-DD	VLM-DD	SLF-DD	SLM-DD	PPF-DD	PPM-DD	MPF-DD	MPM-DD	LPF-DD	LPM-DD	AF-DD	AM-DD
I-DD	PR	**	**	**	**	**	**	**	**	**	**	**	**	**	**	**	**	**	**	**
II-DD	**		NS	**	NS	*	**	**	**	**	**	*	NS	NS	**	NS	**	**	NS	*
III-DD	**	**		**	NS	NS	**	**	*	**	*	NS	NS	NS	**	NS	**	**	NS	NS
IV-DD	**	**	**		**	**	NS	*	**	*	*	**	**	**	**	**	**	NS	**	**
VEF-DD	**	**	**	**		*	**	**	**	**	**	*	NS	NS	**	NS	**	**	NS	*
VEM-DD	**	**	**	**	**		*	NS	NS	NS	NS	NS	*	*	**	NS	**	**	NS	NS
VMF-DD	**	**	**	**	**	*		NS	NS	NS	NS	*	**	**	**	**	**	*	**	*
VMM-DD	**	**	**	**	**	**	NS		NS	NS	NS	NS	**	**	**	**	**	**	**	NS
VLF-DD	**	**	**	**	**	**	**	NS		NS	NS	NS	**	**	**	*	**	**	*	NS
VLM-DD	**	**	**	**	**	*	**	**	**		NS	NS	**	**	**	**	**	*	**	NS
SLF-DD	**	**	**	**	**	**	**	**	*	**		NS	**	**	**	*	**	**	*	NS
SLM-DD	**	**	**	**	**	*	NS	NS	**	**	**		NS	NS	**	NS	**	**	NS	NS
PPF-DD	**	**	**	**	**	NS	NS	**	**	**	**	NS		NS	**	NS	**	**	NS	NS
PPM-DD	**	**	**	**	*	**	**	**	**	*	**	**	**		**	NS	**	**	NS	NS
MPF-DD	**	**	**	**	**	NS	NS	*	**	**	**	NS	NS	**		**	**	**	**	**
MPM-DD	**	**	**	**	**	**	**	**	**	NS	**	**	**	*	**		**	**	NS	NS
LPF-DD	**	**	**	**	**	**	**	NS	NS	**	**	**	**	**	**	**		**	**	**
LPM-DD	**	**	**	**	**	*	NS	NS	**	**	**	NS	NS	**	NS	**	*		**	**
AF-DD	**	**	**	**	**	**	**	**	NS	**	NS	**	**	**	**	**	NS	**		NS
AM-DD	**	**	**	**	**	**	**	**	**	**	NS	**	**	**	**	**	**	**	*	OE
	PR	Lower diagonal															Upper diagonal			OE
CD 5%	0.3461																			0.4234
CD 1%	0.4894																			0.5986

The mean percent population of Oenocytoids were compared for difference between the two types of generations, viz., non-diapausing and diapausing using Student's t-test (n = 15). When the population of Oenocytoids was compared in between the life stages of both the generations it was observed that the mean percent population of Oenocytoids differed significantly at $p < 0.01$ for II instar, III instar and IV instar larvae. The difference was also significant ($p < 0.01$) in between VEM-NDD & VEM-DD; VMF-NDD and VMF-DD; VLF-NDD and VLF-DD; VLM-NDD and VLM-DD; SLM-NDD and SLM-DD; PPF-NDD and PPF-DD; PPM-NDD and PPM-DD; MPF-NDD and MPF-DD; LPF-NDD and LPF-DD; LPM-NDD and LPM-DD, AM-NDD & AM-DD. In case of VEF-NDD and VEF-DD and SLF-NDD and SLF-DD, the difference was observed to be significant at $p < 0.05$. In rest of the life stages, mean percentage population did not significantly differ in between non-diapausing and diapausing generations (Table 4.13).

4.4.1.3. Spherulocytes

The Spherulocytes were significantly higher than Prohaemocytes and Oenocytoids in all the instars and developmental stages (Table 4.14). In NDD I larval instar the percent Spherulocytes were 7.3000 ± 2.3991 % and population jumped to 16.6333 ± 4.2366 % in II instar NDD, 19.8986 ± 5.0618 % in III instar NDD and 20.0628 ± 5.1010 % in IV instar NDD. The ratio decreased in VEF-NDD (14.8667 ± 3.9016 %) and VEM-NDD (12.9333 ± 3.2755 %) which then increased in VMF-NDD (23.7000 ± 5.9811 %) and VMM-NDD (24.1667 ± 6.1250 %), and there after fall in Spherulocytes was regularly noticed in haemolymph of both the sexes as it came down to 5.7000 ± 1.7531 % in MPF-NDD and 7.7333 ± 2.7354 %

Table 4.13. Percent Oenocytoids in different stages of NDD & DD generations of A. mylitta

SN	Age	Mean	±	SE	t Stat	
1	I-NDD	0.5667	±	0.1311	-0.8987	NS
2	I-DD	0.6667	±	0.0861		
3	II-NDD	1.2333	±	0.1311	-4.5826	**
4	II-DD	1.9333	±	0.1311		
5	III-NDD	0.9333	±	0.1311	-13.6699	**
6	III-DD	2.2333	±	0.1311		
7	IV-NDD	4.5325	±	0.2866	4.3553	**
8	IV-DD	3.3333	±	0.2802		
9	VEF-NDD	1.6333	±	0.1940	-2.1282	*
10	VEF-DD	2.0000	±	0.1155		
11	VEM-NDD	1.6667	±	0.1089	-2.9006	**
12	VEM-DD	2.5000	±	0.3333		
13	VMF-NDD	2.0667	±	0.1985	-4.3235	**
14	VMF-DD	2.9667	±	0.1241		
15	VMM-NDD	2.7667	±	0.1870	-0.2543	NS
16	VMM-DD	2.8333	±	0.2639		
17	VLF-NDD	3.8333	±	0.2073	5.5340	**
18	VLF-DD	2.6667	±	0.1721		
19	VLM-NDD	4.1333	±	0.1695	6.1414	**
20	VLM-DD	2.8667	±	0.1822		
21	SLF-NDD	3.8000	±	0.5068	2.5227	*
22	SLF-DD	2.7667	±	0.1615		
23	SLM-NDD	3.2667	±	0.1747	12.2202	**

24 SLM-DD	2.4667	±	0.1695		
25 PPF-NDD	2.8667	±	0.2051	2.7792	**
26 PPF-DD	2.0667	±	0.3582		
27 PPM-NDD	4.8000	±	1.0930	3.3605	**
28 PPM-DD	2.0667	±	0.1471		
29 MPF-NDD	1.8333	±	0.1277	3.1186	**
30 MPF-DD	1.2667	±	0.2198		
31 MPM-NDD	1.8000	±	0.4705	-1.3408	NS
32 MPM-DD	2.2333	±	0.2896		
33 LPF-NDD	2.4333	±	0.5845	-3.3123	**
34 LPF-DD	4.2667	±	0.4314		
35 LPM-NDD	1.4333	±	0.4466	-4.3205	**
36 LPM-DD	3.4333	±	0.3822		
37 AF-NDD	2.6333	±	0.5010	0.8729	NS
38 AF-DD	2.2333	±	0.3993		
39 AM-NDD	0.5000	±	0.3127	-7.3020	**
40 AM-DD	2.4667	±	0.1241		

n = 15 (Abbreviations used are as per section 3.1.2.)

in MPM-NDD. A slight increase in percent population of Spherulocytes was recorded in LPF-NDD (12.6000 ± 3.4345 %) and LPM-NDD (9.1667 ± 3.3951 %), though their population was recorded to be lowest in AF-NDD (5.3333 ± 1.9102 %) and AM-NDD (3.3333 ± 1.6620 %) - Table 4.14.

The mean percent values were subjected for statistical analysis and CD at 5 % was 0.9532 at 1 % 1.3478. Critical difference was observed to be significant in almost all the instar and stages, except in between III instar NDD and IV instar NDD, VMF-NDD and VMM-NDD, PPF-NDD and PPM-NDD. The level of Spherulocyte percent was significantly different between the sexes of the same generation (Table 4.15).

Spherulocytes in DD generation were recorded to be 8.8000 ± 2.3935 % in I instar DD, 18.1333 ± 4.5382% in II instar DD, 18.6333 ± 4.8298 % in IV instar DD and slightly lowered in VEF-DD (15.4667 ± 4.1290%) and VEM-DD (14.2000 ± 3.6157%). It was highest in VMF-DD (23.0333 ± 6.0668 %) and VLM-DD (21.9333 ± 5.5090 %), thereafter it's percentage gradually came down with advancement of life stages, except in LPF-DD (12.3000 ± 3.2497 %) and LPM-DD (13.1333 ± 3.3310 %) and it came down significantly in AF-DD (5.1000 ± 1.3888 %) and AM-DD (4.4667 ± 1.4795%) - Table 4.14.

One way ANOVA was worked out to test significant mean differences among the percentage of Spherulocytes of different life stages of DD generation. The critical difference was observed to 0.6526 at $p < 0.05$ and 0.9228 at $p < 0.01$. The decrease or increase in population was significant ($P < 0.01$) from one stage to other (Table 4.15).

The mean percent of Spherulocytes were compared to record the difference between two generations, viz., non-diapausing and diapausing applying Student's t-Test (n = 15). The mean values were observed to be significant ($p < 0.01$) in most of the cases, except in between III instar NDD & DD, VEF-NDD & VEF-DD, VMF-NDD & VMF-DD, VLM-NDD

& VLM- DD, SLF-NDD & SLF-DD, PPF-NDD & PPF-DD, PPM-NDD & PPM-DD, MPM-NDD & MPM-DD, LPF-NDD & LPF-DD, AF-NDD & AF-DD. Difference was observed to be significant ($p < 0.05$) in between AM-NDD & AM-DD (Table 4.14).

Table 4.14. Percent Spherulocytes in different stages of NDD & DD generations of A. mylitta

SN	Age	Mean	±	SE	t Stat	
1	I-NDD	7.3000	±	2.3991	-2.8299	**
2	I-DD	8.8000	±	2.3935		
3	II-NDD	16.6333	±	4.2366	-5.9161	**
4	II-DD	18.1333	±	4.5382		
5	III-NDD	19.8986	±	5.0618	0.0896	NS
6	III-DD	19.8667	±	5.0306		
7	IV-NDD	20.0628	±	5.1010	3.6183	**
8	IV-DD	18.6333	±	4.8298		
9	VEF-NDD	14.8667	±	3.9016	-1.1310	NS
10	VEF-DD	15.4667	±	4.1290		
11	VEM-NDD	12.9333	±	3.2755	-4.8316	**
12	VEM-DD	14.2000	±	3.6157		
13	VMF-NDD	23.7000	±	5.9811	1.1730	NS
14	VMF-DD	23.0333	±	6.0668		
15	VMM-NDD	24.1667	±	6.1250	7.2161	**
16	VMM-DD	21.4667	±	5.4420		
17	VLF-NDD	17.2000	±	4.5257	-3.3001	**
18	VLF-DD	18.9333	±	4.7766		
19	VLM-NDD	22.6000	±	5.7562	1.6027	NS
20	VLM-DD	21.9333	±	5.5090		
21	SLF-NDD	15.4333	±	4.1037	1.5625	NS
22	SLF-DD	14.5000	±	3.8393		
23	SLM-NDD	12.9667	±	3.4913	-5.9787	**
24	SLM-DD	14.9667	±	3.7411		
25	PPF-NDD	10.0000	±	3.6185	0.2348	NS
26	PPF-DD	9.8333	±	2.5223		
27	PPM-NDD	10.2000	±	3.2290	0.1930	NS
28	PPM-DD	10.0667	±	2.8906		
29	MPF-NDD	5.7000	±	1.7531	-4.7118	**
30	MPF-DD	7.1333	±	1.9189		
31	MPM-NDD	7.7333	±	2.7354	-0.7540	NS
32	MPM-DD	8.1333	±	2.0817		
33	LPF-NDD	12.6000	±	3.4345	0.8158	NS
34	LPF-DD	12.3000	±	3.2479		
35	LPM-NDD	9.1667	±	3.3951	-5.5510	**
36	LPM-DD	13.1333	±	3.3310		
37	AF-NDD	5.3333	±	1.9102	0.6080	NS
38	AF-DD	5.1000	±	1.3888		
39	AM-NDD	3.3333	±	1.6620	-2.2405	*
40	AM-DD	4.4667	±	1.4795		

n = 15 (Abbreviations used are as per section 3.1.2.)

Table 4.15 : Test of significance of DHC mean using CD at 5% and 1% in each age group in NDD generation.

Age	I-NDD	II-NDD	III-NDD	IV-NDD	VEF-NDD	VEM-NDD	VMF-NDD	VMM-NDD	VLF-NDD	VLM-NDD	SLF-NDD	SLM-NDD	PPF-NDD	PPM-NDD	MPF-NDD	MPM-NDD	LPF-NDD	LPM-NDD	AF-NDD	AM-NDD
I-NDD	SP	**	**	**	**	NS	NS	**	**	NS	NS	NS	**	**	**	**	**	**	*	NS
II-NDD	**		**	**	**	**	**	**	**	**	**	**	**	**	**	**	**	**	**	**
III-NDD	**	**		**	**	**	**	**	**	**	**	**	**	**	**	NS	NS	**	**	**
IV-NDD	**	**	NS		**	**	**	**	**	**	**	**	**	**	NS	**	**	NS	**	**
VEF-NDD	**	**	**	**		**	**	**	NS	**	**	*	**	**	**	**	**	**	*	**
VEM-NDD	**	**	**	**	**		NS	**	NS	*	NS	NS	**	**	**	**	**	**	NS	*
VMF-NDD	**	**	**	**	**	**		**	*	NS	NS	NS	**	**	**	**	**	**	NS	NS
VMM-NDD	**	**	**	**	**	**	NS		**	NS	**	**	NS	*	**	**	**	**	**	NS
VLF-NDD	**	NS	**	**	**	**	**	**		**	*	NS	**	**	**	**	**	**	NS	**
VLM-NDD	**	**	**	**	**	**	*	**	**		NS	**	**	**	**	**	**	**	**	NS
SLF-NDD	**	*	**	**	NS	**	**	**	**	**		NS	**	**	**	**	**	**	NS	NS
SLM-NDD	**	**	**	**	**	NS	**	**	**	**	**		**	**	**	**	**	**	NS	**
PPF-NDD	**	**	**	**	**	**	**	**	**	**	**	**		NS	*	**	**	*	**	**
PPM-NDD	**	**	**	**	**	**	**	**	**	**	**	**	NS		NS	**	**	NS	**	**
MPF-NDD	**	**	**	**	**	**	**	**	**	**	**	**	**	**		**	**	NS	**	**
MPM-NDD	NS	**	**	**	**	**	**	**	**	**	**	**	**	**	**		NS	**	**	**
LPF-NDD	**	**	**	**	**	NS	**	**	**	**	**	NS	**	**	**	**		**	**	**
LPM-NDD	**	**	**	**	**	**	**	**	**	**	**	**	NS	*	**	**	**		**	**
AF-NDD	**	**	**	**	**	**	**	**	**	**	**	**	**	**	NS	**	**	**		**
AM-NDD	**	**	**	**	**	**	**	**	**	**	**	**	**	**	**	**	**	**	**	GR
	SP	Lower diagonal															Upper diagonal			GR
CD 5%	0.9532																			2.8197
CD 1%	1.3478																			3.9870

Table 4.16 : Test of significance of DHC mean using CD at 5 and 1 in each age group in DD generation.

Age	I-DD	II-DD	III-DD	IV-DD	VEF-DD	VEM-DD	VMF-DD	VMM-DD	VLF-DD	VLM-DD	SLF-DD	SLM-DD	PPF-DD	PPM-DD	MPF-DD	MPM-DD	LPF-DD	LPM-DD	AF-DD	AM-DD
I-DD	SP	**	**	**	NS	**	**	NS	NS	NS	NS	NS	NS	*	**	**	**	**	NS	*
II-DD	**		NS	**	**	**	**	**	**	**	**	**	**	**	**	**	**	**	**	**
III-DD	**	**		**	**	**	**	**	**	**	**	**	**	**	**	**	**	**	**	**
IV-DD	**	NS	**		**	**	**	**	**	**	**	**	**	**	*	**	**	*	**	**
VEF-DD	**	**	**	**		**	**	NS	NS	NS	NS	NS	NS	NS	**	**	**	**	NS	NS
VEM-DD	**	**	**	**	**		**	**	**	**	**	**	**	**	**	**	**	**	**	**
VMF-DD	**	**	**	**	**	**		**	**	**	**	*	**	**	**	**	**	**	*	**
VMM-DD	**	**	**	**	**	**	**		NS	NS	NS	NS	NS	*	**	**	**	**	NS	*
VLF-DD	**	*	**	NS	**	**	**	**		NS	NS	NS	NS	NS	**	**	**	**	NS	NS
VLM-DD	**	**	**	**	**	**	**	NS	**		NS	NS	NS	NS	**	**	**	**	NS	NS
SLF-DD	**	**	**	**	**	NS	**	**	**	**		NS	NS	*	**	**	**	**	NS	*
SLM-DD	**	**	**	**	NS	*	**	**	**	**	NS		*	**	**	**	**	**	NS	**
PPF-DD	**	**	**	**	**	**	**	**	**	**	**	**		NS	**	**	**	**	*	NS
PPM-DD	**	**	**	**	**	**	**	**	**	**	**	**	NS		**	**	**	**	*	NS
MPF-DD	**	**	**	**	**	**	**	**	**	**	**	**	**	**		NS	NS	NS	**	**
MPM-DD	*	**	**	**	**	**	**	**	**	**	**	**	**	**	**		NS	NS	**	**
LPF-DD	**	**	**	**	**	**	**	**	**	**	**	**	**	**	**	**		NS	**	**
LPM-DD	**	**	**	**	**	**	**	**	**	**	**	**	**	**	**	**	*		**	**
AF-DD	**	**	**	**	**	**	**	**	**	**	**	**	**	**	**	**	**	**		*
AM-DD	**	**	**	**	**	**	**	**	**	**	**	**	**	**	**	**	**	**	NS	GR
	SP	Lower diagonal															Upper diagonal			GR
CD 5%	0.6526																			2.0499
CD 1%	0.9228																			2.8985

4.4.1.4. Granulocytes

In NDD generation percent population of Granulocytes in I instar was 51.0667 ± 1.7624 %. Their percentage lowered in II instar (28.4667 ± 1.1045 %) and slightly increased in III instar (34.7244 ± 0.5133 %), in IV instar (39.6652 ± 0.8595 %), in VEF (57.6000 ± 0.6805 %) and VEM (52.5000 ± 0.6360 %). The ratio of Granulocytes was slightly lower in VMM (46.9333 ± 1.1389 %) and VMF (51.9000 ± 1.4685 %). However, this ratio further increased in SLF (51.9667 ± 0.9763 %) and SLM (53.8667 ± 1.0944 %). After this age group, their population decreased till late pupal stage in both the sexes. A rise in their percentage was recorded in AF-NDD (53.9667 ± 2.3565 %) and AM-NDD (49.3333 ± 1.8663 %) - Table 4.17.

In order to find out the difference in population to be significant or not from one stage to other stage in NDD generation, ANOVA was worked out and CD was 2.8197 at 5 % and 3.9870 at 1 % (upper diagonal in Table 4.15). The difference was significant in between I-NDD & II-NDD, II-NDD & III-NDD, III-NDD & IV-NDD. The detailed comparisons among different life stage are presented in Table 4.15.

Similar trends of Granulocytes (GR) were recorded in life stages of DD generation. In I-DD larval forms, GR comprised of 50.6333 ± 0.4829 % of haemocytes. The percentage lowered in II-DD (34.1333 ± 0.3229 %), III-DD (34.5667 ± 0.4515 %) and IV-DD (44.4333 ± 0.9306 %). This percentage of GR was recorded to be highest in VEM-DD (58.1000 ± 1.9621 %) and VEF-DD (51.3333 ± 1.2287 %). A slight decrease in population percen-tage was recorded in VMM-DD & VMF-DD and it again elevated in VLF-DD (51.5000 ± 0.7513 %) & VLM-DD (51.0000 ± 2.5396 %). In PPF-DD and PPM-DD, their ratio was almost equal viz., 52.3000 ± 1.2198 % and 52.8333 ± 1.7520 %, respectively. It further lowered in MPF-DD, MPM-DD, LPF-DD and LPM-DD. However, in AF-DD, this percentage was 50.1000 ± 0.9362 % and in AM-DD, it was 52.8667 ± 1.3258 % (Table 4.17).

Within different age groups of DD generation of A. mylitta, mean values of GR were subjected to statistical analysis and CD at 5 % was worked out to be 2.0499 and at 1 % 2.8985. The change in percent population was significant at

1 % level in I-DD & II-DD; III-DD & IV-DD; IV-DD & VEM-DD; VEF-DD, VMF-DD & VMM-DD: PPM-DD & MPF-DD; LPM-DD & AF-DD. The difference was significant at 5 % level in between SLF-DD and SLM-DD, AF-DD and AM-DD (Table 4.16).

The mean percent population of Granulocytes were compared to record the difference between two generations, viz., non-diapausing and diapausing applying Student's t-test ($n = 15$). Differences were recorded to be significant ($p < 0.01$) in between II-NDD & II-DD, IV-NDD & IV-DD, VEF-NDD & VEF-DD, VEM-NDD & VEM-DD, VMM-NDD & VMM-DD, VMF-NDD & VMF-DD, VLF-NDD & VLF-DD, SLM-NDD & SLM-DD, PPF-NDD & PPF-DD, PPM-NDD & PPM-DD, and MPM-NDD & MPM-DD. The difference was significant ($p < 0.05$) in between LPF-NDD & LPF-DD, AF-NDD & AF-DD, and AM-NDD & AM-DD (Table 4.17). Rest of the comparisons were non-significant.

Table 4.17. Pèrcent Granulocytes in different stages of NDD & DD generations of *A. mylitta*

SN	Age	Mean	±	SE	t Stat	
1	I-NDD	51.0667	±	1.7624	0.3381	NS
2	I-DD	50.6333	±	0.4829		
3	II-NDD	28.4667	±	1.1045	-6.9444	**
4	II-DD	34.1333	±	0.3229		
5	III-NDD	34.7244	±	0.5133	0.2955	NS
6	III-DD	34.5667	±	0.4515		
7	IV-NDD	39.6652	±	0.8595	-6.4028	**
8	IV-DD	44.4333	±	0.9306		
9	VEF-NDD	57.6000	±	0.6805	5.9502	**
10	VEF-DD	51.3333	±	1.2287		
11	VEM-NDD	52.5000	±	0.6360	-3.6343	**
12	VEM-DD	58.1000	±	1.9621		
13	VMF-NDD	51.9000	±	1.4685	3.0639	**
14	VMF-DD	47.3667	±	1.0424		
15	VMM-NDD	46.9333	±	1.1389	-2.9488	**
16	VMM-DD	50.3000	±	1.0084		
17	VLF-NDD	55.2333	±	1.0829	4.1260	**
18	VLF-DD	51.5000	±	0.7513		
19	VLM-NDD	49.6000	±	1.0202	-0.7972	NS
20	VLM-DD	51.0000	±	2.5395		
21	SLF-NDD	51.9667	±	0.9763	1.5092	NS
22	SLF-DD	50.7000	±	0.4206		
23	SLM-NDD	53.8667	±	1.0944	3.4438	**
24	SLM-DD	49.5667	±	0.9635		
25	PPF-NDD	44.9667	±	1.0099	-8.7403	**
26	PPF-DD	52.3000	±	1.2198		
27	PPM-NDD	43.8333	±	1.6244	-8.4363	**
28	PPM-DD	52.8333	±	1.7520		
29	MPF-NDD	42.1333	±	1.9849	0.1381	NS
30	MPF-DD	41.9333	±	1.1905		
31	MPM-NDD	34.2000	±	2.2802	-3.9199	**
32	MPM-DD	40.2000	±	0.6175		
33	LPF-NDD	34.5333	±	3.6911	-2.0501	*
34	LPF-DD	40.2667	±	1.4770		
35	LPM-NDD	41.2333	±	2.3179	-0.1958	NS
36	LPM-DD	41.5667	±	0.9089		
37	AF-NDD	53.9667	±	2.3565	2.1453	*
38	AF-DD	50.1000	±	0.9362		
39	AM-NDD	49.3333	±	1.8663	-2.5143	*
40	AM-DD	52.8667	±	1.3258		

n = 15 (Abbreviations used are as per section 3.1.2.)

4.4.1.5. Plasmatocytes

Plasmatocytes comprised of 30.7667 ± 1.9469 percent of the total cell population in I-NDD, 46.2333 ± 1.2839 % in II-NDD, 35.6310 ± 0.5011 % in III-NDD and 30.0272 ± 0.5784 % in IV-NDD age groups of NDD generation. Among V instar male and female larval stages, highest population was recorded in VEM-NDD (30.4667 ± 0.6682 %) and 22.4000 ± 0.9711 % in VLF-NDD. The percent population of PLs was generally higher in pupal stages and adults when compared with their female counterparts. In NDD generation, PLs were recorded to be highest in adults of both the sexes (Table 4.18).

Critical difference within the NDD stages was worked out to be 2.7626 at 5 % and 3.9062 at 1 %. Differences in terms of mean decrease or increase were significant in almost all the stages at 1 %, except in between VMM-NDD & VLF-NDD, VLF-NDD & VLM-NDD, SLF-NDD & SLM-NDD, PPF-NDD & PPM-NDD, LPM-NDD & AF-NDD where differences were observed to be non-significant. The differences were significant at 5 % in between VMF-NDD & VMM-NDD, MPF-NDD & MPM-NDD (Table 4.19).

PLs comprised of 30.5333 ± 0.4344 % in I-DD, 38.3667 ± 0.2941 % in II-DD, 32.0000 ± 0.4667 % in III-DD, 28.1000 ± 0.7519 % in IV-DD. Fluctuating trend was recorded in later stages of V instar, early, middle and late pupae. Population of PLs attained its peak during adult stages of both the sexes (Table 4.18).

One way ANOVA was carried out in between PL percent values of DD life stages. Critical difference between the percent availability of PLs in DD life stages was worked out, which was 2.1056 at 5 % and 2.9772 at 1 %. The decrease or increase of mean population percentage in different stages was recorded to be non-significant in between VMF-DD & VMM-DD, VMM-DD & VLF-DD, SLF-DD & SLM-DD, MPF-DD & MPM-DD, LPF-DD & LPM-DD, AF-DD & AM-DD. Differences were significant at 5 % in between IV-DD & VEF-DD and PPF-DD & PPM-DD. Rest of the comparisons were significant at 1 %.

Table 4.18. Percent Plasmatocytes in different stages of NDD & DD generations of *A. mylitta*

SN	Age	Mean	±	SE	t Stat	
1.	I-NDD	30.7667	±	1.9469	0.1697	NS
2.	I-DD	30.5333	±	0.4344		
3.	II-NDD	46.2333	±	1.2839	9.0058	**
4.	II-DD	38.3667	±	0.2941		
5.	III-NDD	35.6310	±	0.5011	8.2469	**
6.	III-DD	32.0000	±	0.4667		
7.	IV-NDD	30.0272	±	0.5784	2.9131	**
8.	IV-DD	28.1000	±	0.7519		
9.	VEF-NDD	22.3333	±	0.7679	-3.1459	**
10.	VEF-DD	25.9000	±	1.2758		
11.	VEM-NDD	30.4667	±	0.6682	7.0352	**
12.	VEM-DD	19.6667	±	1.9364		
13.	VMF-NDD	20.9000	±	1.5058	-2.2510	*

14.	VMF-DD	24.6000	±	1.0375		
15.	VMM-NDD	24.1667	±	1.1901	0.7351	NS
16.	VMM-DD	23.4000	±	1.2862		
17.	VLF-NDD	22.4000	±	0.9711	-3.5707	**
18.	VLF-DD	25.1133	±	0.8728		
19.	VLM-NDD	22.5667	±	1.0453	0.7101	NS
20.	VLM-DD	21.2800	±	2.3602		
21.	SLF-NDD	27.8333	±	1.4902	-2.0072	*
22.	SLF-DD	30.5667	±	0.7479		
23.	SLM-NDD	29.0000	±	1.0914	-1.1780	NS
24.	SLM-DD	30.5000	±	1.0132		
25.	PPF-NDD	34.5000	±	1.1643	7.5153	**
26.	PPF-DD	28.9000	±	1.2723		
27.	PPM-NDD	35.6667	±	1.6285	7.5946	**
28.	PPM-DD	26.7000	±	1.7985		
29.	MPF-NDD	18.8333	±	2.0824	1.2723	NS
30.	MPF-DD	16.9000	±	0.5546		
31.	MPM-NDD	22.4333	±	2.5913	3.1818	**
32.	MPM-DD	16.6000	±	0.7665		
33.	LPF-NDD	26.4333	±	2.2330	-2.2038	*
34.	LPF-DD	30.5667	±	1.5873		
35.	LPM-NDD	34.5000	±	3.0558	2.2946	*
36.	LPM-DD	29.1333	±	1.4794		
37.	AF-NDD	36.9667	±	2.0292	-2.4272	*
38.	AF-DD	41.3667	±	0.9809		
39.	AM-NDD	46.6333	±	1.7185	4.6038	**
40.	AM-DD	39.3000	±	1.4419		

n = 15 (Abbreviations used are as per section 3.1.2.)

The mean percent population of Plasmatocytes were compared to record the difference between two generations, viz., non-diapausing and diapausing applying Student's t-test (n = 15). The differences were found to be significant ($p < 0.01$) in almost all the stages except in between VMF-NDD & VMF-DD, SLF-NDD & SLF-DD, LPF-NDD & LPF-DD, LPM-NDD & LPM-DD, AF-NDD & AF-DD where the differences were significant at $P < 0.05$. The difference was observed to be non-significant in between I-NDD & I-DD, VMM-NDD & VMM-DD, SLM-NDD & SLM-DD and MPF-NDD & MPM-DD (Tables 4.18).

4.4.1.6. Mitotic cells

Cell increase occurs partly through growth and differentiation of haemocytes and partly through mitotic division of other cells. During course of present study, counts were also done of the visible mitotic cells. In I-NDD, the percent mitotic cells were 1.9667 ± 0.2786 %, in II-NDD they were 1.6000 ± 0.2512 %, thereafter they were below 1 % up to

Table 4.19 : Test of significance of DHC mean using CD at 5% and 1% in each age group in NDD generation.

Age	I-NDD	II-NDD	III-NDD	IV-NDD	VEF-NDD	VEM-NDD	VMF-NDD	VMM-NDD	VLF-NDD	VLM-NDD	SLF-NDD	SLM-NDD	PPF-NDD	PPM-NDD	MPF-NDD	MPM-NDD	LPF-NDD	LPM-NDD	AF-NDD	AM-NDD
I-NDD	PL																			
II-NDD	**																			
III-NDD	**	**																		
IV-NDD	NS	**	**																	
VEF-NDD	**	**	**	**																
VEM-NDD	NS	**	**	NS	**															
VMF-NDD	**	**	**	**	NS	**														
VMM-NDD	**	**	**	**	NS	**	*													
VLF-NDD	**	**	**	**	NS	**	NS	NS												
VLM-NDD	**	**	**	**	NS	**	NS	NS	NS											
SLF-NDD	*	**	**	NS	**	NS	**	*	**	**										
SLM-NDD	NS	**	**	NS	**	NS	**	**	**	**	NS									
PPF-NDD	*	**	NS	**	**	**	**	**	**	**	**	**								
PPM-NDD	**	**	NS	**	**	**	**	**	**	**	**	**	NS							
MPF-NDD	**	**	**	**	*	**	NS	**	*	*	**	**	**	**						
MPM-NDD	**	**	**	**	NS	**	NS	NS	NS	NS	**	**	**	**	*					
LPF-NDD	**	**	**	*	**	**	**	NS	**	*	NS	NS	**	**	**	**				
LPM-NDD	*	**	NS	**	**	**	**	**	**	**	**	**	NS	NS	**	**	**			
AF-NDD	**	**	NS	**	**	**	**	**	**	**	**	**	NS	NS	**	**	**	NS		
AM-NDD	**	NS	**	**	**	**	**	**	**	**	**	**	**	**	**	**	**	**	**	
	PL																			
CD 5%	2.7626																			
CD 1%	3.9062																			

Table 4.20 : Test of significance of DHC mean using CD at 5% and 1% in each age group in DD generation.

Age	I-DD	II-DD	III-DD	IV-DD	VEF-DD	VEM-DD	VMF-DD	VMM-DD	VLF-DD	VLM-DD	SLF-DD	SLM-DD	PPF-DD	PPM-DD	MPF-DD	MPM-DD	LPF-DD	LPM-DD	AF-DD	AM-DD
I-DD	PL																			
II-DD	**																			
III-DD	NS	**																		
IV-DD	*	**	**																	
VEF-DD	**	**	**	*																
VEM-DD	**	**	**	**	**															
VMF-DD	**	**	**	**	NS	**														
VMM-DD	**	**	**	**	*	**	NS													
VLF-DD	**	**	**	**	NS	**	NS	NS												
VLM-DD	**	**	**	**	**	NS	**	*	**											
SLF-DD	NS	**	NS	*	**	**	**	**	**	**										
SLM-DD	NS	**	NS	*	**	**	**	**	**	**	NS									
PPF-DD	NS	**	**	NS	**	**	**	**	**	**	NS	NS								
PPM-DD	**	**	**	NS	NS	**	NS	**	NS	**	**	**	*							
MPF-DD	**	**	**	**	**	*	**	**	**	**	**	**	**	**						
MPM-DD	**	**	**	**	**	**	**	**	**	**	**	**	**	**	NS					
LPF-DD	NS	**	NS	*	**	**	**	**	**	**	NS	NS	NS	**	**	**				
LPM-DD	NS	**	*	NS	**	**	**	**	**	**	NS	NS	NS	*	**	**	NS			
AF-DD	**	**	**	**	**	**	**	**	**	**	**	**	**	**	**	**	**	**		
AM-DD	**	NS	**	**	**	**	**	**	**	**	**	**	**	**	**	**	**	**	NS	
	PL																			
CD 5%	2.1056																			
CD 1%	2.9772																			

SLM-NDD and SLF-NDD stage. A jump in percent mitotic cells was recorded in PPF-NDD (3.7667 ± 0.2297 %) and PPM-NDD (3.4000 ± 0.2844 %) and thereafter their percentage dwindled down to 0.0333 ± 0.0455 % in AF-NDD and no mitotic cells were observed in AM-NDD (Table-4.21).

Table 4.21. Percent mitotic cells in different stages of NDD & DD generations of *A. mylitta*

SN	Age	Mean	±	SE	t Stat	
1	I-NDD	1.9667	±	0.2786	1.3484	NS
2	I-DD	1.6333	±	0.1695		
3	II-NDD	1.6000	±	0.2512	2.8062	**
4	II-DD	1.0000	±	0.1155		
5	III-NDD	0.0933	±	0.0686	-0.5439	NS
6	III-DD	0.1333	±	0.0807		
7	IV-NDD	0.8665	±	0.1695	2.6934	**
8	IV-DD	0.4333	±	0.1311		
9	VEF-NDD	0.1667	±	0.1089	-2.3577	*
10	VEF-DD	0.4667	±	0.1409		
11	VEM-NDD	0.1000	±	0.0989	-4.4037	**
12	VEM-DD	1.0333	±	0.2622		
13	VMF-NDD	0.0333	±	0.0455	0.0000	NS
14	VMF-DD	0.0333	±	0.0455		
15	VMM-NDD	0.0667	±	0.0621	-1.4676	NS
16	VMM-DD	0.2000	±	0.0894		
17	VLF-NDD	0.1467	±	0.0903	1.4419	NS
18	VLF-DD	0.0333	±	0.0455		
19	VLM-NDD	0.0800	±	0.0757	-0.2379	NS
20	VLM-DD	0.1000	±	0.0730		
21	SLF-NDD	0.1333	±	0.1047	-3.2139	**
22	SLF-DD	0.5000	±	0.1155		
23	SLM-NDD	0.1000	±	0.0989	-4.0256	**
24	SLM-DD	0.5333	±	0.1241		
25	PPF-NDD	3.7667	±	0.2297	1.8991	*
26	PPF-DD	3.3333	±	0.2553		
27	PPM-NDD	3.4000	±	0.2844	1.4338	NS
28	PPM-DD	3.0333	±	0.1241		
29	MPF-NDD	0.1000	±	0.0989	-2.7500	**
30	MPF-DD	0.4667	±	0.1241		
31	MPM-NDD	0.2000	±	0.1300	-2.9562	**
32	MPM-DD	0.6667	±	0.1089		
33	LPF-NDD	0.1667	±	0.1440	-8.8056	**
34	LPF-DD	1.3667	±	0.1047		
35	LPM-NDD	0.0333	±	0.0455	-4.7402	**
36	LPM-DD	1.0000	±	0.2749		
37	AF-NDD	0.0333	±	0.0455	1.0000	NS
38	AF-DD	0.0000	±	0.0000		
39	AM-NDD	0.0000	±	0.0000	-1.0000	NS
40	AM-DD	0.0333	±	0.0455		

n = 15; Abbreviations used are as per section 3.1.2.

Once the above values put for statistical analysis to find out the differences between the stages, the CD was 0.2472 at 5% and 0.3495 at 1%. The decrease or increase was significant at 1% in between I-NDD & II-NDD; II-NDD & III-NDD; III-NDD & IV-NDD; IV-NDD & VEF-NDD; SLM-NDD & PPF-NDD; PPF-NDD & PPM-NDD and PPM-NDD & MPF-NDD (Table 4.22). Rest of the mean differences were non-significant.

In DD life stages, the percent mitotic cells were highest in I-DD (1.6333 ± 0.1695 %) and in II-DD they were recorded to be 1.0000 ± 0.1155 % and below 1.0000 % in III-DD and IV-DD. A rise in their percentage was recorded in VEM-DD (1.0333 ± 0.2622 %), in PPF-DD (3.3333 ± 0.2553 %) and PPM-DD (3.0333 ± 0.1241 %) and a decrease in percentage was recorded in LPF-DD (0.1667 ± 0.1440 %) and LPM-DD (1.0000 ± 0.2749 %). No mitotic cells were recorded in AF-DD (Table 4.21).

Critical difference was worked out to be 0.4257 at 5 % and 0.6020 at 1 % between the percent mean values of different stages of DD generation. The difference in mean percent mitotic cells was significant at 1 % in between I-DD & II-DD, II-DD & III-DD, VEM-DD & VMF-DD, SLM-DD & PPF-DD, PPM-DD & MPF-DD, MPM-DD & LPF-DD, LPF-DD & LPM-DD, LPM-DD & AF-DD. This difference was significant at 5 % in between VEF-DD and VEM-DD. Rest of the comparisons was non-significant (Table 4.23).

The mean percent population of mitotic cells were compared to record the difference between two generations, viz., non-diapausing and diapausing applying Student's t-test (n = 15). The difference was significant at 5 % in between VEF-NDD & VEF-DD, PPF-NDD & PPF-DD, however, the same was significant at 1 % in between II-NDD & II-DD, IV-NDD & IV-DD, VEM- NDD & VEM-DD, SLF-NDD & SLF-DD, SLM-NDD & SLM-DD, MPF-NDD & MPF-DD, MPM-NDD & MPM-DD, LPF-NDD & LPF-DD, LPM-NDD & LPM-DD. Rest of the comparisons were non-significant (Table 4.21).

4.4.1.7. Degenerated cells:

The traces of degenerated cells were also observed during course of study. They were almost negligible in NDD life stages but their concentration was very high in MPF-NDD (30.2000 ± 2.4701 %), MPM-NDD (32.6000 ± 2.5159 %), LPF-NDD (22.8000 ± 4.2315 %) and in LPM-NDD (13.1667 ± 2.3605 %) - (Table 4.24).

To find out the difference between mean percent degenerated cells of NDD generation different stages, the percent data was subjected for statistical analysis and CD at 5 % was 2.2871 and at 1 % 3.2339. The difference was found to be significant at 1 % in between PPM-NDD and MPF-NDD, MPM-NDD & LPF-NDD, LPF-NDD & LPM-NDD, LPM-NDD & AF-NDD. The difference was significant at 5 % in between MPF-NDD and MPM-NDD. Rest of the comparisons were non-significant (Table 4.22).

In DD life stages, degenerated cells were recorded to be 0.2333 ± 0.0911 % in I instar DD and it fluctuated around this level up to prepupal stage. A jump in the level of degenerated cells was observed in MPF-DD (30.3333 ± 1.0701 %), MPM-DD (29.3667 ± 0.5516 %). The level was observed to be 9.8333 ± 0.5235 % in LPF-DD and 9.8667 ± 0.1940 % in LPM-DD. The level was recorded to be lowest in adults of both the sexes (Table 4.24).

Table 4.22 : Test of significance of DHC mean using CD at 5% and 1% in each age group in NDD generation.

Age	I-NDD	II-NDD	III-NDD	IV-NDD	VEF-NDD	VEM-NDD	VMF-NDD	VMM-NDD	VLF-NDD	VLM-NDD	SLF-NDD	SLM-NDD	PPF-NDD	PPM-NDD	MPF-NDD	MPM-NDD	LPF-NDD	LPM-NDD	AF-NDD	AM-NDD
I-NDD	MC	NS	NS	NS	NS	NS	NS	NS	NS	NS	NS	NS	NS	NS	**	**	**	**	NS	NS
II-NDD	**		NS	NS	NS	NS	NS	NS	NS	NS	NS	NS	NS	NS	**	**	**	**	NS	NS
III-NDD	**	**		NS	NS	NS	NS	NS	NS	NS	NS	NS	NS	NS	**	**	**	**	NS	NS
IV-NDD	**	**	**		NS	NS	NS	NS	NS	NS	NS	NS	NS	NS	**	**	**	**	NS	NS
VEF-NDD	**	**	NS	**		NS	NS	NS	NS	NS	NS	NS	NS	NS	**	**	**	**	NS	NS
VEM-NDD	**	**	NS	**	NS		NS	NS	NS	NS	NS	NS	NS	NS	**	**	**	**	NS	NS
VMF-NDD	**	**	NS	**	NS	NS		NS	NS	NS	NS	NS	NS	NS	**	**	**	**	NS	NS
VMM-NDD	**	**	NS	**	NS	NS	NS		NS	NS	NS	NS	NS	NS	**	**	**	**	NS	NS
VLF-NDD	**	**	NS	**	NS	NS	NS	NS		NS	NS	NS	NS	NS	**	**	**	**	NS	NS
VLM-NDD	**	**	NS	**	NS	NS	NS	NS	NS		NS	NS	NS	NS	**	**	**	**	NS	NS
SLF-NDD	**	**	NS	**	NS	NS	NS	NS	NS	NS		NS	NS	NS	**	**	**	**	NS	NS
SLM-NDD	**	**	NS	**	NS	NS	NS	NS	NS	NS	NS		NS	NS	**	**	**	**	NS	NS
PPF-NDD	**	**	**	**	**	**	**	**	**	**	**	**		NS	**	**	**	**	NS	NS
PPM-NDD	**	**	**	**	**	**	**	**	**	**	**	**	**		**	**	**	**	NS	NS
MPF-NDD	**	**	NS	**	NS	NS	NS	NS	NS	NS	NS	NS	**	**		*	**	**	**	**
MPM-NDD	**	**	NS	**	NS	NS	NS	NS	NS	NS	NS	NS	**	**	NS		**	**	**	**
LPF-NDD	**	**	NS	**	NS	NS	NS	NS	NS	NS	NS	NS	**	**	NS	NS		**	**	**
LPM-NDD	**	**	NS	**	NS	NS	NS	NS	NS	NS	NS	NS	**	**	NS	NS	NS		**	**
AF-NDD	**	**	NS	**	NS	NS	NS	NS	NS	NS	NS	NS	**	**	NS	NS	NS	NS		NS
AM-NDD	**	**	NS	**	NS	NS	NS	NS	NS	NS	NS	NS	**	**	NS	NS	NS	NS	NS	De
	MC		Lower diagonal												Upper diagonal				Deg	
CD 5%	0.2472																		2.2871	
CD 1%	0.3495																		3.2339	

Table 4.23 : Test of significance of DHC mean using CD at 5% and 1% in each age group in DD generation.

Age	I-DD	II-DD	III-DD	IV-DD	VEF-DD	VEM-DD	VMF-DD	VMM-DD	VLF-DD	VLM-DD	SLF-DD	SLM-DD	PPF-DD	PPM-DD	MPF-DD	MPM-DD	LPF-DD	LPM-DD	AF-DD	AM-DD
I-DD	MC	NS	*	*	**	**	NS	NS	NS	NS	NS	NS	**	**	**	**	**	**	NS	NS
II-DD	**		NS	*	**	**	NS	NS	NS	NS	NS	NS	**	**	**	**	**	**	NS	NS
III-DD	**	**		NS	NS	**	*	NS	*	*	*	*	*	**	**	**	**	**	*	*
IV-DD	**	*	NS		NS	**	**	NS	*	**	**	**	NS	**	**	**	**	**	*	*
VEF-DD	**	*	NS	NS		**	**	**	**	**	**	**	NS	**	**	**	**	**	**	**
VEM-DD	*	NS	**	*	*		**	**	**	**	**	**	**	NS	**	**	**	**	**	**
VMF-DD	**	**	NS	NS	*	**		NS	NS	NS	NS	NS	**	**	**	**	**	**	NS	NS
VMM-DD	**	**	NS	NS	NS	**	NS		NS	NS	NS	NS	**	**	**	**	**	**	NS	NS
VLF-DD	**	**	NS	NS	*	**	NS	NS		NS	NS	NS	**	**	**	**	**	**	NS	NS
VLM-DD	**	**	NS	NS	NS	**	NS	NS	NS		NS	NS	**	**	**	**	**	**	NS	NS
SLF-DD	**	*	NS	NS	NS	*	*	NS	*	NS		NS	**	**	**	**	**	**	NS	NS
SLM-DD	**	*	NS	NS	NS	*	*	NS	*	*	NS		**	**	**	**	**	**	NS	NS
PPF-DD	**	**	**	**	**	**	**	**	**	**	**	**		*	**	**	**	**	**	**
PPM-DD	**	**	**	**	**	**	**	**	**	**	**	**	NS		**	**	**	**	**	**
MPF-DD	*	*	NS	NS	NS	*	*	NS	*	NS	NS	NS	**	**		**	**	**	**	**
MPM-DD	**	NS	*	NS	NS	NS	**	*	**	*	NS	NS	**	**	NS		**	**	**	**
LPF-DD	NS	**	**	**	**	**	**	**	**	**	**	**	**	**	**	**		NS	**	**
LPM-DD	**	NS	**	*	*	NS	**	**	**	**	*	*	**	**	*	NS	**		**	**
AF-DD	**	**	NS	*	*	**	NS	NS	NS	NS	*	*	**	**	*	**	**	**		NS
AM-DD	**	**	NS	NS	*	**	NS	NS	NS	NS	*	*	**	**	*	**	**	**	NS	Deg
	MC																			Deg
CD 5%	0.4257																			0.54645
CD 1%	0.6020																			0.77267

To find out the difference between mean percent degenerated cells of DD generation of different stages, the percent data was subjected for statistical analysis and CD at 5 % was 0.54645 and at 1 % 0.77267. The mean difference was significant at 1 % in between VEF-DD & VEM-DD, VEM-DD & VMF-DD, SLM-DD & PPF-DD, PPM-DD and MPF-DD, MPF-DD and MPM-DD, MPM-DD & LPF-DD, LPM-DD & AF-DD. The mean difference was significant at 5 % in PPF-DD & PPM-DD. Others comparisons were non-significant (Table 4.23).

Table 4.24. Percent degenerated cells in different stages of NDD & DD generations of *A. mylitta*

SN	Age	Mean	±	SE	t Stat	
1	I-NDD	0.1667	±	0.0861	-1.4676	NS
2	I-DD	0.2333	±	0.0911		
3	II-NDD	0.0000	±	0.0000	-4.0000	**
4	II-DD	0.2667	±	0.0911		
5	III-NDD	0.1867	±	0.1057	-4.5691	**
6	III-DD	0.7667	±	0.1615		
7	IV-NDD	0.3133	±	0.1738	-2.4313	*
8	IV-DD	0.8667	±	0.2051		
9	VEF-NDD	0.0333	±	0.0455	-5.0082	**
10	VEF-DD	1.3000	±	0.3386		
11	VEM-NDD	0.0667	±	0.0621	-14.2097	**
12	VEM-DD	2.2000	±	0.1738		
13	VMF-NDD	0.1333	±	0.0807	1.0000	NS
14	VMF-DD	0.0667	±	0.0621		
15	VMM-NDD	0.0333	±	0.0455	-4.5163	**
16	VMM-DD	0.4667	±	0.1241		
17	VLF-NDD	0.0533	±	0.0514	-1.3817	NS
18	VLF-DD	0.1533	±	0.0810		
19	VLM-NDD	0.0200	±	0.0273	-1.0000	NS
20	VLM-DD	0.0533	±	0.0514		
21	SLF-NDD	0.1000	±	0.0730	0.5641	NS
22	SLF-DD	0.0667	±	0.0621		
23	SLM-NDD	0.0333	±	0.0455	-1.0000	NS
24	SLM-DD	0.0667	±	0.0621		
25	PPF-NDD	0.9667	±	0.1695	-2.8627	**
26	PPF-DD	1.3667	±	0.1047		
27	PPM-NDD	1.2667	±	0.2297	-4.3770	**
28	PPM-DD	2.1333	±	0.1559		
29	MPF-NDD	30.2000	±	2.4701	-0.0629	NS
30	MPF-DD	30.3333	±	1.0701		
31	MPM-NDD	32.6000	±	2.5159	1.7237	NS
32	MPM-DD	29.3667	±	0.5516		
33	LPF-NDD	22.8000	±	4.2315	4.1636	**
34	LPF-DD	9.8333	±	0.5235		
35	LPM-NDD	13.1667	±	2.3605	1.8658	*
36	LPM-DD	9.8667	±	0.1940		
37	AF-NDD	0.0667	±	0.0911	-1.0000	NS
38	AF-DD	0.1000	±	0.0989		
39	AM-NDD	0.1000	±	0.0989	-0.5641	NS
40	AM-DD	0.1333	±	0.1047		

n = 15; Abbreviations used are as per section 3.1.2.

Student's t-test was applied to work out significant difference between mean values of NDD and DD life stages. The difference was significant at 5 % in between IV- NDD & IV-DD, LPM-NDD & LPM-DD and at 1 % in between II-NDD & II-DD, III-NDD & III-DD, VEF-NDD & VEF-DD, VEM-NDD & VEM-DD, VMM-NDD & VMM-DD, PPF-NDD & PPF-DD, PPM-NDD & PPM-DD and LPF-NDD & LPF-DD (Table 4.24).

4.4.2. DHC between the sexes of NDD and DD generations

4.4.2.1. Prohaemocytes

Student's t-test was calculated to find out difference in means between the two sexes of non-diapausing and diapausing generations. In NDD generation, significant difference in mean percent PR was observed at $p < 0.05$ in between VMF-NDD and VMM-NDD. Mean values differed significantly between VEF-NDD & VEM-NDD, PPF-NDD & PPM-NDD, LPF-NDD & LPM-NDD and AF-NDD & AM-NDD at $p < 0.01$. Females had generally higher percentage of PRs than males.

In DD generation too, females had higher mean percentage of PRs in early and mid-V instar larval stages (VEF-DD and VMF-DD) than males (VEM-DD and VLM-DD), but this trend reversed from late V instar larval and pupal stages (i.e., in between VLF-DD & VLM-DD, SLF-DD & SLM-DD, PPF-DD & PPM-DD, MPF-DD & MPM-DD and LPF-DD & LPM-DD). However, again adult females were having more PRs. The Difference was significant ($p < 0.01$), except in between AF-DD and AM-DD which was significant at $p < 0.05$. In late pupal stage of both the sexes the difference was observed to be non-significant (Table 4.25).

Table 4.25. Percent Prohaemocytes of females and males in different stages of NDD & DD generations of *A. mylitta*

SN	Stage	NDD Mean	±	SE	t Stat		DD Mean	±	SE	t Stat	
1	VEF	3.367	±	0.199	4.322	**	3.533	±	0.147	8.047	**
2	VEM	2.267	±	0.111			2.300	±	0.150		
3	VMF	1.267	±	0.207	-2.553	*	1.933	±	0.122	2.736	**
4	VMM	1.867	±	0.118			1.333	±	0.200		
5	VLF	1.133	±	0.128	0.745	NS	1.600	±	0.136	-6.468	**
6	VLM	1.000	±	0.160			2.767	±	0.111		
7	SLF	0.733	±	0.099	-0.323	NS	0.900	±	0.104	-6.831	**
8	SLM	0.767	±	0.122			1.900	±	0.104		
9	PPF	2.933	±	0.150	11.421	**	2.200	±	0.261	-4.490	**
10	PPF	0.833	±	0.157			3.167	±	0.157		
11	MPF	1.033	±	0.094	3.523	**	1.400	±	0.090	-1.705	NS
12	MPM	0.467	±	0.147			1.867	±	0.260		
13	LPF	1.200	±	0.132	0.702	NS	1.967	±	0.094	-5.000	**
14	LPM	1.033	±	0.155			2.800	±	0.141		
15	AF	1.000	±	0.124	6.441	**	1.100	±	0.161	2.048	*
16	AM	0.100	±	0.075			0.733	±	0.069		

n=15 Abbreviations used are as per section 3.1.2.

4.4.2.2. Oenocytoids

In NDD generation, male (VEM-NDD to VLM-NDD) had higher mean percentage of OEs than females up to late V instar (VEM-NDD to VLM-NDD). Spinning larvae females (SLF-NDD) contained higher mean percentage of OEs than males (SLM-NDD). In prepupal stage, male (PPM-NDD) had higher OEs than female (PPF-NDD), the trend was reversed in mid and late pupal stages (MPF-NDD and LPF-NDD) and adults (AF-NDD). Student's t-test indicated that differences were significant ($p < 0.05$) in between VLF-NDD & VLM-NDD and AF-NDD & AM-NDD. Difference at $p < 0.01$ was found to be significant in between VMF-NDD & VMM-NDD, and AF-NDD & AM-NDD.

In DD generation, difference in mean percent of OEs was found to be non-significant up to late age V instar (VLF-DD & VLM-DD), in pre-pupal stage (PPF-DD & PPM-DD) and in adults (AF-DD & AM-DD). Mean differed significantly ($p < 0.05$) in between SLF-DD & SLM-DD, LPF-DD & LPM-DD and at $p < 0.01$ in between MPF-DD & MPM-DD. Pre-adult females had higher OE, though the trend was reversed in adult stage (Table 4.26).

Table 4.26. Percent Oenocytoids of females and males in different stages of NDD & DD generations of *A. mylitta*

SN	Stage	NDD				DD				
		Mean	± SE	t Stat		Mean	±	SE	t Stat	
1	VEF	1.633	± 0.147	-0.168	NS	2.000	±	0.087	-1.662	NS
2	VEM	1.667	± 0.082			2.500	±	0.253		
3	VMF	2.067	± 0.150	-3.309	**	2.967	±	0.094	0.619	NS
4	VMM	2.767	± 0.142			2.833	±	0.200		
5	VLF	3.833	± 0.157	-2.553	*	2.667	±	0.130	-1.103	NS
6	VLM	4.133	± 0.128			2.867	±	0.138		
7	SLF	3.800	± 0.384	1.394	NS	2.767	±	0.122	1.790	*
8	SLM	3.267	± 0.132			2.467	±	0.128		
9	PPF	2.867	± 0.155	-2.343	*	2.067	±	0.271	0.000	NS
10	PPF	4.800	± 0.828			2.067	±	0.111		
11	MPF	1.833	± 0.097	0.094	NS	1.267	±	0.166	-4.740	**
12	MPM	1.800	± 0.356			2.233	±	0.219		
13	LPF	2.433	± 0.443	2.060	*	4.267	±	0.327	1.804	*
14	LPM	1.433	± 0.338			3.433	±	0.290		
15	AF	2.633	± 0.380	4.757	**	2.233	±	0.302	-0.757	NS
16	AM	0.500	± 0.237			2.467	±	0.094		

n=15 Abbreviations used are as per section 3.1.2.

4.4.2.3. Spherulocytes

In NDD generation, VEF-NDD, VLM-NDD, SLF-NDD, MPM-NDD, LPF-NDD and AF-NDD had higher SPs than their opposite counterparts. The difference was significant ($p < 0.01$) except in between VMF-NDD & VMM-NDD, PPF-NDD & PPM-NDD where no significant difference was observed.

Table 4.27. Percent Spherulocytes of females and males in different stages of NDD & DD generations of *A. mylitta*

SN	Stage	NDD					DD				
		Mean	± SE	t Stat			Mean	±	SE	t Stat	
1	VEF	14.867	± 0.348	5.209	**		15.467	±	0.427	2.785	**
2	VEM	12.933	± 0.159				14.200	±	0.213		
3	VMF	23.700	± 0.261	-1.308	NS		23.033	±	0.570	2.324	*
4	VMM	24.167	± 0.310				21.467	±	0.269		
5	VLF	17.200	± 0.391	-9.774	**		18.933	±	0.219	-10.703	**
6	VLM	22.600	± 0.368				21.933	±	0.262		
7	SLF	15.433	± 0.413	4.664	**		14.500	±	0.410	-1.273	NS
8	SLM	12.967	± 0.359				14.967	±	0.107		
9	PPF	10.000	± 0.728	-0.256	NS		9.833	±	0.157	-0.613	NS
10	PPF	10.200	± 0.569				10.067	±	0.410		
11	MPF	5.700	± 0.289	-3.607	**		7.133	±	0.205	-3.944	**
12	MPM	7.733	± 0.551				8.133	±	0.185		
13	LPF	12.600	± 0.389	3.459	**		12.300	±	0.289	-2.555	*
14	LPM	9.167	± 0.730				13.133	±	0.163		
15	AF	5.333	± 0.406	3.651	**		5.100	±	0.161	1.969	*
16	AM	3.333	± 0.415				4.467	±	0.269		

n=15 (Abbreviations used are as per section 3.1.2.)

In DD generation, the spherulocytes were significantly higher) in early ($p<0.01$) and mid stage ($p<0.05$) fifth instar larvae though the spherulocytes were higher (p<o.o1) in VLM-DD, MPM-DD and MPF-DD stages than respective stages of females The differences were significant ($p < 0.05$) in between VMF-DD & VMM-DD, LPF-DD & LPM-DD, AF-NDD & AM-NDD and at $p < 0.01$ in between VLF-DD & VLM-DD, MPF-DD & MPM-DD (Table 4.27).

4.4.2.4. Granulocytes

In NDD generation, GRs were higher in VEF-NDD, VMF-NDD, VLF-NDD, SLM-NDD, PPF-NDD, MPF-NDD, LPM-NDD and AF-NDD than their respective opposite sex stages. The differences were significant ($p < 0.05$) between LPF-NDD & LPM-NDD, AF-NDD & AM-NDD and at $p < 0.01$ between VEF-NDD & VEM-NDD, VMF-NDD & VMM-NDD, MPF-NDD & MPM-NDD, and VLF-NDD & VLM-NDD. Rest of the comparisons were non-significant.

In DD generation, GRs were higher in VEM-DD, VMM-DD, VLF-DD, SLF-DD, PPM-DD, MPF-DD, LPM-DD and AM-DD stages than their respective opposite sex stages. The difference was significant ($p < 0.01$) between VEF-DD & VEM-DD, VMF-DD & VMM-DD, and AF-DD & AM-DD. However, in remaining stages differences were observed to be non-significant (Table 4.28).

Table 4.28. Percent Granulocytes of females and males in different stages of NDD & DD generations of *A. mylitta*.

SN	Stage	NDD					DD				
		Mean	± SE	t Stat			Mean	±	SE	t Stat	
1	VEF	57.600	± 0.516	6.830	**		51.333	±	0.931	-3.786	**
2	VEM	52.500	± 0.482				58.100	±	1.487		
3	VMF	51.900	± 1.113	4.161	**		47.367	±	0.790	-3.080	**
4	VMM	46.933	± 0.863				50.300	±	0.764		
5	VLF	55.233	± 0.820	4.838	**		51.500	±	0.569	0.257	NS
6	VLM	49.600	± 0.773				51.000	±	1.924		
7	SLF	51.967	± 0.740	-1.440	NS		50.700	±	0.319	1.459	NS
8	SLM	53.867	± 0.829				49.567	±	0.730		
9	PPF	44.967	± 0.765	0.763	NS		52.300	±	0.924	-0.386	NS
10	PPF	43.833	± 1.231				52.833	±	1.327		
11	MPF	42.133	± 1.504	3.170	**		41.933	±	0.902	1.664	NS
12	MPM	34.200	± 1.727				40.200	±	0.468		
13	LPF	34.533	± 2.796	-1.952	*		40.267	±	1.119	-0.978	NS
14	LPM	41.233	± 1.756				41.567	±	0.689		
15	AF	53.967	± 1.785	1.979	*		50.100	±	0.709	-2.985	**
16	AM	49.333	± 1.414				52.867	±	1.004		

n=15 (Abbreviations used are as per section 3.1.2.)

4.4.2.5. Plasmatocytes

In larval stages of NDD generations PLs were higher in all stages of female larvae from early fifth instar to spinning stage. Similar trend was also observed in pupal and adult stages. The differences were significant (P <0.01) in between VEF-NDD & VEM-NDD, LPF-NDD & LPM-NDD, AF-NDD & AM-NDD and at p<0.01, in between VMF-NDD & VMM-NDD (Table 4.29).

In DD generation also, the different stage larvae, pupae and adult females had higher Plasmatocytes than their respective opposite stages of males. The difference was found to be significant (P < 0.01) in between VEF-DD & VEM-DD and at p<0.05, in between VLF-DD & VLM-DD, AF-DD & AM-DD - Table 4.29.

4.4.2.6. Mitotic cells

In NDD generation, no marked difference in percent population of mitotic cells was observed in between the respective stages of two sexes. Similar trend was also observed in DD generation where differences were non-significant in almost all the stages except in between VEF-DD & VEM-DD, VMF-DD & VMM-DD, and MPF-DD & MPM-DD where significant difference was observed (p<0.05) - Table 4.30.

Table 4.29. Percent Plasmatocytes of females and males in different stages of NDD & DD generations of *A. mylitta*

SN	Stage	NDD				DD			
		Mean	± SE	t Stat		Mean	± SE	t Stat	
1	VEF	22.333	± 0.582	-9.506	**	25.900	± 0.967	3.667	**
2	VEM	30.467	± 0.506			19.667	± 1.467		
3	VMF	20.900	± 1.141	-2.556	*	24.600	± 0.786	1.279	NS
4	VMM	24.167	± 0.902			23.400	± 0.974		
5	VLF	22.400	± 0.736	-0.166	NS	25.113	± 0.661	1.989	*
6	VLM	22.567	± 0.792			21.280	± 1.788		
7	SLF	27.833	± 1.129	-0.730	NS	30.567	± 0.567	0.067	NS
8	SLM	29.000	± 0.827			30.500	± 0.768		
9	PPF	34.500	± 0.882	-1.290	NS	28.900	± 0.964	1.557	NS
10	PPF	35.667	± 1.234			26.700	± 1.363		
11	MPF	18.833	± 1.578	-1.359	NS	16.900	± 0.420	0.461	NS
12	MPM	22.433	± 1.963			16.600	± 0.581		
13	LPF	26.433	± 1.692	-2.674	**	30.567	± 1.203	0.841	NS
14	LPM	34.500	± 2.315			29.133	± 1.121		
15	AF	36.967	± 1.537	-4.798	**	41.367	± 0.743	1.779	*
16	AM	46.633	± 1.302			39.300	± 1.092		

n=15 (Abbreviations used are as per section 3.1.2.)

Table 4.30. Percent mitotic cells of females and males in different stages of NDD & DD generations of *A. mylitta*

SN	Stage	NDD				DD			
		Mean	± SE	t Stat		Mean	± SE	t Stat	
1	VEF	0.167	± 0.082	1.000	NS	0.467	± 0.107	-2.605	*
2	VEM	0.100	± 0.075			1.033	± 0.199		
3	VMF	0.033	± 0.035	-1.000	NS	0.033	± 0.035	-2.092	*
4	VMM	0.067	± 0.047			0.200	± 0.068		
5	VLF	0.147	± 0.068	1.000	NS	0.033	± 0.035	-1.000	NS
6	VLM	0.080	± 0.057			0.100	± 0.055		
7	SLF	0.133	± 0.079	0.564	NS	0.500	± 0.087	-0.292	NS
8	SLM	0.100	± 0.075			0.533	± 0.094		
9	PPF	3.767	± 0.174	1.408	NS	3.333	± 0.193	1.718	NS
10	PPF	3.400	± 0.215			3.033	± 0.094		
11	MPF	0.100	± 0.075	-1.146	NS	0.467	± 0.094	-1.871	*
12	MPM	0.200	± 0.098			0.667	± 0.082		
13	LPF	0.167	± 0.109	1.740	NS	1.367	± 0.079	1.661	NS
14	LPM	0.033	± 0.035			1.000	± 0.208		
15	AF	0.033	± 0.035	1.000	NS	0.000	± 0.000	-1.000	NS
16	AM	0.000	± 0.000			0.033	± 0.035		

n=15 (Abbreviations used are as per section 3.1.2.)

4.4.2.7. Degenerated cells

In NDD generation presence of degenerated cells was observed to be non-significant in between different stages of two sexes except in between LPF-NDD & LPM-NDD where difference was significant ($P < 0.01$). In DD generation in early, mid and late V instars male had significant higher ($p < 0.01$ or $p < 0.05$) percentage of degenerated cells than females; this was also true in pre-pupal stage. In VLM-DD degenerated cells were significantly higher ($P<0.05$) than VLF-DD. In remaining stages the difference was non-significant - Table 4.31.

Table 4.31. Percent degenerated cells of females and males in different stages of NDD & DD generations of *A. mylitta*

SN	Stage	NDD					DD				
		Mean	±	SE	t Stat		Mean	±	SE	t Stat	
1	VEF	0.033	±	0.035	-1.000	NS	1.300	±	0.257	-3.674	**
2	VEM	0.067	±	0.047			2.200	±	0.132		
3	VMF	0.133	±	0.061	1.382	NS	0.067	±	0.047	-5.527	**
4	VMM	0.033	±	0.035			0.467	±	0.094		
5	VLF	0.053	±	0.039	1.000	NS	0.153	±	0.061	1.871	*
6	VLM	0.020	±	0.021			0.053	±	0.039		
7	SLF	0.100	±	0.055	1.468	NS	0.067	±	0.047	0.000	NS
8	SLM	0.033	±	0.035			0.067	±	0.047		
9	PPF	0.967	±	0.128	1.057	NS	1.367	±	0.079	-4.766	**
10	PPF	0.797	±	0.083			2.133	±	0.118		
11	MPF	30.200	±	1.871	-0.872	NS	30.333	±	0.811	1.189	NS
12	MPM	32.600	±	1.906			29.367	±	0.418		
13	LPF	22.800	±	3.206	2.825	**	9.833	±	0.397	-0.093	NS
14	LPM	13.167	±	1.788			9.867	±	0.147		
15	AF	0.067	±	0.069	-1.000	NS	0.100	±	0.075	-0.564	NS
16	AM	0.100	±	0.075			0.133	±	0.079		

n=15 (Abbreviations used are as per section 3.1.2.)

Summery data showing test of significance among means of differential haemocyte counts using CD at 5% and 1% in each age group of both NDD and DD generations for all types of haemocytes are given in Table 4.32 & 4.33.

Table 4.32. Summary of Differential Haemocyte Counts of non-diapausing generation of *Antheraea mylitta*.

Age	Prohaemocytes			Oenocytes			Spherulocytes			Granulocytes			Plasmatocytes			Mitotic cells			Degenerated cells		
	Mean	±	SE	Mean	±	SE	Mean	±	SE	Mean	±	SE	Mean	±	SE	Mean	±	SE	Mean	±	SE
I-NDD	8.1667	±	0.1517	0.5667	±	0.0959	7.3000	±	0.4572	51.0667	±	1.2899	30.7667	±	1.4250	1.9667	±	0.2039	0.1667	±	0.0630
II-NDD	5.8333	±	0.1931	1.2333	±	0.0959	16.6333	±	0.2261	28.4667	±	0.8084	46.2333	±	0.9397	1.6000	±	0.1839	0.0000	±	0.0000
III-NDD	8.5328	±	0.2039	0.9333	±	0.0959	19.8986	±	0.2677	34.7244	±	0.3757	35.6310	±	0.3667	0.0933	±	0.0502	0.1867	±	0.0774
IV-NDD	4.5325	±	0.2098	4.5325	±	0.2098	20.0628	±	0.2760	39.6652	±	0.6291	30.0272	±	0.4233	0.8665	±	0.1241	0.3133	±	0.1272
VEF-NDD	3.3667	±	0.1919	1.6333	±	0.1420	14.8667	±	0.3362	57.6000	±	0.4981	22.3333	±	0.5620	0.1667	±	0.0797	0.0333	±	0.0333
VEM-NDD	2.2667	±	0.1076	1.6667	±	0.0797	12.9333	±	0.1533	52.5000	±	0.4655	30.4667	±	0.4891	0.1000	±	0.0724	0.0667	±	0.0454
VMF-NDD	1.2667	±	0.2004	2.0667	±	0.1453	23.7000	±	0.2526	51.9000	±	1.0748	20.9000	±	1.1022	0.0333	±	0.0333	0.1333	±	0.0591
VMM-NDD	1.8667	±	0.1141	2.7667	±	0.1369	24.1667	±	0.2995	46.9333	±	0.8336	24.1667	±	0.8711	0.0667	±	0.0454	0.0333	±	0.0333
VLF-NDD	1.1333	±	0.1241	3.8333	±	0.1517	17.2000	±	0.3773	55.2333	±	0.7926	22.4000	±	0.7108	0.1467	±	0.0661	0.0533	±	0.0376
VLM-NDD	1.0000	±	0.1543	4.1333	±	0.1241	22.6000	±	0.3559	49.6000	±	0.7467	22.5667	±	0.7651	0.0800	±	0.0554	0.0200	±	0.0200
SLF-NDD	0.7333	±	0.0959	3.8000	±	0.3710	15.4333	±	0.3990	51.9667	±	0.7146	27.8333	±	1.0907	0.1333	±	0.0766	0.1000	±	0.0535
SLM-NDD	0.7667	±	0.1182	3.2667	±	0.1279	12.9667	±	0.3466	53.8667	±	0.8010	29.0000	±	0.7988	0.1000	±	0.0724	0.0333	±	0.0333
PPF-NDD	2.9333	±	0.1453	2.8667	±	0.1501	10.0000	±	0.7037	44.9667	±	0.7392	34.5000	±	0.8522	3.7667	±	0.1681	0.9667	±	0.1241
PPM-NDD	0.8333	±	0.1517	4.8000	±	0.8000	10.2000	±	0.5495	43.8333	±	1.1889	35.6667	±	1.1919	3.4000	±	0.2082	1.2667	±	0.1681
MPF-NDD	1.2000	±	0.1272	1.8333	±	0.0934	5.7000	±	0.2795	42.1333	±	1.4528	18.8333	±	1.5241	0.1000	±	0.0724	30.2000	±	1.8079
MPM-NDD	1.0333	±	0.1501	1.8000	±	0.3443	7.7333	±	0.5320	34.2000	±	1.6689	22.4333	±	1.8967	0.2000	±	0.0951	32.6000	±	1.8415
LPF-NDD	1.0333	±	0.0909	2.4333	±	0.4278	12.6000	±	0.3754	34.5333	±	2.7016	26.4333	±	1.6344	0.1667	±	0.1054	22.8000	±	3.0972
LPM-NDD	0.4667	±	0.1420	1.4333	±	0.3268	9.1667	±	0.7049	41.2333	±	1.6965	34.5000	±	2.2366	0.0333	±	0.0333	13.1667	±	1.7277
AF-NDD	1.0000	±	0.1195	2.6333	±	0.3667	5.3333	±	0.3924	53.9667	±	1.7248	36.9667	±	1.4852	0.0333	±	0.0333	0.0667	±	0.0667
AM-NDD	0.1000	±	0.0724	0.5000	±	0.2289	3.3333	±	0.4014	49.3333	±	1.3660	46.6333	±	1.2578	0.0000	±	0.0000	0.1000	±	0.0724
CD 5%	0.3453			0.6574			0.9532			2.8197			2.7626			0.2472			2.2871		
CD 1%	0.4883			0.9295			1.3478			3.9870			3.9062			0.3495			3.2339		

n = 15

Table 4.33. Summary of Differential Haemocyte Counts of diapausing generation of *Antheraea mylitta*.

Age	Prohaemocytes			Oenocytes			Spherulocytes			Granulocytes			Plasmatocytes			Mitotic cells			Degenerated cells		
	Mean	±	SE	Mean	±	SE	Mean	±	SE	Mean	±	SE	Mean	±	SE	Mean	±	SE	Mean	±	SE
I-DD	7.5667	±	0.2175	0.6667	±	0.0630	8.8000	±	0.2526	50.6333	±	0.3534	30.5333	±	0.3180	1.6333	±	0.1241	0.1667	±	0.0630
II-DD	6.1667	±	0.1260	1.9333	±	0.0959	18.1333	±	0.0909	34.1333	±	0.2364	38.3667	±	0.2153	1.0000	±	0.0845	0.2667	±	0.0667
III-DD	10.4333	±	0.1369	2.2333	±	0.0959	19.8667	±	0.2153	34.5667	±	0.3305	32.0000	±	0.3416	0.1333	±	0.0591	0.7667	±	0.1182
IV-DD	4.2000	±	0.0816	3.3333	±	0.2051	18.6333	±	0.3857	44.4333	±	0.6812	28.1000	±	0.5503	0.4333	±	0.0959	0.8667	±	0.1501
VEF-DD	3.5333	±	0.1420	2.0000	±	0.0845	15.4667	±	0.4125	51.3333	±	0.8993	25.9000	±	0.9338	0.4667	±	0.1031	1.3000	±	0.2478
VEM-DD	2.3000	±	0.1447	2.5000	±	0.2440	14.2000	±	0.2059	58.1000	±	1.4361	19.6667	±	1.4173	1.0333	±	0.1919	2.2000	±	0.1272
VMF-DD	1.9333	±	0.1182	2.9667	±	0.0909	23.0333	±	0.5509	47.3667	±	0.7629	24.6000	±	0.7594	0.0333	±	0.0333	0.0667	±	0.0454
VMM-DD	1.6000	±	0.1309	2.8333	±	0.1931	21.4667	±	0.2603	50.3000	±	0.7381	23.4000	±	0.9414	0.2000	±	0.0655	0.4667	±	0.0909
VLF-DD	1.3333	±	0.1931	2.6667	±	0.1260	18.9333	±	0.2119	51.5000	±	0.5499	25.1133	±	0.6388	0.0333	±	0.0333	0.1533	±	0.0593
VLM-DD	2.7667	±	0.1076	2.8667	±	0.1333	21.9333	±	0.2529	51.0000	±	1.8587	21.2800	±	1.7275	0.1000	±	0.0535	0.0533	±	0.0376
SLF-DD	0.9000	±	0.1000	2.7667	±	0.1182	14.5000	±	0.3964	50.7000	±	0.3078	30.5667	±	0.5474	0.5000	±	0.0845	0.0667	±	0.0454
SLM-DD	1.9000	±	0.1000	2.4667	±	0.1241	14.9667	±	0.1031	49.5667	±	0.7052	30.5000	±	0.7416	0.5333	±	0.0909	0.0667	±	0.0454
PPF-DD	2.2000	±	0.2526	2.0667	±	0.2622	9.8333	±	0.1517	52.3000	±	0.8928	28.9000	±	0.9312	3.3333	±	0.1869	1.3667	±	0.0766
PPM-DD	3.1667	±	0.1517	2.0667	±	0.1076	10.0667	±	0.3960	52.8333	±	1.2824	26.7000	±	1.3164	3.0333	±	0.0909	2.1333	±	0.1141
MPF-DD	1.9667	±	0.0909	1.2667	±	0.1609	7.1333	±	0.1980	41.9333	±	0.8713	16.9000	±	0.4059	0.4667	±	0.0909	30.3333	±	0.7833
MPM-DD	2.8000	±	0.1363	2.2333	±	0.2119	8.1333	±	0.1791	40.2000	±	0.4520	16.6000	±	0.5610	0.6667	±	0.0797	29.3667	±	0.4038
LPF-DD	1.4000	±	0.0873	4.2667	±	0.3157	12.3000	±	0.2795	40.2667	±	1.0811	30.5667	±	1.1618	2.0333	±	0.6805	9.6667	±	0.3673
LPM-DD	1.8667	±	0.2510	3.4333	±	0.2797	13.1333	±	0.1579	41.5667	±	0.6652	29.1333	±	1.0828	1.0000	±	0.2012	9.8667	±	0.1420
AF-DD	1.1000	±	0.1558	2.2333	±	0.2922	5.1000	±	0.1558	50.1000	±	0.6852	41.3667	±	0.7179	0.0000	±	0.0000	0.1000	±	0.0724
AM-DD	0.7333	±	0.0667	2.4667	±	0.0909	4.4667	±	0.2603	52.8667	±	0.9704	39.3000	±	1.0554	0.0333	±	0.0333	0.1333	±	0.0766
CD 5%	0.3461			0.4234			0.6526			2.0499			2.1056			0.4257			0.5465		
CD 1%	0.4894			0.5986			0.9228			2.8985			2.9772			0.6020			0.7727		

n = 15

4.5. Histochemical studies

4.5.1. Histochemical studies on NDD and DD generations

With the help of staining/histochemical technique qualitative presence of general protein, nucleic acid DNA (Feulgen-Schiff's method), nucleic acid DNA & RNA (methyl-pyronin Y method), PA/S substances, glycogen, lipids and alkaline phosphatases were estimated in the cytoplasm and nucleus of haemocytes of different life stages of non-diapausing (NDD) and diapausing (DD) generations of *A. mylitta*.

4.5.1.1. General protein

The qualitative presence of general proteins in the cytoplasm and nuclei of different haemocytes in different stages of NDD and DD generations were observed. The findings are presented in Table-4.34.

In NDD generation, the presence of general protein was strong (++) in the cytoplasm of Prohaemocytes starting from I-NDD to SLM-NDD, thereafter, the reaction was observed to be weak (+) in other advance stages of growth in pupae till emergence of adults. However, in the nuclei of these haemocytes very strong (+++) presence of general protein was recorded from first instar to spinning larval stage of both the sexes. In rest of the advance stages of growth under study the reaction was observed to be strong (++) till emergence of adults.

Similar pattern of reaction for the presence of protein was also observed in DD generation except with a difference of weak presence of protein in cytoplasm of Prohaemocytes of spinning larvae and strong (++) presence of protein in the nucleus of spinning larvae. In cytoplasm of Oenocytoids (OEs), weak (+) presence of general protein was observed in all the stages of both NDD and DD generations. In nucleus of OEs the strong (++) presence of protein was observed from first to late fifth instar larvae of both the generations. With the advancement of growth, in other stages, protein was available in traces (±) in both the generations.

In the cytoplasm of Spherulocytes strong (++) presence of general protein was observed from first to late fifth instar male and female larvae. Its presence was weak (+) in spinning larvae of both the sexes of NDD and DD generation. Thereafter, reaction was again observed to be strong (++) till emergence of adults in both the generations.

Nucleus of spherulocytes showed very strong (+++) presence of general protein from first to late fifth instar larval stage of both the sexes in NDD and DD generation. Thereafter, strong (++) presence of proteins was observed till emergence of adults in both the sexes.

Granulocyte cytoplasm also showed strong (++) presence of general protein from first to late fifth instar larval stages of both the sexes in both the generations. Though the reaction was observed to be weak (+) in spinning larvae and also in late pupal stages of both the sexes in NDD and DD. The reaction was observed to be strong (++) in early, middle aged pupae and in adults of both the sexes in both the generations.

Very strong (+++) presence of general protein was recorded in the nucleus of Granulocytes up to late fifth instar larval stage of both the sexes. In all the remaining stages the reaction was observed to be strong (++) till emergence of adults.

In the cytoplasm of Plasmatocyte strong (++) presence of general protein was observed in all the stages, whereas, its nucleus showed very strong (+++) presence of general proteins from I instar to spinning stage larvae of both the sexes in both the generations thereafter strong (++) presence in remaining stages till the emergence of adult.

Table 4.34. Presence of general protein through histochemical tests in cytoplasm and nuclei of different haemocytes in NDD and DD generations of *A. mylitta.*

SN	Age	Name of the Haemocyte									
		PRs		OEs		SPs		GRs		PLs	
		Cyto.	Nucl.	Cyto.	Nucl.	Cyto.	Nucl.	Cyto.	Nucl.	Cyto.	Nucl.
1	I-NDD	++	+++	+	++	++	+++	++	+++	++	+++
2	I-DD	++	+++	+	++	++	+++	++	+++	++	+++
3	II-NDD	++	+++	+	++	++	+++	++	+++	++	+++
4	II-DD	++	+++	+	++	++	+++	++	+++	++	+++
5	III-NDD	++	+++	+	++	++	+++	++	+++	++	+++
6	III-DD	++	+++	+	++	++	+++	++	+++	++	+++
7	IV-NDD	++	+++	+	++	++	+++	++	+++	++	+++
8	IV-DD	++	+++	+	++	++	+++	++	+++	++	+++
9	VEF-NDD	++	+++	+	++	++	+++	++	+++	++	+++
10	VEF-DD	++	+++	+	++	++	+++	++	+++	++	+++
11	VEM-NDD	++	+++	+	++	++	+++	++	+++	++	+++
12	VEM-DD	++	+++	+	++	++	+++	++	+++	++	+++
13	VMF-NDD	++	+++	+	++	++	+++	++	+++	++	+++
14	VMF-DD	++	+++	+	++	++	+++	++	+++	++	+++
15	VMM-NDD	++	+++	+	++	++	+++	++	+++	++	+++
16	VMM-DD	++	+++	+	++	++	+++	++	+++	++	+++
17	VLF-NDD	++	+++	+	++	++	+++	++	+++	++	+++
18	VLF-DD	++	+++	+	++	++	+++	++	+++	++	+++
19	VLM-NDD	++	+++	+	++	++	+++	++	+++	++	+++
20	VLM-DD	++	+++	+	++	++	+++	++	+++	++	+++
21	SLF-NDD	++	+++	+	±	+	++	+	++	++	+++
22	SLF-DD	+	++	+	±	+	++	+	++	++	+++
23	SLM-NDD	++	+++	+	±	+	++	+	++	++	+++
24	SLM-DD	+	++	+	±	+	++	+	++	++	+++
25	PPF-NDD	+	++	+	±	++	++	++	++	++	++
26	PPF-DD	+	++	+	±	++	++	++	++	++	++
27	PPM-NDD	+	++	+	±	++	++	++	++	++	++
28	PPM-DD	+	++	+	±	++	++	++	++	++	++
29	MPF-NDD	+	++	+	±	++	++	++	++	++	++
30	MPF-DD	+	++	+	±	++	++	++	++	++	++
31	MPM-NDD	+	++	+	±	++	++	++	++	++	++
32	MPM-DD	+	++	+	±	++	++	++	++	++	++
33	LPF-NDD	+	++	+	±	++	++	+	++	++	++
34	LPF-DD	+	++	+	±	++	++	+	++	++	++
35	LPM-NDD	+	++	+	±	++	++	+	++	++	++
36	LPM-DD	+	++	+	±	++	++	+	++	++	++
37	AF-NDD	+	++	+	±	++	++	++	++	++	++
38	AF-DD	+	++	+	±	++	++	++	++	++	++
39	AM-NDD	+	++	+	±	++	++	++	++	++	++
40	AM-DD	+	++	+	±	++	++	++	++	++	++

ABBREVIATIONS/SYMBOLS USED: PRs " Prohaemocytes; OEs " Oenocytoids; SPs " Spherulocytes; GRs " Granulocytes; PLs " Plasmatocytes; Cyto. " Cytoplasm; Nucl. " Nucleus; +++ " Very strong; ++ " Strong; + " Weak; ± "Traces; -- " Absent.

4.5.1.2. Nucleic acids

4.5.1.2.1. DNA (Feulgen-Schiff's Method)

DNA was absent (-) in the cytoplasm of all the haemocytes of both non-diapausing and diapausing generations, which is a natural phenomenon because DNA are only available in the nuclei of cells as is evident from Table 4.35.

In the nuclei of Prohaemocytes, very strong (+++) presence of DNA was observed up to spinning larval stages of both the sexes of NDD generations. Thereafter, its presence was observed to be strong (++) till emergence of adults of NDD. In DD generations, the presence of DNA was recorded to be similar to NDD generations except in spinning larvae of both the sexes where reaction was observed to be strong (++).

Strong (++) presence of DNA was observed in the nuclei of Oenocytoid of both the generations from first to late fifth instar larval stages. Thereafter, it was recorded to be in traces till emergence of adults.

The nucleus of Spherulocytes and Granulocytes showed very strong (+++) presence of DNA from first instar to late fifth instar larval stages, thereafter, reaction was observed to be strong (++) till emergence of adults in both NDD and DD generations.

Very strong (+++) presence of DNA was observed in the nuclei of plasmatocytes up to spinning larval stages of both the sexes in both the generations. Though in pupal and adult stages the reaction was observed to be strong (++).

Table 4.35. Presence of DNA through histochemical tests (Feulgen-Schiff's method) in cytoplasm and nuclei of different haemocytes in NDD and DD generations of *A. mylitta*.

SN	Age	Name of the Haemocyte									
		PRs		OEs		SPs		GRs		PLs	
		Cyto.	Nucl.	Cyto.	Nucl.	Cyto.	Nucl.	Cyto.	Nucl.	Cyto.	Nucl.
1	I-NDD	-	+++	-	++	-	+++	-	+++	-	+++
2	I-DD	-	+++	-	++	-	+++	-	+++	-	+++
3	II-NDD	-	+++	-	++	-	+++	-	+++	-	+++
4	II-DD	-	+++	-	++	-	+++	-	+++	-	+++
5	III-NDD	-	+++	-	++	-	+++	-	+++	-	+++
6	III-DD	-	+++	-	++	-	+++	-	+++	-	+++
7	IV-NDD	-	+++	-	++	-	+++	-	+++	-	+++
8	IV-DD	-	+++	-	++	-	+++	-	+++	-	+++
9	VEF-NDD	-	+++	-	++	-	+++	-	+++	-	+++
10	VEF-DD	-	+++	-	++	-	+++	-	+++	-	+++
11	VEM-NDD	-	+++	-	++	-	+++	-	+++	-	+++
12	VEM-DD	-	+++	-	++	-	+++	-	+++	-	+++
13	VMF-NDD	-	+++	-	++	-	+++	-	+++	-	+++
14	VMF-DD	-	+++	-	++	-	+++	-	+++	-	+++
15	VMM-NDD	-	+++	-	++	-	+++	-	+++	-	+++
16	VMM-DD	-	+++	-	++	-	+++	-	+++	-	+++

conti.

17	VLF-NDD	-	+++	-	++	-	+++	-	+++	-	+++
18	VLF-DD	-	+++	-	++	-	+++	-	+++	-	+++
19	VLM-NDD	-	+++	-	++	-	+++	-	+++	-	+++
20	VLM-DD	-	+++	-	++	-	+++	-	+++	-	+++
21	SLF-NDD	-	+++	-	±	-	++	-	++	-	+++
22	SLF-DD	-	++	-	±	-	++	-	++	-	+++
23	SLM-NDD	-	+++	-	±	-	++	-	++	-	+++
24	SLM-DD	-	++	-	±	-	++	-	++	-	+++
25	PPF-NDD	-	++	-	±	-	++	-	++	-	++
26	PPF-DD	-	++	-	±	-	++	-	++	-	++
27	PPM-NDD	-	++	-	±	-	++	-	++	-	++
28	PPM-DD	-	++	-	±	-	++	-	++	-	++
29	MPF-NDD	-	++	-	±	-	++	-	++	-	++
30	MPF-DD	-	++	-	±	-	++	-	++	-	++
31	MPM-NDD	-	++	-	±	-	++	-	++	-	++
32	MPM-DD	-	++	-	±	-	++	-	++	-	++
33	LPF-NDD	-	++	-	±	-	++	-	++	-	++
34	LPF-DD	-	++	-	±	-	++	-	++	-	++
35	LPM-NDD	-	++	-	±	-	++	-	++	-	++
36	LPM-DD	-	++	-	±	-	++	-	++	-	++
37	AF-NDD	-	++	-	±	-	++	-	++	-	++
38	AF-DD	-	++	-	±	-	++	-	++	-	++
39	AM-NDD	-	++	-	±	-	++	-	++	-	++
40	AM-DD	-	++	-	±	-	++	-	++	-	++

ABBREVIATIONS/SYMBOLS USED: PRs " Prohaemocytes; OEs " Oenocytoids; SPs " Spherulocytes; GRs " Granulocytes; PLs " Plasmatocytes; Cyto. " Cytoplasm; Nucl. " Nucleus; +++ " Very strong; ++ " Strong; + " Weak; ± "Traces; -- " Absent.

4.5.1.2.2. DNA and RNA (methyl-pyronin Y method)

In Non-diapausing generations, in the cytoplasm of Prohaemocytes the presence of nucleic acid was strong (++) from first instar to spinning larval stages of the sexes. In all other stages it was observed to be weak (+). In the nucleic acid of these haemocytes, very strong (+++) reaction of nucleic acids (DNA & RNA) was observed from first instar larvae to spinning larval stages of both the sexes, thereafter, the reaction was strong (++) till emergence of adults.

In diapausing generation also, presence of nucleic acids was observed to be strong (++) in the cytoplasm and very strong (+++) in the nuclei of Prohaemocytes from first to late fifth instar larval stages. In rest of the stages, its presence was observed to be weak (+) in the cytoplasm and strong (++) in the nuclei from spinning larval stage till emergence of adults.

Cytoplasm of Oenocytoids of both NDD and DD generations showed weak (+) presence of Nucleic acids in all the stages. Though the strong (++) presence of nucleic acids was observed in the nuclei of Oenocytoids from first to late fifth instar stages, thereafter it was available in traces (±).

In the cytoplasm of Spherulocytes of both NDD and DD generations, strong (++) presence of these acids was observed from first instar up to late fifth instar stages and again from prepupal stage to emergence of adults, however in spinning larval stages, its presence was weak (+). In the nuclei, presence of nucleic acid was observed to be very strong (+++) from first instar to late fifth instar, thereafter its presence was recorded to be strong (++) till emergence of adults.

In the cytoplasm and nuclei of Granulocytes, the reaction was observed to similar to that of Spherulocytes except in the cytoplasm of late pupal stages of both the sexes and generations where reaction was observed to be weak (+).

In the cytoplasm of Plasmatocytes the reaction was observed to be strong (++) in all the stages. However, in their nuclei, the reaction was observed to be very strong (+++) from first instar to spinning larval stage, thereafter, it appeared to be strong (++) till emergence of adults (Table 4.36).

Table 4.36. Presence of DNA & RNA through histochemical tests (methyl-pyronin Y method) in cytoplasm and nuclei of different haemocytes in NDD and DD generations of *A. mylitta*.

SN	Age	Name of the Haemocyte									
		PRs		OEs		SPs		GRs		PLs	
		Cyto.	Nucl.	Cyto.	Nucl.	Cyto.	Nucl.	Cyto.	Nucl.	Cyto.	Nucl.
1	I-NDD	++	+++	+	++	++	+++	++	+++	++	+++
2	I-DD	++	+++	+	++	++	+++	++	+++	++	+++
3	II-NDD	++	+++	+	++	++	+++	++	+++	++	+++
4	II-DD	++	+++	+	++	++	+++	++	+++	++	+++
5	III-NDD	++	+++	+	++	++	+++	++	+++	++	+++
6	III-DD	++	+++	+	++	++	+++	++	+++	++	+++
7	IV-NDD	++	+++	+	++	++	+++	++	+++	++	+++
8	IV-DD	++	+++	+	++	++	+++	++	+++	++	+++
9	VEF-NDD	++	+++	+	++	++	+++	++	+++	++	+++
10	VEF-DD	++	+++	+	++	++	+++	++	+++	++	+++
11	VEM-NDD	++	+++	+	++	++	+++	++	+++	++	+++
12	VEM-DD	++	+++	+	++	++	+++	++	+++	++	+++
13	VMF-NDD	++	+++	+	++	++	+++	++	+++	++	+++
14	VMF-DD	++	+++	+	++	++	+++	++	+++	++	+++
15	VMM-NDD	++	+++	+	++	++	+++	++	+++	++	+++
16	VMM-DD	++	+++	+	++	++	+++	++	+++	++	+++
17	VLF-NDD	++	+++	+	++	++	+++	++	+++	++	+++
18	VLF-DD	++	+++	+	++	++	+++	++	+++	++	+++
19	VLM-NDD	++	+++	+	++	++	+++	++	+++	++	+++
20	VLM-DD	++	+++	+	++	++	+++	++	+++	++	+++
21	SLF-NDD	++	+++	+	±	+	++	+	++	++	+++
22	SLF-DD	+	++	+	±	+	++	+	++	++	+++
23	SLM-NDD	++	+++	+	±	+	++	+	++	++	+++

24SLM-DD	+	++	+	±	+	++	+	++	++	+++
25PPF-NDD	+	++	+	±	++	++	++	++	++	++
26PPF-DD	+	++	+	±	++	++	++	++	++	++
27PPM-NDD	+	++	+	±	++	++	++	++	++	++
28PPM-DD	+	++	+	±	++	++	++	++	++	++
29MPF-NDD	+	++	+	±	++	++	++	++	++	++
30MPF-DD	+	++	+	±	++	++	++	++	++	++
31MPM-NDD	+	++	+	±	++	++	++	++	++	++
32MPM-DD	+	++	+	±	++	++	++	++	++	++
33LPF-NDD	+	++	+	±	++	++	+	++	++	++
34LPF-DD	+	++	+	±	++	++	+	++	++	++
35LPM-NDD	+	++	+	±	++	++	+	++	++	++
36LPM-DD	+	++	+	±	++	++	+	++	++	++
37AF-NDD	+	++	+	±	++	++	++	++	++	++
38AF-DD	+	++	+	±	++	++	++	++	++	++
39AM-NDD	+	++	+	±	++	++	++	++	++	++
40AM-DD	+	++	+	±	++	++	++	++	++	++

ABBREVIATIONS/SYMBOLS USED: PRs " Prohaemocytes; OEs " Oenocytoids; SPs " Spherulocytes; GRs " Granulocytes; PLs " Plasmatocytes; Cyto. " Cytoplasm; Nucl. " Nucleus; +++ " Very strong; ++ " Strong; + " Weak; ± "Traces; -- " Absent.

4.5.1.3. Presence of PA/S substances

Presence of PA/S substances was not found in the nuclei of the different haemocytes. The test results for PA/S substances in the cytoplasm of different haemocytes are shown in Table-4.37.

In the cytoplasm of Prohaemocytes, in NDD generation very strong (+++) presence of PA/S substances was observed from first instar to spinning larval stage, thereafter, its presence was observed to be strong (++), till emergence of adults.

In DD generations, from first to late fifth larval stages only strong (++) presence of PA/S substances was observed. Its presence was weak (+) in spinning larvae, late pupae and strong (++) in remaining stages.

In the cytoplasm of Oenocytoids, the reaction was observed to be strong (++) from I instar to late fifth instars, thereafter it was available in traces (±) till emergence of adults in both the sexes and generations.

In the cytoplasm of Spherulocytes and Plasmatocytes, very strong (+++) presence of these substances was observed from I to late V instar larvae of NDD generations, thereafter, their presence was strong (++) till emergence of adults.

In DD generation, in the cytoplasm of Spherulocytes and Plasmatocytes strong (++) presence of PA/S substances was recorded from first instar larval stage to adults of both the sexes, except in the cytoplasm of Spherulocytes of spinning male and female larvae, where weak (+) presence of these substances was recorded.

In the cytoplasm of Granulocytes, very strong (+++) reaction of PA/S substances was recorded from first instar to late fifth instar stage, thereafter, their presence was recorded to be only strong (++) till emergence of adults in both the generations and sexes.

Table 4.37. Presence of PA/S substances through histochemical tests in cytoplasm and nuclei of different haemocytes in NDD and DD generations of *A. mylitta.*

SN	Age	Name of the Haemocyte									
		PRs		OEs		SPs		GRs		PLs	
		Cyto.	Nucl.	Cyto.	Nucl.	Cyto.	Nucl.	Cyto.	Nucl.	Cyto.	Nucl.
1	I-NDD	+++	-	++	-	+++	-	+++	-	+++	-
1	I-DD	++	-	++	-	++	-	+++	-	++	-
2	II-NDD	+++	-	++	-	+++	-	+++	-	+++	-
2	II-DD	++	-	++	-	++	-	+++	-	++	-
3	III-NDD	+++	-	++	-	+++	-	+++	-	+++	-
3	III-DD	++	-	++	-	++	-	+++	-	++	-
4	IV-NDD	+++	-	++	-	+++	-	+++	-	+++	-
4	IV-DD	++	-	++	-	++	-	+++	-	++	-
5	VEF-NDD	+++	-	++	-	+++	-	+++	-	+++	-
5	VEF-DD	++	-	++	-	++	-	+++	-	++	-
6	VEM-NDD	+++	-	++	-	+++	-	+++	-	+++	-
6	VEM-DD	++	-	++	-	++	-	+++	-	++	-
7	VMF-NDD	+++	-	++	-	+++	-	+++	-	+++	-
7	VMF-DD	++	-	++	-	++	-	+++	-	++	-
8	VMM-NDD	+++	-	++	-	+++	-	+++	-	+++	-
8	VMM-DD	++	-	++	-	++	-	+++	-	++	-
9	VLF-NDD	+++	-	++	-	+++	-	+++	-	+++	-
9	VLF-DD	++	-	++	-	++	-	+++	-	++	-
10	VLM-NDD	+++	-	++	-	+++	-	+++	-	+++	-
10	VLM-DD	++	-	++	-	++	-	+++	-	++	-
11	SLF-NDD	+++	-	±	-	++	-	++	-	++	-
11	SLF-DD	+	-	±	-	+	-	++	-	++	-
12	SLM-NDD	+++	-	±	-	++	-	++	-	++	-
12	SLM-DD	+	-	±	-	+	-	++	-	++	-
13	PPF-NDD	++	-	±	-	++	-	++	-	++	-
13	PPF-DD	++	-	±	-	++	-	++	-	++	-
14	PPM-NDD	++	-	±	-	++	-	++	-	++	-
14	PPM-DD	++	-	±	-	++	-	++	-	++	-
15	MPF-NDD	++	-	±	-	++	-	++	-	++	-
15	MPF-DD	++	-	±	-	++	-	++	-	++	-
16	MPM-NDD	++	-	±	-	++	-	++	-	++	-
16	MPM-DD	++	-	±	-	++	-	++	-	++	-
17	LPF-NDD	++	-	±	-	++	-	++	-	++	-
17	LPF-DD	+	-	±	-	++	-	++	-	++	-
18	LPM-NDD	++	-	±	-	++	-	++	-	++	-
18	LPM-DD	+	-	±	-	++	-	++	-	++	-
19	AF-NDD	++	-	±	-	++	-	++	-	++	-
19	AF-DD	++	-	±	-	++	-	++	-	++	-
20	AM-NDD	++	-	±	-	++	-	++	-	++	-
20	AM-DD	++	-	±	-	++	-	++	-	++	-

ABBREVIATIONS/SYMBOLS USED: PRs " Prohaemocytes; OEs " Oenocytoids; SPs " Spherulocytes; GRs " Granulocytes; PLs " Plasmatocytes; Cyto. " Cytoplasm; Nucl. " Nucleus; +++ " Very strong; ++ " Strong; + " Weak; ± "Traces; -- " Absent.

The inclusions of carbohydrate rich substances in Spherulocytes gave intense red to orange fluorescence. Nuclei of all haemocytes gave yellowish fluorescence (Fig 4.18 a & b).

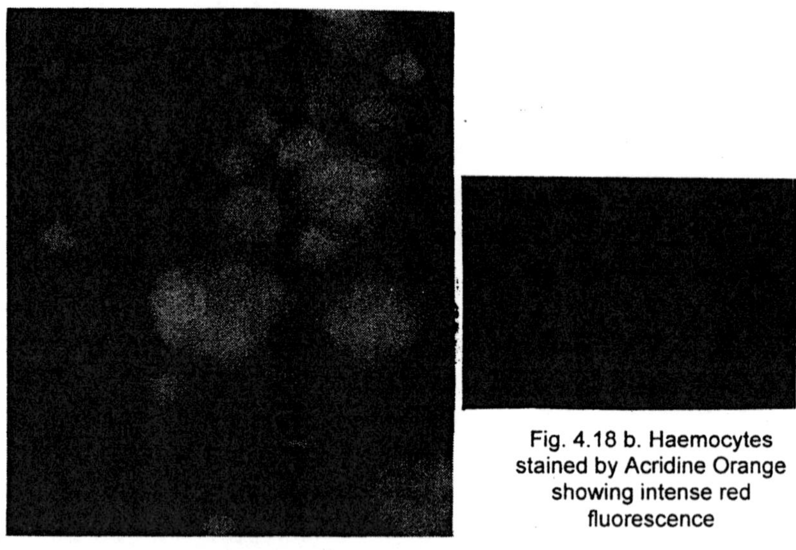
Fig. 4.18 b. Haemocytes stained by Acridine Orange showing intense red fluorescence

Fig. 4.18a. : Haemocytes stained by Acridine Orange showing various colours of fluorescence

Fig. 4.18b. : Haemocytes stained by Acridine Orange showing intense red fluorescence.

4.5.1.4. Bound lipids

Qualitative presence of bound lipids was not observed in the nuclei of different haemocytes and the same is presented in Table-4.38. In NDD generation, strong (++) presence of bound lipids was observed in the cytoplasm of Prohaemocytes from first instar to spinning larvae, thereafter reaction was weak (+) till the emergence of adults. In case of DD generation presence was observed to be strong (++) from first instar to late fifth instar and weak (+) reaction was recorded from spinning larval stage to emergence of adults.

In the cytoplasm of Oenocytoids of NDD generation weak (+) presence of bound lipids was recorded from first instar to late fifth instar larvae, thereafter, it was available in traces (±) till emergence of adults whereas in case of DD generation, their cytoplasm showed strong (++) presence of bound lipids from first to late fifth instar larval stages and thereafter it was also available in traces (±) like NDD generation.

In NDD generation, the cytoplasm of Spherulocytes showed very strong (+++) presence of bound lipids from first instar to late fifth instar larval stage. Thereafter, it was strong (++) till emergence of adults. In DD generation, from first to late fifth instar strong (++) presence of bound lipids was recorded which was weak (+) in spinning larval stage, strong (++) in different stages of pupae and again weak (+) in adults of both the sexes.

Table 4.38. Presence of bound lipids through histochemical tests in cytoplasm and nuclei of different haemocytes in NDD and DD generations of *A. mylitta*.

SN	Age	Name of the Haemocyte									
		PRs		OEs		SPs		GRs		PLs	
		Cyto.	Nucl.	Cyto.	Nucl.	Cyto.	Nucl.	Cyto.	Nucl.	Cyto.	Nucl.
1	I-NDD	++	-	+	-	+++	-	++	-	+	-
2	I-DD	++	-	++	-	++	-	++	-	+	-
3	II-NDD	++	-	+	-	+++	-	++	-	+	-
4	II-DD	++	-	++	-	++	-	++	-	+	-
5	III-NDD	++	-	+	-	+++	-	++	-	+	-
6	III-DD	++	-	++	-	++	-	++	-	+	-
7	IV-NDD	++	-	+	-	+++	-	++	-	+	-
8	IV-DD	++	-	++	-	++	-	++	-	+	-
9	VEF-NDD	++	-	+	-	+++	-	++	-	+	-
10	VEF-DD	++	-	++	-	++	-	++	-	+	-
11	VEM-NDD	++	-	+	-	+++	-	++	-	+	-
12	VEM-DD	++	-	++	-	++	-	++	-	+	-
13	VMF-NDD	++	-	+	-	+++	-	++	-	+	-
14	VMF-DD	++	-	++	-	++	-	++	-	+	-
15	VMM-NDD	++	-	+	-	+++	-	++	-	+	-
16	VMM-DD	++	-	++	-	++	-	++	-	+	-
17	VLF-NDD	++	-	+	-	+++	-	++	-	+	-
18	VLF-DD	++	-	++	-	++	-	++	-	+	-
19	VLM-NDD	++	-	+	-	+++	-	++	-	+	-
20	VLM-DD	++	-	++	-	++	-	++	-	+	-
21	SLF-NDD	++	-	±	-	++	-	++	-	±	-
22	SLF-DD	+	-	+	-	+	-	++	-	±	-
23	SLM-NDD	++	-	±	-	++	-	++	-	±	-
24	SLM-DD	+	-	+	-	+	-	++	-	±	-
25	PPF-NDD	+	-	±	-	++	-	++	-	±	-
26	PPF-DD	+	-	+	-	++	-	+	-	±	-
27	PPM-NDD	+	-	±	-	++	-	++	-	±	-
28	PPM-DD	+	-	+	-	++	-	+	-	±	-
29	MPF-NDD	+	-	±	-	++	-	+	-	±	-
30	MPF-DD	+	-	+	-	++	-	+	-	±	-
31	MPM-NDD	+	-	±	-	++	-	+	-	±	-
32	MPM-DD	+	-	+	-	++	-	+	-	±	-
33	LPF-NDD	+	-	±	-	++	-	+	-	±	-
34	LPF-DD	+	-	+	-	++	-	+	-	±	-
35	LPM-NDD	+	-	±	-	++	-	+	-	±	-
36	LPM-DD	+	-	+	-	++	-	+	-	±	-
37	AF-NDD	+	-	±	-	++	-	+	-	±	-
38	AF-DD	+	-	+	-	+	-	+	-	±	-
39	AM-NDD	+	-	±	-	++	-	+	-	±	-
40	AM-DD	+	-	+	-	+	-	+	-	±	-

ABBREVIATIONS/SYMBOLS USED: PRs " Prohaemocytes; OEs " Oenocytoids; SPs " Spherulocytes; GRs " Granulocytes; PLs " Plasmatocytes; Cyto. " Cytoplasm; Nucl. " Nucleus; +++ " Very strong; ++ " Strong; + " Weak; ± "Traces; -- " Absent.

The cytoplasm of Granulocytes showed strong (++) reaction from first instar larvae to mid pupal stages of NDD generation and in later stages reaction was observed to be weak (+) till the emergence of adults. In DD generation presence of bound lipids was similar like NDD generation from first instar to spinning larval stage, thereafter, the reaction was observed to be weak (+) till emergence of adults.

In the cytoplasm of Plasmatocytes of both the generations weak (+) presence of bound lipids was observed till late fifth instar larval stage, thereafter, bound lipids were available in traces (±) till emergence of adults.

4.5.1.5. Glycogen

In the nuclei of all the haemocytes of both NDD and DD generation, no traces of glycogen could be observed. The presence of glycogen in the cytoplasm of different haemocytes is presented in Table - 4.39. In NDD generation, in the cytoplasm of Prohaemocytes, very strong (+++) presence of glycogen was recorded from first instar to spinning larval stages. In later stages the presence was recorded to be strong (++) till emergence of adults. In DD generation, strong (++) presence was observed in almost all the stages taken up for study except in spinning larvae and late pupae where presence of glycogen was weak (+).

In the cytoplasm of Oenocytoids of both the generations strong (++) presence of glycogen was recorded from first instar to late fifth instar larval stages, thereafter, it was available in traces (±) till emergence of adults.

Cytoplasm of Spherulocytes, showed weak (+) presence of glycogen from first instar to late fifth instar larvae of NDD and up to mid fifth instar larvae of DD generation. In remaining stages, it was available in traces (±) till emergence of adults in both the generations.

Table 4.39. Presence of glycogen through histochemical tests in cytoplasm and nuclei of different haemocytes in NDD and DD generations of *A. mylitta*.

SN	Age	Name of the Haemocyte									
		PRs		OEs		SPs		GRs		PLs	
		Cyto.	Nucl.	Cyto.	Nucl.	Cyto.	Nucl.	Cyto.	Nucl.	Cyto.	Nucl.
1	I-NDD	+++	-	++	-	+	-	+++	-	+	-
2	I-DD	++	-	++	-	+	-	+++	-	+	-
3	II-NDD	+++	-	++	-	+	-	+++	-	+	-
4	II-DD	++	-	++	-	+	-	+++	-	+	-
5	III-NDD	+++	-	++	-	+	-	+++	-	+	-
6	III-DD	++	-	++	-	+	-	+++	-	+	-
7	IV-NDD	+++	-	++	-	+	-	+++	-	+	-
8	IV-DD	++	-	++	-	+	-	+++	-	+	-
9	VEF-NDD	+++	-	++	-	+	-	+++	-	+	-
10	VEF-DD	++	-	++	-	+	-	+++	-	+	-
11	VEM-NDD	+++	-	++	-	+	-	+++	-	+	-
12	VEM-DD	++	-	++	-	+	-	+++	-	+	-

13	VMF-NDD	+++	-	++	-	+	-	+++	-	+	-
14	VMF-DD	++	-	++	-	+	-	+++	-	+	-
15	VMM-NDD	+++	-	++	-	+	-	+++	-	+	-
16	VMM-DD	++	-	++	-	+	-	+++	-	+	-
17	VLF-NDD	+++	-	++	-	+	-	+++	-	+	-
18	VLF-DD	++	-	++	-	±	-	+++	-	±	-
19	VLM-NDD	+++	-	++	-	+	-	+++	-	±	-
20	VLM-DD	++	-	++	-	±	-	+++	-	±	-
21	SLF-NDD	+++	-	±	-	±	-	++	-	±	-
22	SLF-DD	+	-	±	-	±	-	++	-	±	-
23	SLM-NDD	+++	-	±	-	±	-	++	-	±	-
24	SLM-DD	+	-	±	-	±	-	++	-	±	-
25	PPF-NDD	++	-	±	-	±	-	++	-	±	-
26	PPF-DD	++	-	±	-	±	-	++	-	±	-
27	PPM-NDD	++	-	±	-	±	-	++	-	±	-
28	PPM-DD	++	-	±	-	±	-	++	-	±	-
29	MPF-NDD	++	-	±	-	±	-	++	-	±	-
30	MPF-DD	++	-	±	-	±	-	++	-	±	-
31	MPM-NDD	++	-	±	-	±	-	++	-	±	-
32	MPM-DD	++	-	±	-	±	-	++	-	±	-
33	LPF-NDD	++	-	±	-	±	-	++	-	±	-
34	LPF-DD	+	-	±	-	±	-	++	-	±	-
35	LPM-NDD	++	-	±	-	±	-	++	-	±	-
36	LPM-DD	+	-	±	-	±	-	++	-	±	-
37	AF-NDD	++	-	±	-	±	-	++	-	±	-
38	AF-DD	++	-	±	-	±	-	++	-	±	-
39	AM-NDD	++	-	±	-	±	-	++	-	±	-
40	AM-DD	++	-	±	-	±	-	++	-	±	-

ABBREVIATIONS/SYMBOLS USED: PRs " Prohaemocytes; OEs " Oenocytoids; SPs " Spherulocytes; GRs " Granulocytes; PLs " Plasmatocytes; Cyto. " Cytoplasm; Nucl. " Nucleus; +++ " Very strong; ++ " Strong; + " Weak; ± "Traces; -- " Absent.

Cytoplasm of Plasmatocytes showed weak (+) presence of glycogen from first instar to mid-fifth instar larval stages in both the generations, thereafter, it was available in traces (±) -Table 4.39.

4.5.1.6. Alkaline phosphatases

No traces of alkaline phosphatases were observed in the nuclei of all the haemocytes. Its presence in cytoplasm of different haemocytes is presented in Table-4.40. In the cytoplasm of Prohaemocytes, presence of alkaline phosphatases was observed in traces (±) from first to fifth instar larval stages in both non-diapausing and diapausing generations.

In cytoplasm of Spherulocytes and Granulocytes in both the generations, weak (+) presence of alkaline phosphatases was observed from first instar larval stage to mid-pupal stage.

Other haemocytes viz., Oenocytoids and Plasmatocytes showed no traces (-) of alkaline phosphatases in their cytoplasm (Table 4.40).

Table 4.40. Presence of alkaline phosphatases through histochemical tests in cytoplasm and nuclei of different haemocytes in NDD and DD generations of *A. mylitta*.

SN	Age	Name of the Haemocyte									
		PRs		OEs		SPs		GRs		PLs	
		Cyto.	Nucl.	Cyto.	Nucl.	Cyto.	Nucl.	Cyto.	Nucl.	Cyto.	Nucl.
1	I-NDD	±	-	-	-	+	-	+	-	-	-
2	I-DD	±	-	-	-	+	-	+	-	-	-
3	II-NDD	±	-	-	-	+	-	+	-	-	-
4	II-DD	±	-	-	-	+	-	+	-	-	-
5	III-NDD	±	-	-	-	+	-	+	-	-	-
6	III-DD	±	-	-	-	+	-	+	-	-	-
7	IV-NDD	±	-	-	-	+	-	+	-	-	-
8	IV-DD	±	-	-	-	+	-	+	-	-	-
9	VEF-NDD	±	-	-	-	+	-	+	-	-	-
10	VEF-DD	±	-	-	-	+	-	+	-	-	-
11	VEM-NDD	±	-	-	-	+	-	+	-	-	-
12	VEM-DD	±	-	-	-	+	-	+	-	-	-
13	VMF-NDD	±	-	-	-	+	-	+	-	-	-
14	VMF-DD	±	-	-	-	+	-	+	-	-	-
15	VMM-NDD	±	-	-	-	+	-	+	-	-	-
16	VMM-DD	±	-	-	-	+	-	+	-	-	-
17	VLF-NDD	±	-	-	-	+	-	+	-	-	-
18	VLF-DD	±	-	-	-	+	-	+	-	-	-
19	VLM-NDD	±	-	-	-	+	-	+	-	-	-
20	VLM-DD	±	-	-	-	+	-	+	-	-	-
21	SLF-NDD	±	-	-	-	+	-	+	-	-	-
22	SLF-DD	±	-	-	-	+	-	+	-	-	-
23	SLM-NDD	±	-	-	-	+	-	+	-	-	-
24	SLM-DD	±	-	-	-	+	-	+	-	-	-
25	PPF-NDD	-	-	-	-	+	-	+	-	-	-
26	PPF-DD	-	-	-	-	+	-	+	-	-	-
27	PPM-NDD	-	-	-	-	+	-	+	-	-	-
28	PPM-DD	-	-	-	-	+	-	+	-	-	-
29	MPF-NDD	-	-	-	-	+	-	+	-	-	-
30	MPF-DD	-	-	-	-	+	-	+	-	-	-
31	MPM-NDD	-	-	-	-	+	-	+	-	-	-
32	MPM-DD	-	-	-	-	+	-	+	-	-	-
33	LPF-NDD	-	-	-	-	-	-	-	-	-	-
34	LPF-DD	-	-	-	-	-	-	-	-	-	-
35	LPM-NDD	-	-	-	-	-	-	-	-	-	-
36	LPM-DD	-	-	-	-	-	-	-	-	-	-
37	AF-NDD	-	-	-	-	-	-	-	-	-	-
38	AF-DD	-	-	-	-	-	-	-	-	-	-
39	AM-NDD	-	-	-	-	-	-	-	-	-	-
40	AM-DD	-	-	-	-	-	-	-	-	-	-

ABBREVIATIONS/SYMBOLS USED: PRs " Prohaemocytes; OEs " Oenocytoids; SPs " Spherulocytes; GRs " Granulocytes; PLs " Plasmatocytes; Cyto. " Cytoplasm; Nucl. " Nucleus; +++ " Very strong; ++ " Strong; + " Weak; ± "Traces; -- " Absent.

4.5.2. Histochemical studies between sexes of NDD and DD generations

Observations on the differences in qualitative presence of general protein, DNA, nucleic acids (DNA & RNA), PA/S substances, bound lipids, glycogen and alkaline phosphatases between females and males of non-diapausing and diapausing generations were recorded and are presented in Tables - 4.41 to 4.54.

4.5.2.1. General protein

Reaction with mercury Bromophenol blue indicated strong (++) presence of general protein in cytoplasm of Prohaemocytes of both the sexes of NDD generations form early fifth larval stage to spinning larval stage, thereafter it presence was weak (+) till emergence of adults. In case of DD generation strong (++) presence was only recorded up to late fifth instars of both the sexes and in later stages the reaction was observed to be weak (+). Contrary to this, the reaction was very strong (+++) in nucleus from early fifth instar larval stage up to spinning larval stage in NDD generations and till late fifth instar in DD generations, and in the remaining stages the reaction was observed to be strong (++) till emergence of adults, in both the sexes and generations.

Table 4.41. Sex-wise presence of general protein through histochemical test in cytoplasm & nuclei of different haemocytes in NDD generation of *A. mylitta*

SN	Age	Name of the Haemocyte									
		PRs		OEs		SPs		GRs		PLs	
		Cyto.	Nucl.	Cyto.	Nucl.	Cyto.	Nucl.	Cyto.	Nucl.	Cyto.	Nucl.
1	VEF-NDD	++	+++	+	++	++	+++	++	+++	++	+++
2	VEM-NDD	++	+++	+	++	++	+++	++	+++	++	+++
3	VMF-NDD	++	+++	+	++	++	+++	++	+++	++	+++
4	VMM-NDD	++	+++	+	++	++	+++	++	+++	++	+++
5	VLF-NDD	++	+++	+	++	++	+++	++	+++	++	+++
6	VLM-NDD	++	+++	+	++	++	+++	++	+++	++	+++
7	SLF-NDD	++	+++	+	±	+	++	+	++	++	+++
8	SLM-NDD	++	+++	+	±	+	++	+	++	++	+++
9	PPF-NDD	+	++	+	±	++	++	++	++	++	++
10	PPM-NDD	+	++	+	±	++	++	++	++	++	++
11	MPF-NDD	+	++	+	±	++	++	++	++	++	++
12	MPM-NDD	+	++	+	±	++	++	++	++	++	++
13	LPF-NDD	+	++	+	±	++	++	+	++	++	++
14	LPM-NDD	+	++	+	±	++	++	+	++	++	++
15	AF-NDD	+	++	+	±	++	++	++	++	++	++
16	AM-NDD	+	++	+	±	++	++	++	++	++	++

ABBREVIATIONS/SYMBOLS USED: PRs " Prohaemocytes; OEs " Oenocytoids; SPs " Spherulocytes; GRs " Granulocytes; PLs " Plasmatocytes; Cyto. " Cytoplasm; Nucl. " Nucleus; +++ " Very strong; ++ " Strong; + " Weak; ± "Traces; -- " Absent.

In the cytoplasm of Oenocytoids weak (+) reaction was noticed in all the stages of both the sexes of NDD and DD generations, however, in nuclei, strong (++) presence was recorded from early fifth instar to late fifth instar in both the sexes, thereafter, it was available in traces (±) till emergence of adults.

In the cytoplasm of Spherulocytes strong (++) reaction from early fifth to late fifth instar larval stage in both NDD and DD generations and weak (+) presence in spinning larvae of both sexes was observed. Thereafter, in cytoplasm of remaining stages strong (++) reaction was observed. In the nuclei of Spherulocytes, very strong (+++) reaction was recorded from early fifth to late fifth instar and thereafter, reaction showed strong (++) presence of protein till emergence of adults in both the sexes and generations.

In cytoplasm of Granulocytes, the reaction was strong (++) up from early fifth instar to late V instar larvae, in pre and mid pupal stages and also in adults of both the sexes and generations, however, weak (+) presence was recorded in spinning larvae and late pupae. Their nuclei showed very strong (+++) presence of protein starting from early fifth instar to late fifth instar of both the sexes and generations, thereafter strong (++) presence of protein was observed till emergence of adults.

Table 4.42. Sex-wise presence of general protein through histochemical test in cytoplasm & nuclei of different haemocytes in DD generation of *A. mylitta*

SN	Age	Name of the Haemocyte									
		PRs		OEs		SPs		GRs		PLs	
		Cyto.	Nucl.	Cyto.	Nucl.	Cyto.	Nucl.	Cyto.	Nucl.	Cyto.	Nucl.
1	VEF-DD	++	+++	+	++	++	+++	++	+++	++	+++
2	VEM-DD	++	+++	+	++	++	+++	++	+++	++	+++
3	VMF-DD	++	+++	+	++	++	+++	++	+++	++	+++
4	VMM-DD	++	+++	+	++	++	+++	++	+++	++	+++
5	VLF-DD	++	+++	+	++	++	+++	++	+++	++	+++
6	VLM-DD	++	+++	+	++	++	+++	++	+++	++	+++
7	SLF-DD	+	++	+	±	+	++	+	++	++	+++
8	SLM-DD	+	++	+	±	+	++	+	++	++	+++
9	PPF-DD	+	++	+	±	++	++	++	++	++	++
10	PPM-DD	+	++	+	±	++	++	++	++	++	++
11	MPF-DD	+	++	+	±	++	++	++	++	++	++
12	MPM-DD	+	++	+	±	++	++	++	++	++	++
13	LPF-DD	+	++	+	±	++	++	+	++	++	++
14	LPM-DD	+	++	+	±	++	++	+	++	++	++
15	AF-DD	+	++	+	±	++	++	++	++	++	++
16	AM-DD	+	++	+	±	++	++	++	++	++	++

ABBREVIATIONS/SYMBOLS USED: PRs " Prohaemocytes; OEs " Oenocytoids; SPs " Spherulocytes; GRs " Granulocytes; PLs " Plasmatocytes; Cyto. " Cytoplasm; Nucl. " Nucleus; +++ " Very strong; ++ " Strong; + " Weak; ± "Traces; -- " Absent.

In the cytoplasm of Plasmatocytes, strong (++) presence of protein was recorded in all the stages of both the sexes and generations. However, in case of nuclei, very strong (+++)

presence of protein was recorded up to spinning larval stage thereafter, it showed strong (++) reaction from prepupal stages till emergence of adults (Table 4.41 & 4.42).

4.5.2.2. Nucleic acid

4.5.2.2.1. DNA (Feulgen-Schiff's Method)

Only nuclei of all the haemocytes showed the presence of nucleic acid DNA. The details of the observations are presented in Tables 4.43 & 4.44.

Nucleus of Prohaemocytes showed very strong (+++) presence of DNA from early fifth instar to spinning larval stages of both the sexes in NDD generation and late fifth instar of both the sexes in DD generation. The remaining stages of both the generations had strong (++) presence of DNA till emergence of adults.

Oenocytoids nuclei showed strong (++) presence of DNA from early fifth to late fifth instar larval stage, thereafter, it was available in traces (±) till emergence of adults in both the generations.

The nuclei of Spherulocytes and Granulocytes showed very strong (+++) presence of DNA from early fifth instar to late fifth instar larval stage, thereafter, presence of DNA was strong (++) in remaining stages till emergence of adults of both the sexes in NDD and DD generations. In nuclei of Plasmatocytes, very strong (+++) presence of DNA was observed from early fifth instar larval stage to spinning larval stage of both the sexes in NDD and DD generations. Thereafter, the presence of DNA was observed to be strong (++) till emergence of adults.

Table 4.43. Sex-wise presence of DNA through histochemical test (Feulgen-Schiff's method) in cytoplasm & nuclei of different haemocytes in NDD generation of *A. mylitta*

SN	Age	Name of the Haemocyte									
		PRs		OEs		SPs		GRs		PLs	
		Cyto.	Nucl.	Cyto.	Nucl.	Cyto.	Nucl.	Cyto.	Nucl.	Cyto.	Nucl.
1	VEF-NDD	-	+++	-	++	-	+++	-	+++	-	+++
2	VEM-NDD	-	+++	-	++	-	+++	-	+++	-	+++
3	VMF-NDD	-	+++	-	++	-	+++	-	+++	-	+++
4	VMM-NDD	-	+++	-	++	-	+++	-	+++	-	+++
5	VLF-NDD	-	+++	-	++	-	+++	-	+++	-	+++
6	VLM-NDD	-	+++	-	++	-	+++	-	+++	-	+++
7	SLF-NDD	-	+++	-	±	-	++	-	++	-	+++
8	SLM-NDD	-	+++	-	±	-	++	-	++	-	+++
9	PPF-NDD	-	++	-	±	-	++	-	++	-	++
10	PPM-NDD	-	++	-	±	-	++	-	++	-	++
11	MPF-NDD	-	++	-	±	-	++	-	++	-	++
12	MPM-NDD	-	++	-	±	-	++	-	++	-	++
13	LPF-NDD	-	++	-	±	-	++	-	++	-	++
14	LPM-NDD	-	++	-	±	-	++	-	++	-	++
15	AF-NDD	-	++	-	±	-	++	-	++	-	++
16	AM-NDD	-	++	-	±	-	++	-	++	-	++

ABBREVIATIONS/SYMBOLS USED: As in Table 4.44

Table 4.44. Sex-wise presence of DNA through histochemical test (Feulgen-Schiff's method) in cytoplasm & nuclei of different haemocytes in DD generation of *A. mylitta*

SNAge	Name of the Haemocyte									
	PRs		OEs		SPs		GRs		PLs	
	Cyto.	Nucl.	Cyto.	Nucl.	Cyto.	Nucl.	Cyto.	Nucl.	Cyto.	Nucl.
1 VEF-DD	-	+++	-	++	-	+++	-	+++	-	+++
2 VEM-DD	-	+++	-	++	-	+++	-	+++	-	+++
3 VMF-DD	-	+++	-	++	-	+++	-	+++	-	+++
4 VMM-DD	-	+++	-	++	-	+++	-	+++	-	+++
5 VLF-DD	-	+++	-	++	-	+++	-	+++	-	+++
6 VLM-DD	-	+++	-	++	-	+++	-	+++	-	+++
7 SLF-DD	-	++	-	±	-	++	-	++	-	+++
8 SLM-DD	-	++	-	±	-	++	-	++	-	+++
9 PPF-DD	-	++	-	±	-	++	-	++	-	++
10 PPM-DD	-	++	-	±	-	++	-	++	-	++
11 MPF-DD	-	++	-	±	-	++	-	++	-	++
12 MPM-DD	-	++	-	±	-	++	-	++	-	++
13 LPF-DD	-	++	-	±	-	++	-	++	-	++
14 LPM-DD	-	++	-	±	-	++	-	++	-	++
15 AF-DD	-	++	-	±	-	++	-	++	-	++
16 AM-DD	-	++	-	±	-	++	-	++	-	++

ABBREVIATIONS/SYMBOLS USED: PRs " Prohaemocytes; OEs " Oenocytoids; SPs " Spherulocytes; GRs " Granulocytes; PLs " Plasmatocytes; Cyto." Cytoplasm; Nucl. " Nucleus; +++ " Very strong; ++ " Strong; + " Weak; ± "Traces; -- " Absent.

4.5.2.2.2. DNA and RNA (methyl-pyronin Y method)

Histochemical tests were carried out in order to record the presence of DNA & RNA through Pyronin green + Methyl green tests, in cytoplasm and nucleus of different haemocytes in both the sexes of NDD and DD generations and presented in Tables 4.45 & 4.46.

The cytoplasm of Prohaemocytes showed strong (++) reaction from early fifth to spinning larvae in NDD generation and up to late fifth instar larval stage in DD generation of both the sexes. In remaining stages, reaction showed weak (+) presence of DNA and RNA till emergence of adults in both the sexes. The nuclei of Prohaemocytes showed very strong (+++) reaction from early fifth to spinning larvae in NDD generation and up to late fifth instar larval stage in DD generation of both the sexes. In remaing stages reaction showed strong (++) presence of DNA and RNA, till emergence of adults, in both the sexes and generations.

In cytoplasm of Oenocytoids weak (+) presence of DNA and RNA was observed in all the stages of both the sexes and generations. In nuclei of Oenocytoids, reaction was observed to be strong (++) from first instar to late fifth instar, thereafter it was available in traces (±) in both the sexes and generation.

In cytoplasm of Spherulocytes, strong (++) reaction from early fifth instar to late fifth instars larvae, weak (+) presence in spinning larval stage and again strong (++) presence was observed till emergence of adults in both the sexes of both generations.

In the cytoplasm of Granulocytes weak (+) presence of these acids was recorded in late pupal stages of both the sexes and generations. In remaining stages, the presence of these acids was strong (++) or similar to that of Spherulocytes.

In Plasmatocyte cytoplasm in all the stages, reaction was very strong (++) in all the stages of both the sexes and generations. However, nucleus of Plasmatocytes showed very strong (+++) reaction from early fifth instar to spinning larval stage, thereafter, only strong (++) presence of these acids was observed till emergence of adults of both the sexes and generations.

Table 4.45. Sex-wise presence of DNA & RNA through histochemical test (methyl-pyronin Y method) in cytoplasm & nuclei of different haemocytes in NDD generation of *A. mylitta*

SN	Age	Name of the Haemocyte									
		PRs		OEs		SPs		GRs		PLs	
		Cyto.	Nucl.	Cyto.	Nucl.	Cyto.	Nucl.	Cyto.	Nucl.	Cyto.	Nucl.
1	VEF-NDD	++	+++	+	++	++	+++	++	+++	++	+++
2	VEM-NDD	++	+++	+	++	++	+++	++	+++	++	+++
3	VMF-NDD	++	+++	+	++	++	+++	++	+++	++	+++
4	VMM-NDD	++	+++	+	++	++	+++	++	+++	++	+++
5	VLF-NDD	++	+++	+	++	++	+++	++	+++	++	+++
6	VLM-NDD	++	+++	+	++	++	+++	++	+++	++	+++
7	SLF-NDD	++	+++	+	±	+	++	+	++	++	+++
8	SLM-NDD	++	+++	+	±	+	++	+	++	++	+++
9	PPF-NDD	+	++	+	±	++	++	++	++	++	++
10	PPM-NDD	+	++	+	±	++	++	++	++	++	++
11	MPF-NDD	+	++	+	±	++	++	++	++	++	++
12	MPM-NDD	+	++	+	±	++	++	++	++	++	++
13	LPF-NDD	+	++	+	±	++	++	+	++	++	++
14	LPM-NDD	+	++	+	±	++	++	+	++	++	++
15	AF-NDD	+	++	+	±	++	++	++	++	++	++
16	AM-NDD	+	++	+	±	++	++	++	++	++	++

ABBREVIATIONS/SYMBOLS USED: PRs " Prohaemocytes; OEs " Oenocytoids; SPs " Spherulocytes; GRs " Granulocytes; PLs " Plasmatocytes; Cyto. " Cytoplasm; Nucl." Nucleus;+++ " Very strong; ++ " Strong; + " Weak; ± "Traces; -- " Absent.

4.5.2.3. Presence of PA/S substances

In between the sexes of non-diapausing and diapausing generation, presence of PA/S substances in haemocytes was observed histochemically after reaction with Periodic acid and Schiff's reagent which is presented in Tables 4.47 & 4.48. No traces of PA/S substances were found in the nuclei of different haemocytes.

Table 4.46. Sex-wise presence of DNA & RNA through histochemical test (methyl-pyronin Y method) in cytoplasm & nuclei of different haemocytes in DD generation of *A. mylitta*.

SN	Age	Name of the Haemocyte									
		PRs		OEs		SPs		GRs		PLs	
		Cyto.	Nucl.	Cyto.	Nucl.	Cyto.	Nucl.	Cyto.	Nucl.	Cyto.	Nucl.
1	VEF-DD	++	+++	+	++	++	+++	++	+++	++	+++
2	VEM-DD	++	+++	+	++	++	+++	++	+++	++	+++
3	VMF-DD	++	+++	+	++	++	+++	++	+++	++	+++
4	VMM-DD	++	+++	+	++	++	+++	++	+++	++	+++
5	VLF-DD	++	+++	+	++	++	+++	++	+++	++	+++
6	VLM-DD	++	+++	+	++	++	+++	++	+++	++	+++
7	SLF-DD	+	++	+	±	+	++	+	++	++	+++
8	SLM-DD	+	++	+	±	+	++	+	++	++	+++
9	PPF-DD	+	++	+	±	++	++	++	++	++	++
10	PPM-DD	+	++	+	±	++	++	++	++	++	++
11	MPF-DD	+	++	+	±	++	++	++	++	++	++
12	MPM-DD	+	++	+	±	++	++	++	++	++	++
13	LPF-DD	+	++	+	±	++	++	+	++	++	++
14	LPM-DD	+	++	+	±	++	++	+	++	++	++
15	AF-DD	+	++	+	±	++	++	++	++	++	++
16	AM-DD	+	++	+	±	++	++	++	++	++	++

ABBREVIATIONS/SYMBOLS USED: PRs " Prohaemocytes; OEs " Oenocytoids; SPs " Spherulocytes; GRs " Granulocytes; PLs " Plasmatocytes; Cyto. " Cytoplasm; Nucl." Nucleus;+++ " Very strong; ++ " Strong; + " Weak; ± "Traces; -- " Absent.

In NDD generation in cytoplasm of Prohaemocytes from early fifth to spinning larvae reaction was very strong (+++) and thereafter strong (++) in remaining stages. However, in DD lot, reaction was strong (++) up to late V instar, weak (+) in spinning larvae and late pupal stage, and again strong (++) in remaining stages of DD in both the sexes.

In the cytoplasm of Oenocytoids, reaction was strong (++) until late fifth instar in both the sexes of NDD & DD, and it was in traces (±) in remaining stages.

In the cytoplasm of Spherulocytes of different stages of NDD very strong (+++) reaction was observed from early fifth to late V instar larvae, thereafter, it was strong (++) in remaining stages in both the sexes, till emergence of adults. In DD, reaction was strong (++) from early fifth to late fifth instar, weak (+) in spinning larvae, and again strong (++) in remaining stages of both the sexes, till emergence of adults.

In the cytoplasm of Granulocytes, reaction was very strong (+++) from early fifth instar to late fifth instar, thereafter, the reaction was strong (++) in both the sexes and generation.

In the cytoplasm of Plasmatocytes in NDD generation reaction was very strong (+++) and strong (++) in DD generation from early fifth instar to late fifth instar, however, in remaining stages of both the generations, the reaction was strong (++).

Table 4.47. Sex-wise presence of PA/S substances through histochemical test in cytoplasm & nuclei of different haemocytes in NDD generation of *A. mylitta*

SN	Age	PRs		OEs		SPs		GRs		PLs	
		Cyto.	Nucl.	Cyto.	Nucl.	Cyto.	Nucl.	Cyto.	Nucl.	Cyto.	Nucl.
1	VEF-NDD	+++	-	++	-	+++	-	+++	-	+++	-
2	VEM-NDD	+++	-	++	-	+++	-	+++	-	+++	-
3	VMF-NDD	+++	-	++	-	+++	-	+++	-	+++	-
4	VMM-NDD	+++	-	++	-	+++	-	+++	-	+++	-
5	VLF-NDD	+++	-	++	-	+++	-	+++	-	+++	-
6	VLM-NDD	+++	-	++	-	+++	-	+++	-	+++	-
7	SLF-NDD	+++	-	±	-	++	-	++	-	++	-
8	SLM-NDD	+++	-	±	-	++	-	++	-	++	-
9	PPF-NDD	++	-	±	-	++	-	++	-	++	-
10	PPM-NDD	++	-	±	-	++	-	++	-	++	-
11	MPF-NDD	++	-	±	-	++	-	++	-	++	-
12	MPM-NDD	++	-	±	-	++	-	++	-	++	-
13	LPF-NDD	++	-	±	-	++	-	++	-	++	-
14	LPM-NDD	++	-	±	-	++	-	++	-	++	-
15	AF-NDD	++	-	±	-	++	-	++	-	++	-
16	AM-NDD	++	-	±	-	++	-	++	-	++	-

Name of the Haemocyte (column group heading for PRs–PLs).

ABBREVIATIONS/SYMBOLS USED: PRs " Prohaemocytes; OEs " Oenocytoids; SPs " Spherulocytes; GRs " Granulocytes; PLs " Plasmatocytes; Cyto. " Cytoplasm; Nucl. " Nucleus; +++ " Very strong; ++ " Strong; + " Weak; ± "Traces; -- " Absent.

Table 4.48. Sex-wise presence of PA/S substances through histochemical test in cytoplasm & nuclei of different haemocytes in DD generation of *A. mylitta*

SN	Age	PRs		OEs		SPs		GRs		PLs	
		Cyto.	Nucl.	Cyto.	Nucl.	Cyto.	Nucl.	Cyto.	Nucl.	Cyto.	Nucl.
1	VEF-DD	++	-	++	-	++	-	+++	-	++	-
2	VEM-DD	++	-	++	-	++	-	+++	-	++	-
3	VMF-DD	++	-	++	-	++	-	+++	-	++	-
4	VMM-DD	++	-	++	-	++	-	+++	-	++	-
5	VLF-DD	++	-	++	-	++	-	+++	-	++	-
6	VLM-DD	++	-	++	-	++	-	+++	-	++	-
7	SLF-DD	+	-	±	-	+	-	++	-	++	-
8	SLM-DD	+	-	±	-	+	-	++	-	++	-
9	PPF-DD	++	-	±	-	++	-	++	-	++	-
10	PPM-DD	++	-	±	-	++	-	++	-	++	-
11	MPF-DD	++	-	±	-	++	-	++	-	++	-
12	MPM-DD	++	-	±	-	++	-	++	-	++	-
13	LPF-DD	+	-	±	-	++	-	++	-	++	-
14	LPM-DD	+	-	±	-	++	-	++	-	++	-
15	AF-DD	++	-	±	-	++	-	++	-	++	-
16	AM-DD	++	-	±	-	++	-	++	-	++	-

Name of the Haemocyte (column group heading for PRs–PLs).

ABBREVIATIONS/SYMBOLS USED: PRs " Prohaemocytes; OEs " Oenocytoids; SPs " Spherulocytes; GRs " Granulocytes; PLs " Plasmatocytes; Cyto. " Cytoplasm; Nucl. " Nucleus; +++ " Very strong; ++ " Strong; + " Weak; ± "Traces; -- " Absent.

4.5.2.4. Bound lipids

The presence of bound lipids in the haemocytes of different life stages of both the sexes of NDD and DD generations was recorded with the help of Acetone Sudan Black B reaction tests and the findings are presented in Tables 4.49 & 4.50. The presence of bound lipids was not observed in the nuclei of different haemocytes.

The bound lipids present in cytoplasm of Prohaemocytes showed strong (++) reaction from early fifth instar to spinning larval stage in NDD and up to late fifth instar in DD generation in both the sexes. Thereafter, the presence was weak (+) in remaining stages, till emergence of adults.

Table 4.49. Sex-wise presence of bound lipids through histochemical test in cytoplasm & nuclei of different haemocytes in NDD generation of *A. mylitta*

SN	Age	Name of the Haemocyte									
		PRs		OEs		SPs		GRs		PLs	
		Cyto.	Nucl.	Cyto.	Nucl.	Cyto.	Nucl.	Cyto.	Nucl.	Cyto.	Nucl.
1	VEF-NDD	++	-	+	-	+++	-	++	-	+	-
2	VEM-NDD	++	-	+	-	+++	-	++	-	+	-
3	VMF-NDD	++	-	+	-	+++	-	++	-	+	-
4	VMM-NDD	++	-	+	-	+++	-	++	-	+	-
5	VLF-NDD	++	-	+	-	+++	-	++	-	+	-
6	VLM-NDD	++	-	+	-	+++	-	++	-	+	-
7	SLF-NDD	++	-	±	-	++	-	++	-	±	-
8	SLM-NDD	++	-	±	-	++	-	++	-	±	-
9	PPF-NDD	+	-	±	-	++	-	++	-	±	-
10	PPM-NDD	+	-	±	-	++	-	++	-	±	-
11	MPF-NDD	+	-	±	-	++	-	+	-	±	-
12	MPM-NDD	+	-	±	-	++	-	+	-	±	-
13	LPF-NDD	+	-	±	-	++	-	+	-	±	-
14	LPM-NDD	+	-	±	-	++	-	+	-	±	-
15	AF-NDD	+	-	±	-	++	-	+	-	±	-
16	AM-NDD	+	-	±	-	++	-	+	-	±	-

ABBREVIATIONS/SYMBOLS USED: PRs " Prohaemocytes; OEs " Oenocytoids; SPs " Spherulocytes; GRs " Granulocytes; PLs " Plasmatocytes; Cyto. " Cytoplasm; Nucl. " Nucleus; +++ " Very strong; ++ " Strong; + " Weak; ± "Traces, -- " Absent.

In the cytoplasm of Oenocytoids of both the sexes of NDD generation, weak (+) presence of bound lipids was observed from early fifth instar to late fifth instar and in remaining stages, it was available in traces (±). In DD generation, strong (++) presence of bound lipids was recorded from early to late fifth instar larval stage and in remaining stages of both the sexes, its presence was observed to be weak (+) till emergence of adults.

In Spherulocytes of NDD generation, very strong (+++) presence of bound lipid was recorded from early fifth instar to late fifth instar larvae and strong (++) in remaining stages of both the sexes. Comparatively, in DD generation, presence of bound lipid was obserced to be strong (++) from early fifth to late fifth instar larval stage, weak (+) in spinning larvae and again strong (++) from prepupal to late pupal stages and thereafter weak (+) in adults of both the sexes.

Table 4.50. Sex-wise presence of bound lipids through histochemical test in cytoplasm & nuclei of different haemocytes in DD generation of *A. mylitta*

SN	Age	Name of the Haemocyte									
		PRs		OEs		SPs		GRs		PLs	
		Cyto.	Nucl.	Cyto.	Nucl.	Cyto.	Nucl.	Cyto.	Nucl.	Cyto.	Nucl.
1	VEF-DD	++	-	++	-	++	-	++	-	+	-
2	VEM-DD	++	-	++	-	++	-	++	-	+	-
3	VMF-DD	++	-	++	-	++	-	++	-	+	-
4	VMM-DD	++	-	++	-	++	-	++	-	+	-
5	VLF-DD	++	-	++	-	++	-	++	-	+	-
6	VLM-DD	++	-	++	-	++	-	++	-	+	-
7	SLF-DD	+	-	+	-	+	-	++	-	±	-
8	SLM-DD	+	-	+	-	+	-	++	-	±	-
9	PPF-DD	+	-	+	-	++	-	+	-	±	-
10	PPM-DD	+	-	+	-	++	-	+	-	±	-
11	MPF-DD	+	-	+	-	++	-	+	-	±	-
12	MPM-DD	+	-	+	-	++	-	+	-	±	-
13	LPF-DD	+	-	+	-	++	-	+	-	±	-
14	LPM-DD	+	-	+	-	++	-	+	-	±	-
15	AF-DD	+	-	+	-	+	-	+	-	±	-
16	AM-DD	+	-	+	-	+	-	+	-	±	-

ABBREVIATIONS/SYMBOLS USED: PRs " Prohaemocytes; OEs " Oenocytoids; SPs " Spherulocytes; GRs " Granulocytes; PLs " Plasmatocytes; Cyto. " Cytoplasm; Nucl. " Nucleus; +++ " Very strong; ++ " Strong; + " Weak; ± "Traces; -- " Absent.

In Granulocytes of NDD and DD generations, the Acetone Sudan Black B reaction was strong (++) until spinning larval stage. It was strong (++) in prepupal stages of NDD, however, in remaining stages of NDD and DD, the presence of bound lipids was weak (+) in both the sexes.

The Plasmatocytes of both the generation weak (+) reaction was observed until late fifth instar, thereafter, it was available in traces (±).

4.5.2.5. Glycogen

The presence of glycogen in cytoplasm and nuclei of different haemocytes of different stages of NDD and DD generation in both the sexes was observed with the help of Best's ammonium carmine tests which is presented in (Tables 4.51 & 4.52). It is clear from the observations that no traces of glycogen were found in the nuclei of haemocytes.

In the cytoplasm of Prohaemocytes strong (++) presence of glycogen was observed from early fifth instar to late fifth instar larvae, pre and mid pupae and in adults; weak (+) presence in spinning larvae and late pupal stage of both the sexes of DD generation.

Table 4.51. Sex-wise presence of glycogen through histochemical test in cytoplasm & nuclei of different haemocytes in NDD generation of *A. mylitta*

SN	Age	PRs		OEs		SPs		GRs		PLs	
		Cyto.	Nucl.	Cyto.	Nucl.	Cyto.	Nucl.	Cyto.	Nucl.	Cyto.	Nucl.
1	VEF-NDD	+++	-	++	-	+	-	+++	-	+	-
2	VEM-NDD	+++	-	++	-	+	-	+++	-	+	-
3	VMF-NDD	+++	-	++	-	+	-	+++	-	+	-
4	VMM-NDD	+++	-	++	-	+	-	+++	-	+	-
5	VLF-NDD	+++	-	++	-	+	-	+++	-	±	-
6	VLM-NDD	+++	-	++	-	+	-	+++	-	±	-
7	SLF-NDD	+++	-	±	-	±	-	++	-	±	-
8	SLM-NDD	+++	-	±	-	±	-	++	-	±	-
9	PPF-NDD	++	-	±	-	±	-	++	-	±	-
10	PPM-NDD	++	-	±	-	±	-	++	-	±	-
11	MPF-NDD	++	-	±	-	±	-	++	-	±	-
12	MPM-NDD	++	-	±	-	±	-	++	-	±	-
13	LPF-NDD	++	-	±	-	±	-	++	-	±	-
14	LPM-NDD	++	-	±	-	±	-	++	-	±	-
15	AF-NDD	++	-	±	-	±	-	++	-	±	-
16	AM-NDD	++	-	±	-	±	-	++	-	±	-

ABBREVIATIONS/SYMBOLS USED: PRs " Prohaemocytes; OEs " Oenocytoids; SPs " Spherulocytes; GRs " Granulocytes; PLs " Plasmatocytes; Cyto. " Cytoplasm; Nucl. " Nucleus; +++ " Very strong; ++ " Strong; + " Weak; ± "Traces; -- " Absent.

In case of DD generation the presence was observed to be very strong (+++) from early fifth to spinning larval stage of both the sexes and thereafter the presence was recorded to be strong (++), till emergence of adults.

In cytoplasm of Oenocytoids of both the sexes and generations, strong (++) presence was recorded from early fifth instar to late fifth instar larval stage, thereafter, it was observed to be in traces (±).

In the cytoplasm of spherulocytes, weak (+) presence of glycogen was recorded from early fifth instar to mid fifth instar larval stage of both the sexes and generations, thereafter, it was observed in traces (±) till emergence of adults. Besides, weak (+) presence was also observed in late fifth instar in both the sexes of NDD generation.

In cytoplasm of Granulocytes reaction was observed to be very strong (+++) from early fifth instar to late fifth instar larval stage and strong (++) in remaining stages till emergence of adults in both the sexes and generations.

In cytoplasm of Plasmatocytes, weak (+) reaction was observed from early fifth to mid fifth instars larval stage of both the sexes and generations. Thereafter, it was observed in traces (±) till emergence of adults.

Table 4.52. Sex-wise presence of glycogen through histochemical test in cytoplasm & nuclei of different haemocytes in DD generation of *A. mylitta*

SN	Age	PRs Cyto.	PRs Nucl.	OEs Cyto.	OEs Nucl.	SPs Cyto.	SPs Nucl.	GRs Cyto.	GRs Nucl.	PLs Cyto.	PLs Nucl.
1	VEF-DD	++	-	++	-	+	-	+++	-	+	-
2	VEM-DD	++	-	++	-	+	-	+++	-	+	-
3	VMF-DD	++	-	++	-	+	-	+++	-	+	-
4	VMM-DD	++	-	++	-	+	-	+++	-	+	-
5	VLF-DD	++	-	++	-	±	-	+++	-	±	-
6	VLM-DD	++	-	++	-	±	-	+++	-	±	-
7	SLF-DD	+	-	±	-	±	-	++	-	±	-
8	SLM-DD	+	-	±	-	±	-	++	-	±	-
9	PPF-DD	++	-	±	-	±	-	++	-	±	-
10	PPM-DD	++	-	±	-	±	-	++	-	±	-
11	MPF-DD	++	-	±	-	±	-	++	-	±	-
12	MPM-DD	++	-	±	-	±	-	++	-	±	-
13	LPF-DD	+	-	±	-	±	-	++	-	±	-
14	LPM-DD	+	-	±	-	±	-	++	-	±	-
15	AF-DD	++	-	±	-	±	-	++	-	±	-
16	AM-DD	++	-	±	-	±	-	++	-	±	-

ABBREVIATIONS/SYMBOLS USED: PRs " Prohaemocytes; OEs " Oenocytoids; SPs " Spherulocytes; GRs " Granulocytes; PLs " Plasmatocytes; Cyto. " Cytoplasm; Nucl. " Nucleus; +++ " Very strong; ++ " Strong; + " Weak; ± "Traces; -- " Absent.

4.5.2.6. Alkaline phosphatases

In none of the nuclei of Prohaemocytes, Oenocytoids, Spherulocytes, Granulocytes and Plasmatocytes in any stage of both the sexes of NDD and DD generations Cobalt sulphide reaction showed presence of alkaline phosphatases. The observations are presented in Tables 4.53. & 4.54.

In cytoplasm of Prohaemocytes, it was available in traces (±) up to spinning larval stage of both the sexes and generations. In remaining stages, it was absent (-).

In cytoplasm of Oenocytoids and Plasmatocytes, no traces of alkaline phosphatases were observed in both the sexes and generations.

In cytoplasm of Spherulocytes and Granulocytes, weak (+) presence was observed up to mid-pupal stage, thereafter, it was absent (-) in both the sexes and generations.

Table 4.53. Sex-wise presence of alkaline phosphatases through histochemical test in cytoplasm & nuclei of different haemocytes in NDD generation of *A. mylitta*

SNAge	Name of the Haemocyte									
	PRs		OEs		SPs		GRs		PLs	
	Cyto.	Nucl.	Cyto.	Nucl.	Cyto.	Nucl.	Cyto.	Nucl.	Cyto.	Nucl.
1 VEF-NDD	±	-	-	-	+	-	+	-	-	-
2 VEM-NDD	±	-	-	-	+	-	+	-	-	-
3 VMF-NDD	±	-	-	-	+	-	+	-	-	-
4 VMM-NDD	±	-	-	-	+	-	+	-	-	-
5 VLF-NDD	±	-	-	-	+	-	+	-	-	-
6 VLM-NDD	±	-	-	-	+	-	+	-	-	-
7 SLF-NDD	±	-	-	-	+	-	+	-	-	-
8 SLM-NDD	±	-	-	-	+	-	+	-	-	-
9 PPF-NDD	-	-	-	-	+	-	+	-	-	-
10 PPM-NDD	-	-	-	-	+	-	+	-	-	-
11 MPF-NDD	-	-	-	-	+	-	+	-	-	-
12 MPM-NDD	-	-	-	-	+	-	+	-	-	-
13 LPF-NDD	-	-	-	-	-	-	-	-	-	-
14 LPM-NDD	-	-	-	-	-	-	-	-	-	-
15 AF-NDD	-	-	-	-	-	-	-	-	-	-
16 AM-NDD	-	-	-	-	-	-	-	-	-	-

ABBREVIATIONS/SYMBOLS USED: PRs " Prohaemocytes; OEs " Oenocytoids; SPs " Spherulocytes; GRs " Granulocytes; PLs " Plasmatocytes; Cyto. " Cytoplasm; Nucl. " Nucleus; +++ " Very strong; ++ " Strong; + " Weak; ± "Traces; -- " Absent.

Table 4.54. Sex-wise presence of alkaline phosphatases through histochemical test in cytoplasm & nuclei of different haemocytes in DD generation of *A. mylitta*

SNAge	Name of the Haemocyte									
	PRs		OEs		SPs		GRs		PLs	
	Cyto.	Nucl.	Cyto.	Nucl.	Cyto.	Nucl.	Cyto.	Nucl.	Cyto.	Nucl.
1 VEF-DD	±	-	-	-	+	-	+	-	-	-
2 VEM-DD	±	-	-	-	+	-	+	-	-	-
3 VMF-DD	±	-	-	-	+	-	+	-	-	-
4 VMM-DD	±	-	-	-	+	-	+	-	-	-
5 VLF-DD	±	-	-	-	+	-	+	-	-	-
6 VLM-DD	±	-	-	-	+	-	+	-	-	-
7 SLF-DD	±	-	-	-	+	-	+	-	-	-
8 SLM-DD	±	-	-	-	+	-	+	-	-	-
9 PPF-DD	-	-	-	-	+	-	+	-	-	-
10 PPM-DD	-	-	-	-	+	-	+	-	-	-
11 MPF-DD	-	-	-	-	+	-	+	-	-	-
12 MPM-DD	-	-	-	-	+	-	+	-	-	-
13 LPF-DD	-	-	-	-	-	-	-	-	-	-
14 LPM-DD	-	-	-	-	-	-	-	-	-	-
15 AF-DD	-	-	-	-	-	-	-	-	-	-
16 AM-DD	-	-	-	-	-	-	-	-	-	-

ABBREVIATIONS/SYMBOLS USED: PRs " Prohaemocytes; OEs " Oenocytoids; SPs " Spherulocytes; GRs " Granulocytes; PLs " Plasmatocytes; Cyto. " Cytoplasm; Nucl. " Nucleus; +++ " Very strong; ++ " Strong; + " Weak; ± "Traces; -- " Absent.

4.6. Physical properties and chemical composition of haemolymph of *Antheraea mylitta*

4.6.1. Specific gravity

Under the present study, attempts were made to estimate the specific gravity of haemolymph of different life stages of NDD and DD generations of A. mylitta. Student's t-test was used to determine the means of different life stages of the generations and presented in Table (4.55, 4.56 and 4.57).

The mean specific gravity of I-NDD was 1.082 ± 0.004 and was significantly higher ($p < 0.01$) than I-DD (1.0047 ± 0.0013). Similarly, in second third, fourth and early fifth larval stage of both the sexes had significantly higher specific gravity in NDD generation when compared with respective stage of DD generation. This difference was also significant ($p < 0.01$) in between VMM-NDD (1.0143 ± 0.0008) & VMM-DD (1.0113 ± 0.0009), MPF-NDD (1.0217 ± 0.0009) & MPF-DD (1.0380 ± 0.0014), MPM-NDD (1.0191 ± 0.0007) & MPM-DD (1.0433±0.0013), LPF-NDD (1.0233±0.0007) & LPF-DD (1.0453 ± 0.0013), AF-NDD (1.0550 ± 0.0006) & AF - DD (1.0620 ± 0.0011) and AM - NDD

Table 4.55. Specific gravity of the haemolymph of different stages in NDD & DD generations of *A. mylitta*

SN	Age	Mean	±	SE	t Stat	
1	I-NDD	1.0182	±	0.0004	10.5170	**
2	I-DD	1.0047	±	0.0013		
3	II-NDD	1.0100	±	0.0000	3.0551	**
4	II-DD	1.0060	±	0.0013		
5	III-NDD	1.0134	±	0.0006	5.5159	**
6	III-DD	1.0067	±	0.0013		
7	IV-NDD	1.0159	±	0.0007	3.7668	**
8	IV-DD	1.0093	±	0.0015		
9	VEF-NDD	1.0140	±	0.0008	4.3353	**
10	VEF-DD	1.0093	±	0.0007		
11	VEM-NDD	1.0151	±	0.0007	4.5431	**
12	VEM-DD	1.0080	±	0.0014		
13	VMF-NDD	1.0148	±	0.0008	0.0871	NS
14	VMF-DD	1.0147	±	0.0013		
15	VMM-NDD	1.0143	±	0.0008	2.6834	**
16	VMM-DD	1.0113	±	0.0009		
17	VLF-NDD	1.0153	±	0.0007	1.3284	NS
18	VLF-DD	1.0133	±	0.0013		
19	VLM-NDD	1.0137	±	0.0006	-1.5934	NS
20	VLM-DD	1.0160	±	0.0013		
21	SLF-NDD	1.0139	±	0.0008	-2.2747	*
22	SLF-DD	1.0173	±	0.0012		
23	SLM-NDD	1.0129	±	0.0007	-0.7063	NS

24	SLM-DD	1.0140	±	0.0013		
25	PPF-NDD	1.0189	±	0.0007	-0.3371	NS
26	PPF-DD	1.0193	±	0.0007		
27	PPM-NDD	1.0201	±	0.0009	-0.5478	NS
28	PPM-DD	1.0213	±	0.0019		
29	MPF-NDD	1.0217	±	0.0009	-9.6362	**
30	MPF-DD	1.0380	±	0.0014		
31	MPM-NDD	1.0191	±	0.0007	-16.6356	**
32	MPM-DD	1.0433	±	0.0013		
33	LPF-NDD	1.0233	±	0.0007	-16.8496	**
34	LPF-DD	1.0453	±	0.0013		
35	LPM-NDD	1.0238	±	0.0006	-23.9878	**
36	LPM-DD	1.0560	±	0.0013		
37	AF-NDD	1.0550	±	0.0006	-5.8762	**
38	AF-DD	1.0620	±	0.0011		
39	AM-NDD	1.0307	±	0.0021	-11.9270	**
40	AM-DD	1.0620	±	0.0011		

n = 15 (Abbreviations used are as per section 3.1.2.)

(1.0307 ± 0.0021) & AM-DD (1.0620 ± 0.0011). The difference was observed to be significant ($p < 0.05$) in between SLF-NDD (1.0139 ± 0.0008) and SLF-DD (1.0173 ± 0.0012). Comparisons in between other stages of both the generations were observed to be non-significant.

Table 4.56. Specific gravity of the haemolymph of different sexes in NDD generation of *A. mylitta*

SN	Age	Mean	±	SE	t Stat	
1	VEF-NDD	1.0140	±	0.0008	-1.1923	NS
2	VEM-NDD	1.0151	±	0.0007		
3	VMF-NDD	1.0148	±	0.0008	0.4193	NS
4	VMM-NDD	1.0143	±	0.0008		
5	VLF-NDD	1.0153	±	0.0007	1.5369	NS
6	VLM-NDD	1.0137	±	0.0006		
7	SLF-NDD	1.0139	±	0.0008	0.7902	NS
8	SLM-NDD	1.0129	±	0.0007		
9	PPF-NDD	1.0189	±	0.0007	-1.0836	NS
10	PPM-NDD	1.0201	±	0.0009		
11	MPF-NDD	1.0217	±	0.0009	1.8362	*
12	MPM-NDD	1.0191	±	0.0007		
13	LPF-NDD	1.0233	±	0.0007	-0.4910	NS
14	LPM-NDD	1.0238	±	0.0006		
15	AF-NDD	1.0550	±	0.0006	10.5398	**
16	AM-NDD	1.0307	±	0.0021		

n = 15 (Abbreviations used are as per section 3.1.2.)

Between the sexes of NDD generation, difference in the level of specific gravity was recorded to be non-significant (Student's t-test) except in MPF-NDD (1.0217 ± 0.0009) & MPM-NDD (1.0191 ± 0.0007) where difference was significant ($p < 0.05$) and in between AF-NDD (1.0550 ± 0.0006) and AM-NDD (1.0307 ± 0.0021) where difference was significant at $p < 0.01$.

Table 4.57. Specific gravity of the haemolymph of different sexes in DD generation of *A. mylitta*

SN	Age	Mean	±	SE	t Stat	
1	VEF-DD	1.0093	±	0.0007	0.8069	NS
2	VEM-DD	1.0080	±	0.0014		
3	VMF-DD	1.0147	±	0.0013	2.0917	*
4	VMM-DD	1.0113	±	0.0009		
5	VLF-DD	1.0133	±	0.0013	-1.2929	NS
6	VLM-DD	1.0160	±	0.0013		
7	SLF-DD	1.0173	±	0.0012	2.0917	*
8	SLM-DD	1.0140	±	0.0013		
9	PPF-DD	1.0193	±	0.0007	-1.0000	NS
10	PPM-DD	1.0213	±	0.0019		
11	MPF-DD	1.0380	±	0.0014	-2.7792	**
12	MPM-DD	1.0433	±	0.0013		
13	LPF-DD	1.0453	±	0.0013	-5.1717	**
14	LPM-DD	1.0560	±	0.0013		
15	AF-DD	1.0620	±	0.0011	0.0000	NS
16	AM-DD	1.0620	±	0.0011		

n = 15 (Abbreviations used are as per section 3.1.2.)

Student's t-test between sexes of DD forms revealed that the mean difference in specific gravity values of two sexes was significant ($p < 0.05$) in between VMF-DD (1.0147 ± 0.0013) & VMM-DD (1.0133 ± 0.0009), SLF-DD (1.0173 ± 0.0012) & SLM-DD (1.0140 ± 0.0013) and $p < 0.01$ was in between MPF-DD (1.0380 ± 0.0014) & MPM- DD (1.0433 ± 0.0013), LPF- DD (1.0453 ± 0.0013) & LPM- DD (1.0560 ± 0.0013).

4.6.2. pH value

In general, the haemolymph pH of the life stages ranged between 6.713 to 6.930, which may be considered slightly acidic nearer to neutral (Table 4.58).

No significant difference was recorded among different life stage of the NDD and DD generation and in between the sexes of both generations (Tables-4.59 & 5.60). However, pH of II and III instar NDD and DD larvae had significantly ($p < 0.01$) different mean values of pH. Similarly, mean pH of MPM-NDD & MPM-DD larvae were also different at $p < 0.05$.

Table 4.58. Hydrogen ion concentration (pH) of the haemolymph of different stages in NDD & DD generations of *A. mylitta*

SN	Age	Mean	±	SE	t Stat	
1	I-NDD	6.930	±	0.015	1.705	NS
2	I-DD	6.883	±	0.024		
3	II-NDD	6.910	±	0.017	14.000	**
4	II-DD	6.733	±	0.022		
5	III-NDD	6.883	±	0.012	8.828	**
6	III-DD	6.713	±	0.018		
7	IV-NDD	6.890	±	0.015	-0.269	NS
8	IV-DD	6.897	±	0.020		
9	VEF-NDD	6.897	±	0.014	1.000	NS
10	VEF-DD	6.870	±	0.022		
11	VEM-NDD	6.870	±	0.014	0.000	NS
12	VEM-DD	6.870	±	0.017		
13	VMF-NDD	6.870	±	0.017	0.000	NS
14	VMF-DD	6.870	±	0.020		
15	VMM-NDD	6.883	±	0.016	0.000	NS
16	VMM-DD	6.883	±	0.019		
17	VLF-NDD	6.897	±	0.020	-0.435	NS
18	VLF-DD	6.910	±	0.017		
19	VLM-NDD	6.897	±	0.017	1.000	NS
20	VLM-DD	6.877	±	0.016		
21	SLF-NDD	6.870	±	0.017	-1.169	NS
22	SLF-DD	6.897	±	0.017		
23	SLM-NDD	6.877	±	0.016	0.000	NS
24	SLM-DD	6.877	±	0.013		
25	PPF-NDD	6.897	±	0.022	0.000	NS
26	PPF-DD	6.897	±	0.022		
27	PPM-NDD	6.903	±	0.021	0.459	NS
28	PPM-DD	6.890	±	0.015		
29	MPF-NDD	6.863	±	0.022	0.000	NS
30	MPF-DD	6.863	±	0.014		
31	MPM-NDD	6.930	±	0.018	2.103	*
32	MPM-DD	6.890	±	0.021		
33	LPF-NDD	6.910	±	0.017	0.000	NS
34	LPF-DD	6.910	±	0.020		
35	LPM-NDD	6.883	±	0.021	-0.587	NS
36	LPM-DD	6.903	±	0.019		
37	AF-NDD	6.890	±	0.018	0.202	NS
38	AF-DD	6.883	±	0.021		
39	AM-NDD	6.910	±	0.017	0.899	NS
40	AM-DD	6.890	±	0.015		

n=15 (Abbreviations used are as per section 3.1.2.)

Between sexes, the means were similar in almost all the stages of life cycle studied except in between MPF-NDD (6.863 ± 0.022) and MPM-NDD (6.930 ± 0.018) where the difference was significant (P<0.05).

Table 4.59. Hydrogen ion concentration (pH) in females & males of NDD generation

SN	Age	Mean	±	SE	t Stat	
1	VEF-NDD	6.897	±	0.014	1.468	NS
2	VEM-NDD	6.870	±	0.014		
3	VMF-NDD	6.870	±	0.017	-0.487	NS
4	VMM-NDD	6.883	±	0.016		
5	VLF-NDD	6.897	±	0.020	0.000	NS
6	VLM-NDD	6.897	±	0.017		
7	SLF-NDD	6.870	±	0.017	-0.292	NS
8	SLM-NDD	6.877	±	0.016		
9	PPF-NDD	6.897	±	0.022	-0.202	NS
10	PPM-NDD	6.903	±	0.021		
11	MPF-NDD	6.863	±	0.022	-2.320	*
12	MPM-NDD	6.930	±	0.018		
13	LPF-NDD	6.910	±	0.017	1.075	NS
14	LPM-NDD	6.883	±	0.021		
15	AF-NDD	6.890	±	0.018	-0.899	NS
16	AM-NDD	6.910	±	0.017		

n = 15 (Abbreviations used are as per section 3.1.2.)

Table 4.60. Hydrogen ion concentration (pH) in females and males of DD generation

SN	Age	Mean	±	SE	t Stat	
1	VEF-DD	6.870	±	0.022	0.000	NS
2	VEM-DD	6.870	±	0.017		
3	VMF-DD	6.870	±	0.020	-0.487	NS
4	VMM-DD	6.883	±	0.019		
5	VLF-DD	6.910	±	0.017	1.581	NS
6	VLM-DD	6.877	±	0.016		
7	SLF-DD	6.897	±	0.017	0.899	NS
8	SLM-DD	6.877	±	0.013		
9	PPF-DD	6.897	±	0.022	0.250	NS
10	PPM-DD	6.890	±	0.015		
11	MPF-DD	6.863	±	0.014	-1.000	NS
12	MPM-DD	6.890	±	0.021		
13	LPF-DD	6.910	±	0.020	0.292	NS
14	LPM-DD	6.903	±	0.019		
15	AF-DD	6.883	±	0.021	-0.250	NS
16	AM-DD	6.890	±	0.015		

n = 15 (Abbreviations used are as per section 3.1.2.)

4.6.3. Haemolymph cations

The cations, namely, Na+, K+, Ca++ and Mg++ were recorded in the haemolymph of different stages of both the sexes and generations of A. mylitta. The detailed observations are mentioned in following pages.

4.6.3.1. Sodium (Na+)

Sodium concentration in terms of mEq/l was recorded in a fluctuating manner in different stages of NDD generations as it was 1.4700 ± 0.1134 in I-NDD, 1.3100 ± 0.0532 in II-NDD, 1.3700 ± 0.0545 in III-NDD and 1.2900 ± 0.1048 in IV-NDD. The sodium concentration increased up to 1.6700 ± 0.0632 in early male fifth instar and lowered in mid fifth instar following a rise in late male fifth instar. In prepupal stage of male of NDD, concentration rose up to 1.8100 ± 0.0292. It was recorded to be highest in adults of both the sexes such as in female (5.2100 ± 0.2603) and male (6.8600 ± 0.2207). Fluctuating trend was also recorded in DD generation.

One way ANOVA was applied to observe the difference in between mean Na+ concentration in respective stage of NDD and DD generations. The difference in concentration of Na+ was significant ($p < 0.05$) in between II-NDD & II-DD, VEF-NDD & VEF-DD, VLM-NDD & VLM-DD, PPF-NDD & PPF-DD, PPM-NDD & PPM-DD, LPM-NDD & LPM-DD. Significant difference ($p < 0.01$) was also observed in between VMF-NDD & VMF-DD, SLF-NDD & SLF-DD, SLM-NDD & SLM-DD, AM-NDD & AM-DD (Table 4.61). Adults NDD (females and males) had higher Na+ than adults of DD generations.

Between the different stages of the two sexes of NDD generations, one-way ANOVA was done which indicated that Na+ was significantly higher ($p < 0.01$) in males, particularly, in VEM (1.67 ± 0.063) than VEF NDD (1.2600 ± 0.0502). The trend continued to be the same in between LPM-NDD (1.5900 ± 0.0399) and LPF-NDD (1.1900 ± 0.0399) and adult male (6.8600 ± 0.2207) and female (5.2100 ± 0.2603) (Table 4.62).

Table 4.61. Sodium ion concentration (mEq/l) of the haemolymph of different stages in NDD & DD generations of *A. mylitta*

SN	Age	Mean	±	SE	t Stat	
1	I-NDD	1.4700	±	0.1134	-0.3203	NS
2	I-DD	1.5100	±	0.0823		
3	II-NDD	1.3100	±	0.0532	2.0925	*
4	II-DD	1.1900	±	0.0508		
5	III-NDD	1.3700	±	0.0545	1.7928	NS
6	III-DD	1.2700	±	0.0417		
7	IV-NDD	1.2900	±	0.1048	-0.9491	NS
8	IV-DD	1.4000	±	0.0703		
9	VEF-NDD	1.2600	±	0.0502	-1.8773	*
10	VEF-DD	1.3700	±	0.0274		
11	VEM-NDD	1.6700	±	0.0630	0.3273	NS
12	VEM-DD	1.6500	±	0.0283		
13	VMF-NDD	1.1800	±	0.0344	-2.8900	**

14	VMF-DD	1.3700	±	0.0417		
15	VMM-NDD	1.5200	±	0.0886	0.6882	NS
16	VMM-DD	1.4800	±	0.0625		
17	VLF-NDD	1.3600	±	0.0358	0.4804	NS
18	VLF-DD	1.3400	±	0.0281		
19	VLM-NDD	1.6200	±	0.0378	2.6656	*
20	VLM-DD	1.4700	±	0.0387		
21	SLF-NDD	1.1500	±	0.0360	-3.6979	**
22	SLF-DD	1.4300	±	0.0738		
23	SLM-NDD	1.3600	±	0.0549	-4.2055	**
24	SLM-DD	1.6200	±	0.0540		
25	PPF-NDD	1.3200	±	0.0439	-2.5776	*
26	PPF-DD	1.5100	±	0.0618		
27	PPM-NDD	1.8100	±	0.0292	2.7618	*
28	PPM-DD	1.6400	±	0.0613		
29	MPF-NDD	1.2500	±	0.0423	0.2458	NS
30	MPF-DD	1.2300	±	0.0498		
31	MPM-NDD	1.4800	±	0.0466	1.8020	NS
32	MPM-DD	1.3100	±	0.0728		
33	LPF-NDD	1.1900	±	0.0399	0.5571	NS
34	LPF-DD	1.1500	±	0.0423		
35	LPM-NDD	1.5900	±	0.0483	2.2119	*
36	LPM-DD	1.5000	±	0.0272		
37	AF-NDD	5.2100	±	0.2603	0.6509	NS
38	AF-DD	5.0000	±	0.1707		
39	AM-NDD	6.8600	±	0.2207	2.9215	**
40	AM-DD	5.9900	±	0.2687		

n = 10 (Abbreviations used are as per section 3.1.2.)

In DD generation, the difference between the sexes was significant ($P<0.05$) in between in SLF-DD (1.4300 ± 0.0738) & SLM-DD (1.6200 ± 0.0540) and at $P < 0.01$ was in between VEF-DD & VEM-DD, VMF-DD & VMM-DD, VLF-DD & VLM-DD, LPF-DD & LPM-DD, AF-DD & AM-DD. Comparisons between the means availability of Na+ in the remaining stages of both the sexes was non-significant (Table 4.63).

4.6.3.2. Potassium (K+)

K+ concentration was also recorded in different larval instar, different aged pupae and adults of both NDD and DD of A. mylitta and is shown in Table 4.64 to 4.66. In I-NDD K+ concentration was 51.0500 ± 0.4229 and with little decrease or increase in its level it prevailed till emergence of adults. The highest concentration was recorded to be in adults female NDD (68.7700 ± 0.2178) and male 64.6400 ± 0.3808). Similarly, in DD generation K+ concentration ranged in between 47.4500 ± 0.7275 (MPM-DD) to 51.5100 ± 0.4484 (MPF-DD) and it was recorded to be highest in adults of both the sexes i.e. female (64.0600 ± 0.3656) and male (60.8200 ± 0.2567) -Table 4.64.

Table 4.62. Sodium ion concentration (mEq/l) in females and males of NDD generation in *A. mylitta*

SN	Age	Mean	±	SE	t Stat	
1	VEF-NDD	1.2600	±	0.0502	-5.5623	**
2	VEM-NDD	1.6700	±	0.0630		
3	VMF-NDD	1.1800	±	0.0344	-3.4701	**
4	VMM-NDD	1.5200	±	0.0886		
5	VLF-NDD	1.3600	±	0.0358	-4.6284	**
6	VLM-NDD	1.6200	±	0.0378		
7	SLF-NDD	1.1500	±	0.0360	-3.5839	**
8	SLM-NDD	1.3600	±	0.0549		
9	PPF-NDD	1.3200	±	0.0439	-8.3625	**
10	PPM-NDD	1.8100	±	0.0292		
11	MPF-NDD	1.2500	±	0.0423	-3.8512	**
12	MPM-NDD	1.4800	±	0.0466		
13	LPF-NDD	1.1900	±	0.0399	-7.4421	**
14	LPM-NDD	1.5900	±	0.0483		
15	AF-NDD	5.2100	±	0.2603	-4.7015	**
16	AM-NDD	6.8600	±	0.2207		

n = 10 (Abbreviations used are as per section 3.1.2.)

Table 4.63. Sodium ion concentration (mEq/l) in females and males of DD generation in *A. mylitta*

SN	Age	Mean	±	SE	t Stat	
1	VEF-DD	1.3700	±	0.0274	-7.2029	**
2	VEM-DD	1.6500	±	0.0283		
3	VMF-DD	1.3700	±	0.0417	-3.4980	**
4	VMM-DD	1.4800	±	0.0625		
5	VLF-DD	1.3400	±	0.0281	-3.5455	**
6	VLM-DD	1.4700	±	0.0387		
7	SLF-DD	1.4300	±	0.0738	-2.4327	*
8	SLM-DD	1.6200	±	0.0540		
9	PPF-DD	1.5100	±	0.0618	-1.7086	NS
10	PPM-DD	1.6400	±	0.0613		
11	MPF-DD	1.2300	±	0.0498	-0.9370	NS
12	MPM-DD	1.3100	±	0.0728		
13	LPF-DD	1.1500	±	0.0423	-7.0000	**
14	LPM-DD	1.5000	±	0.0272		
15	AF-DD	5.0000	±	0.1707	-4.6220	**
16	AM-DD	5.9900	±	0.2687		

n = 10 (Abbreviations used are as per section 3.1.2.)

ANOVA (one way) was worked out to find out the difference between means of K+ ion concentration in the sexes of NDD generation. A significant difference ($p < 0.01$) was recorded in between VEF-NDD & VEM-NDD, VMF-NDD & VMM-NDD, VLF-NDD & VLM-NDD, SLF-NDD & SLM-NDD and AF-NDD & AM-NDD. The difference was significant ($P < 0.05$) in between PPF-NDD & PPM-NDD and MPF-NDD & MPM-NDD and the difference was observed to be non-significant in between, LPF-NDD & LPM-NDD. In late pupal and adult stage the K+ concentration was higher in female than male viz; LPF-NDD (48.1800 ± 0.4436), LPM-NDD (48.64 ± 0.1619), AF-NDD (68.7700 ± 0.2178) and AM-NDD (64.6400 ± 0.3808)-Table 4.65.

The difference between the mean concentration of K+ in the two sexes of DD generation was also significant ($P<0.01$) in between VEF-DD & VEM-DD, VLF-DD & VLM-DD, SLF-DD & SLM-DD, MPF-DD & MPM-DD, and AF-DD & AM-DD. The difference was significant ($P<0.05$) in between VMF-DD & VMM-DD, PPF-DD & PPM-DD and non-significant difference was recorded in between LPF-DD and LPM-DD. K+ concentration was more in female (64.06 ± 0.3656) than male (60.82 ± 0.2567) - Table 4.66.

Table 4.64. Potassium ion concentration (mEq/l) of the haemolymph of different stages in NDD & DD generations of *A. mylitta*

SN	Age	Mean	±	SE	t Stat	
1	I-NDD	51.0500	±	0.4229	1.6216	NS
2	I-DD	49.6600	±	0.7073		
3	II-NDD	49.5200	±	0.2017	-2.4385	*
4	II-DD	50.4100	±	0.3476		
5	III-NDD	49.8400	±	0.2700	0.2793	NS
6	III-DD	49.7200	±	0.2841		
7	IV-NDD	50.4500	±	0.3850	-1.2262	NS
8	IV-DD	51.0300	±	0.3903		
9	VEF-NDD	51.2200	±	0.1720	0.7802	NS
10	VEF-DD	50.7800	±	0.4954		
11	VEM-NDD	48.4600	±	0.3785	-0.4647	NS
12	VEM-DD	48.7000	±	0.3407		
13	VMF-NDD	51.5500	±	0.4235	2.4970	*
14	VMF-DD	49.7700	±	0.6356		
15	VMM-NDD	49.0700	±	0.2067	1.3591	NS
16	VMM-DD	48.6900	±	0.1972		
17	VLF-NDD	47.8800	±	0.5335	-2.6494	*
18	VLF-DD	49.4300	±	0.6068		
19	VLM-NDD	45.8200	±	0.2089	-1.9579	*
20	VLM-DD	47.4500	±	0.7275		
21	SLF-NDD	48.1000	±	0.3006	-2.4099	*
22	SLF-DD	49.1300	±	0.3954		
23	SLM-NDD	51.1600	±	0.2038	0.7211	NS
24	SLM-DD	50.8100	±	0.3646		

25	PPF-NDD	50.0000	±	0.2419	-2.2621	*
26	PPF-DD	50.6300	±	0.1988		
27	PPM-NDD	51.2000	±	0.3725	3.2399	**
28	PPM-DD	49.4000	±	0.5509		
29	MPF-NDD	50.5000	±	0.5670	-1.3504	NS
30	MPF-DD	51.5100	±	0.4484		
31	MPM-NDD	48.7800	±	0.2628	-0.9508	NS
32	MPM-DD	49.0800	±	0.2542		
33	LPF-NDD	48.1800	±	0.4436	0.8998	NS
34	LPF-DD	47.6900	±	0.4985		
35	LPM-NDD	48.6400	±	0.1619	1.9418	*
36	LPM-DD	47.5300	±	0.4937		
37	AF-NDD	68.7700	±	0.2178	14.0397	**
38	AF-DD	64.0600	±	0.3656		
39	AM-NDD	64.6400	±	0.3808	7.6103	**
40	AM-DD	60.8200	±	0.2567		

n = 10 (Abbreviations used are as per section 3.1.2.)

Table 4.65. Potassium ion concentration (mEq/l) in females and males of NDD generation in *A. mylitta*

SN	Age	Mean	±	SE	t Stat	
1	VEF-NDD	51.2200	±	0.1720	6.3460	**
2	VEM-NDD	48.4600	±	0.3785		
3	VMF-NDD	51.5500	±	0.4235	5.6119	**
4	VMM-NDD	49.0700	±	0.2067		
5	VLF-NDD	47.8800	±	0.5335	3.8163	**
6	VLM-NDD	45.8200	±	0.2089		
7	SLF-NDD	48.1000	±	0.3006	-11.8474	**
8	SLM-NDD	51.1600	±	0.2038		
9	PPF-NDD	50.0000	±	0.2419	-2.8111	*
10	PPM-NDD	51.2000	±	0.3725		
11	MPF-NDD	50.5000	±	0.5670	2.5892	*
12	MPM-NDD	48.7800	±	0.2628		
13	LPF-NDD	48.1800	±	0.4436	-0.9350	NS
14	LPM-NDD	48.6400	±	0.1619		
15	AF-NDD	68.7700	±	0.2178	9.8629	**
16	AM-NDD	64.6400	±	0.3808		

n = 10 (Abbreviations used are as per section 3.1.2.)

Table 4.66. Potassium ion concentration (mEq/l) in females and males of DD generation in *A. mylitta*

SN	Age	Mean	±	SE	t Stat	
1	VEF-DD	50.7800	±	0.4954	3.4918	**
2	VEM-DD	48.7000	±	0.3407		
3	VMF-DD	49.7700	±	0.6356	1.9173	*
4	VMM-DD	48.6900	±	0.1972		
5	VLF-DD	49.4300	±	0.6068	3.6589	**
6	VLM-DD	47.4500	±	0.7275		
7	SLF-DD	49.1300	±	0.3954	-2.8814	**
8	SLM-DD	50.8100	±	0.3646		
9	PPF-DD	50.6300	±	0.1988	1.9240	*
10	PPM-DD	49.4000	±	0.5509		
11	MPF-DD	51.5100	±	0.4484	4.1687	**
12	MPM-DD	49.0800	±	0.2542		
13	LPF-DD	47.6900	±	0.4985	0.2725	NS
14	LPM-DD	47.5300	±	0.4937		
15	AF-DD	64.0600	±	0.3656	5.9570	**
16	AM-DD	60.8200	±	0.2567		

n = 10 (Abbreviations used are as per section 3.1.2.)

4.6.3.3. Calcium (Ca++)

Ca++ concentration of different stages of NDD and DD generations and also in both the sexes was estimated and is presented in Table- 4.67 to 4.69.

Ca++ concentration was 22.1800 ± 0.474 in I-NDD generation, 21.9800 ± 0.2119 in II-NDD, 23.4600 ± 0.2630 in III-NDD and 21.6500 ± 0.1945 in IV-NDD. The concentration was higher in VMF-NDD (22.0600 ± 0.6887), VMM-NDD (24.0800 ± 0.394), LPF-NDD (22.5000 ± 0.1736) and LPM-NDD (23.3400 ± 0.1326) though it was highest in AF-NDD (27.7800 ± 0.2403) & AM-NDD (30.6900 ± 0.2263). In DD generation Ca+ was 20.3800 ± 0.7648 in I-DD, 21.0600 ± 0.4214 in II-DD, 21.0500 ± 7686 in III-DD and 21.2200 ± 0.3246 in IV-DD. The concentration of Ca+ was observed to be highest in AF-DD (26.2400 ± 0.4608) and AM-DD (26.6700 ± 1.5533).

Student's t-test was implied in between different stages of NDD and DD generations to test the difference in between means. The difference was significant ($p < 0.05$) in between I-NDD and I-DD, SLF-NDD & SLF-DD, SLM- NDD & SLM-DD, PPM-NDD & PPM-DD, MPF-NDD and MPF-DD and AM-NDD and AM-DD. The difference was observed to be significant ($P < 0.01$) in between II-NDD and II-DD, VLF-NDD & VLF-DD. In rest of the comparisons the non-significant difference was observed.

Data were statistically analysed to find out the difference between the mean concentrations of Ca+ of sexes in NDD generation. The difference was non-significant in between VEF-NDD & VEM-NDD, and significant ($p < 0.01$) in between VLF-NDD & VLM-NDD, SLF-NDD & SLM-NDD, PPF-NDD & PPM-NDD, MPF-NDD & MPM-NDD, LPF-NDD & LPM-NDD, AF-NDD & AM-NDD and $p<0.05$ was in between VMF-NDD and VMM-NDD.

In DD generation, the difference was significant ($P < 0.05$) in between VEF-DD & VEM-DD, VMF-DD & VMM-DD, and LPF-DD & LPM-DD. However, in all other stages the difference was non-significant -Table 4.69.

Table 4.67. Calcium ion concentration (mEq/l) of the haemolymph of different stages in NDD & DD generations of *A. mylitta*

SN	Age	Mean	±	SE	t Stat	
1	I-NDD	22.1800	±	0.4740	2.2807	*
2	I-DD	20.3800	±	0.7648		
3	II-NDD	21.9800	±	0.2119	1.7549	NS
4	II-DD	21.0600	±	0.4214		
5	III-NDD	23.4600	±	0.2630	3.6891	**
6	III-DD	21.0500	±	0.7686		
7	IV-NDD	21.6500	±	0.1945	0.9718	NS
8	IV-DD	21.2200	±	0.3246		
9	VEF-NDD	21.8000	±	0.3289	1.8148	NS
10	VEF-DD	20.8900	±	0.4250		
11	VEM-NDD	21.4000	±	0.4640	-1.0881	NS
12	VEM-DD	21.9200	±	0.3318		
13	VMF-NDD	22.0600	±	0.6887	-0.3010	NS
14	VMF-DD	22.3200	±	0.4219		
15	VMM-NDD	24.0800	±	0.3940	0.6343	NS
16	VMM-DD	23.7300	±	0.3367		
17	VLF-NDD	19.9700	±	0.3425	-3.0958	**
18	VLF-DD	23.9000	±	1.3546		
19	VLM-NDD	22.2200	±	0.2797	-1.2066	NS
20	VLM-DD	23.8500	±	1.4509		
21	SLF-NDD	18.7700	±	0.3010	-2.7212	*
22	SLF-DD	20.5700	±	0.6678		
23	SLM-NDD	19.9100	±	0.0637	-2.5809	*
24	SLM-DD	20.8300	±	0.3641		
25	PPF-NDD	19.4500	±	0.3608	-2.8853	**
26	PPF-DD	21.8600	±	0.7489		
27	PPM-NDD	20.7400	±	0.2316	-1.9401	*
28	PPM-DD	21.1700	±	0.1925		
29	MPF-NDD	20.0000	±	0.3023	-2.2239	*
30	MPF-DD	20.9100	±	0.4678		
31	MPM-NDD	22.1800	±	0.1684	1.4562	NS
32	MPM-DD	21.6600	±	0.4086		
33	LPF-NDD	22.5000	±	0.1736	0.4314	NS
34	LPF-DD	22.2400	±	0.5814		
35	LPM-NDD	23.3400	±	0.1326	-1.1825	NS
36	LPM-DD	23.6900	±	0.3053		
37	AF-NDD	27.7800	±	0.2403	1.1176	NS
38	AF-DD	26.2400	±	1.4608		
39	AM-NDD	30.6900	±	0.2263	2.5021	*
40	AM-DD	26.6700	±	1.5533		

n = 10 (Abbreviations used are as per section 3.1.2.)

Table 4.68. Calcium ion concentration (mEq/l) in females and males of NDD generation in *A. mylitta*

SN	Age	Mean	±	SE	t Stat	
1	VEF-NDD	21.8000	±	0.3289	0.6761	NS
2	VEM-NDD	21.4000	±	0.4640		
3	VMF-NDD	22.0600	±	0.6887	-2.6973	*
4	VMM-NDD	24.0800	±	0.3940		
5	VLF-NDD	19.9700	±	0.3425	-15.2273	**
6	VLM-NDD	22.2200	±	0.2797		
7	SLF-NDD	18.7700	±	0.3010	-4.2472	**
8	SLM-NDD	19.9100	±	0.0637		
9	PPF-NDD	19.4500	±	0.3608	-4.2923	**
10	PPM-NDD	20.7400	±	0.2316		
11	MPF-NDD	20.0000	±	0.3023	-6.1534	**
12	MPM-NDD	22.1800	±	0.1684		
13	LPF-NDD	22.5000	±	0.1736	-3.6133	**
14	LPM-NDD	23.3400	±	0.1326		
15	AF-NDD	27.7800	±	0.2403	-11.0791	**
16	AM-NDD	30.6900	±	0.2263		

n = 10 (Abbreviations used are as per section 3.1.2.)

Table 4.69. Calcium ion concentration (mEq/l) in females and males of DD generation in *A. mylitta*

SN	Age	Mean	±	SE	t Stat	
1	VEF-DD	20.8900	±	0.4250	-1.8387	*
2	VEM-DD	21.9200	±	0.3318		
3	VMF-DD	22.3200	±	0.4219	-2.3883	*
4	VMM-DD	23.7300	±	0.3367		
5	VLF-DD	23.9000	±	1.3546	0.1351	NS
6	VLM-DD	23.8500	±	1.4509		
7	SLF-DD	20.5700	±	0.6678	-0.6722	NS
8	SLM-DD	20.8300	±	0.3641		
9	PPF-DD	21.8600	±	0.7489	1.0237	NS
10	PPM-DD	21.1700	±	0.1925		
11	MPF-DD	20.9100	±	0.4678	-1.0330	NS
12	MPM-DD	21.6600	±	0.4086		
13	LPF-DD	22.2400	±	0.5814	-2.4736	*
14	LPM-DD	23.6900	±	0.3053		
15	AF-DD	26.2400	±	1.4608	-1.7685	NS
16	AM-DD	26.6700	±	1.5533		

n = 10 (Abbreviations used are as per section 3.1.2.)

4.6.3.4. Magnesium (Mg++)

The concentration Mg++ in m Eq/l was observed in both NDD and DD generations in different stages of development and also in both the sexes. The data are presented in Tables - 4.70 to 4.72.

Table 4.70. Magnesium ion concentration (mEq/l) of the haemolymph of different stages in NDD & DD generations of *A. mylitta*

SN	Age	Mean	±	SE	t Stat	
1	I-NDD	52.3130	±	0.7145	2.9353	**
2	I-DD	49.4030	±	0.7935		
3	II-NDD	48.7070	±	0.6391	-4.4286	**
4	II-DD	52.5260	±	0.7853		
5	III-NDD	49.7310	±	0.2454	-0.7161	NS
6	III-DD	50.0870	±	0.4686		
7	IV-NDD	49.9240	±	0.1248	-0.6050	NS
8	IV-DD	50.2310	±	0.5764		
9	VEF-NDD	51.3080	±	0.1101	6.2603	**
10	VEF-DD	49.8660	±	0.2251		
11	VEM-NDD	47.8910	±	0.1141	-3.7546	**
12	VEM-DD	48.7600	±	0.1734		
13	VMF-NDD	52.2740	±	0.1732	8.9755	**
14	VMF-DD	49.3950	±	0.3469		
15	VMM-NDD	48.8390	±	0.1255	1.3067	NS
16	VMM-DD	48.5590	±	0.2324		
17	VLF-NDD	69.0470	±	0.4309	9.2288	**
18	VLF-DD	64.3100	±	0.2963		
19	VLM-NDD	65.3380	±	0.3366	10.1518	**
20	VLM-DD	61.1540	±	0.3029		
21	SLF-NDD	47.9040	±	0.3747	-1.3045	NS
22	SLF-DD	49.0360	±	0.8143		
23	SLM-NDD	51.0440	±	0.2226	0.9577	NS
24	SLM-DD	50.1670	±	1.0102		
25	PPF-NDD	33.7310	±	0.2454	-14.9323	**
26	PPF-DD	44.9090	±	0.7441		
27	PPM-NDD	31.9240	±	0.1248	-9.8476	**
28	PPM-DD	41.3290	±	0.9731		
29	MPF-NDD	35.3830	±	0.1386	-47.2085	**
30	MPF-DD	50.0940	±	0.2537		
31	MPM-NDD	29.8910	±	0.1141	-33.8162	**
32	MPM-DD	46.2460	±	0.5763		
33	LPF-NDD	38.2740	±	0.1732	10.3710	**
34	LPF-DD	35.4160	±	0.2478		
35	LPM-NDD	29.8390	±	0.1255	-5.3796	**
36	LPM-DD	32.2280	±	0.4004		
37	AF-NDD	48.0970	±	0.1506	-3.3055	**
38	AF-DD	48.9990	±	0.2168		
39	AM-NDD	46.0670	±	0.1614	-0.8681	NS
40	AM-DD	46.3170	±	0.2056		

n = 10 (Abbreviations used are as per section 3.1.2.)

In NDD generation, the concentration of Mg++ was recorded to be 52.3130 ± 0.7145 in I-NDD, slightly lowered in II-NDD (48.707 ± 0.6391), 49.7310 ± 0.2454 in III-NDD, 49.9240 ± 0.1248 in IV-NDD, 51.3080 ± 0.1101 in VEF-NDD, 47.8910 ± 0.1141 in VEM-NDD, 52.2740 ± 0.1732 in VMF-NDD and 48.839 ± 0.1255 in VMM-NDD. An increase in Mg++ concentration was observed in VLF-NDD (69.047 ± 0.4309) and VLM-NDD (65.3380 ± 0.3366). A decreasing trend was recorded in both the sexes of NDD generations in spinning larvae and pupal stages. However, concentration of Mg++ was recorded to be highest in adults such as in AF-NDD (48.097 ± 0.1506) and AM-NDD (46.067 ± 0.1614).

The concentration of Mg++ was 49.403 ± 0.7935 in I-DD, a bit higher in II-DD (52.5260 ± 0.7853), III-DD (50.0870 ± 0.4686), IV-DD (50.2310 ± 0.5764) and went up to its highest level in VLF-DD (64.310 ± 0.2963) and VLM-DD (61.1540 ± 0.3029), thereafter a decreasing trend was noticed till late pupal stage of both the sexes viz; LPF-DD (31.416 ± 0.2478) and LPM-DD (32.2280 ± 0.4004). Again, higher concentration was recorded in adults such as in females (48.999 ± 0.2168) and males (46.317 ± 0.2056)-Table 4.70.

One way ANOVA was applied and it was observed that the difference in Mg++ concentration in respective stages of NDD and DD generation was significant ($p < 0.01$) in all the stages except in between III-NDD & III-DD, IV-NDD & IV-DD, VMM-NDD & VMM-DD, SLF-NDD and SLF-DD, SLM-NDD & SLM-DD and AM-NDD & AM-DD.

One way ANOVA was also applied to see the difference between the sexes of different stages of NDD generation. The difference was significant ($p < 0.01$) between all the comparisons in between respective male and female larval instars, spinning larvae, pupae and adults. Difference was also found to be significant ($p < 0.01$) in between the sexes of DD generations viz. in between VEF-DD & VEM-DD, VLF-DD & VLM-DD, PPF-DD & PPM-DD, MPF-DD & MPM-DD, LPF-DD & LPM-DD and AF-DD & AM-DD. Though, the difference was significant ($p < 0.05$) in between VMF-DD & VMM-DD in rest of the comparisons difference was observed to be non-significant viz., in SLF-DD and SLM-DD.

Table 4.71. Magnesium ion concentration (mEq/l) in females and males of NDD generation in *A. mylitta*

SN	Age	Mean	±	SE	t Stat	
1	VEF-NDD	51.3080	±	0.1101	22.1027	**
2	VEM-NDD	47.8910	±	0.1141		
3	VMF-NDD	52.2740	±	0.1732	15.1796	**
4	VMM-NDD	48.8390	±	0.1255		
5	VLF-NDD	69.0470	±	0.4309	20.2803	**
6	VLM-NDD	65.3380	±	0.3366		
7	SLF-NDD	47.9040	±	0.3747	-9.0897	**
8	SLM-NDD	51.0440	±	0.2226		
9	PPF-NDD	33.7310	±	0.2454	9.2814	**
10	PPM-NDD	31.9240	±	0.1248		
11	MPF-NDD	35.3830	±	0.1386	31.3709	**
12	MPM-NDD	29.8910	±	0.1141		
13	LPF-NDD	38.2740	±	0.1732	37.2752	**
14	LPM-NDD	29.8390	±	0.1255		
15	AF-NDD	48.0970	±	0.1506	11.9233	**
16	AM-NDD	46.0670	±	0.1614		

n = 10 (Abbreviations used are as per section 3.1.2.)

Table 4.72. Magnesium ion concentration (mEq/l) in females and males of DD generation in *A. mylitta*

SN	Age	Mean	±	SE	t Stat	
1	VEF-DD	49.8660	±	0.2251	4.0460	**
2	VEM-DD	48.7600	±	0.1734		
3	VMF-DD	49.3950	±	0.3469	2.0069	*
4	VMM-DD	48.5590	±	0.2324		
5	VLF-DD	64.3100	±	0.2963	8.6692	**
6	VLM-DD	61.1540	±	0.3029		
7	SLF-DD	49.0360	±	0.8143	-0.7989	NS
8	SLM-DD	50.1670	±	1.0102		
9	PPF-DD	44.9090	±	0.7441	4.9151	**
10	PPM-DD	41.3290	±	0.9731		
11	MPF-DD	50.0940	±	0.2537	6.9458	**
12	MPM-DD	46.2460	±	0.5763		
13	LPF-DD	35.4160	±	0.2478	7.6299	**
14	LPM-DD	32.2280	±	0.4004		
15	AF-DD	48.9990	±	0.2168	8.3745	**
16	AM-DD	46.3170	±	0.2056		

n = 10 (Abbreviations used are as per section 3.1.2.)

4.6.4. Trehalose (mg/ml) in haemolymph

Trehalose is generally present in large amounts in the haemolymph and is reported to be important for metabolic processes and chitin synthesis. Trehalose availability was estimated in the haemolymph of all the stages of NDD and DD generations taken up for the present study and the data thus generated is presented in Table 4.70 to 4.72.

The concentration of trehalose in NDD generation was recorded to be lowest in I-NDD (0.8501 ± 0.0191) which gradually increased through different larval instars VEF-NDD (7.4848 ± 0.0584), VEM-NDD (5.4851 ± 0.0424), different pupal stages (MPF-NDD-31.1720 ± 0.1239, MPM-NDD-27.9328 ± 0.1819). The trehalose concentration decreased during late pupal stage viz. in LPF-NDD (22.8288 ± 0.0632) and LPM-NDD (21.6364 ± 0.1717) and this decrease continued till emergence of adults (AF-NDD - 14.9465 ± 0.0780 and AM-NDD - 12.8615 ± 0.0422).

The concentration of trehalose in DD generation also consistently increased starting from I-DD (1.0502 ± 0.0043) and attained its peak in mid pupal stage viz. in MPF-DD (35.6581 ± 0.1089) and MPM-DD (31.0781 ± 0.1937). Thereafter, trehalose concentration started to decrease and it was recorded to be 34.8714 ± 0.0793 in LPF-DD and 26.9440 ± 0.0621in LPM-DD. The level of trehalose further went down in adults of both the sexes such as, 27.3160 ± 0.2179 in AF-DD and 19.5578 ± 0.3098 in AM-NDD.

When compared the availability of trehalose in NDD generation with that of DD generation, it was higher in I, II, III and IV instars of DD generations. In early fifth instar female and male and also in mid fifth instar female, the concentration was recorded to be higher in NDD generations than DD. In rest all the stages the concentration of trehalose was higher in DD generation. T-test applied indicated that the differences were significant ($p < 0.01$) in all the respective stages of NDD and DD.

The differences were significant ($p < 0.01$) in between sexes of NDD generation viz. in between VEF-NDD & VEM-NDD, VMF-NDD & VMM-NDD, VLF-NDD & VLM-NDD, SLF-NDD & SLM-NDD, PPF-NDD & PPM-NDD, MPF-NDD & MPM-NDD, LPF-NDD & LPM-NDD, AF-NDD & AM-NDD. Late pupae and adult female had high level of trehalose than their male counterparts.

Similarly, the comparison was made in between the different sexes of DD generation. Male fifth larval instar had significant higher-level of trehalose than females ($p < 0.01$). Spinning female larvae had high level of trehalose than spinning male larvae ($p < 0.01$).

Table 4.73. Trehalose concentration (mg/ml) of the haemolymph of different stages in NDD & DD generations of *A. mylitta*

SN	Age	Mean	±	SE	t Stat	
1	I-NDD	0.8501	±	0.0191	-12.1291	**
2	I-DD	1.0502	±	0.0043		
3	II-NDD	1.1439	±	0.0170	-21.7885	**
4	II-DD	1.4614	±	0.0039		
5	III-NDD	1.2725	±	0.0139	-20.4100	**
6	III-DD	1.5534	±	0.0016		
7	IV-NDD	2.3226	±	0.0054	-58.5038	**
8	IV-DD	2.9045	±	0.0076		
9	VEF-NDD	7.4848	±	0.0584	62.2341	**
10	VEF-DD	3.1512	±	0.0101		
11	VEM-NDD	5.4851	±	0.0424	5.8627	**
12	VEM-DD	5.0330	±	0.0338		
13	VMF-NDD	10.2124	±	0.0694	29.4071	**
14	VMF-DD	7.0630	±	0.0321		
15	VMM-NDD	7.4760	±	0.0543	-28.4727	**
16	VMM-DD	10.1750	±	0.0460		
17	VLF-NDD	13.0355	±	0.0650	-7.6351	**
18	VLF-DD	13.5211	±	0.0114		
19	VLM-NDD	9.8987	±	0.1143	-78.7173	**
20	VLM-DD	18.9064	±	0.0355		
21	SLF-NDD	15.8074	±	0.0591	-78.2294	**
22	SLF-DD	22.1808	±	0.0520		
23	SLM-NDD	12.3600	±	0.0561	-65.7962	**

24	SLM-DD	17.9347	±	0.1083		
25	PPF-NDD	27.9124	±	0.1615	-36.0238	**
26	PPF-DD	35.5637	±	0.0982		
27	PPM-NDD	23.6620	±	0.0200	-42.5761	**
28	PPM-DD	30.3947	±	0.1247		
29	MPF-NDD	31.1720	±	0.1239	-20.6082	**
30	MPF-DD	35.6581	±	0.1089		
31	MPM-NDD	27.9328	±	0.1819	-8.1433	**
32	MPM-DD	31.0736	±	0.1937		
33	LPF-NDD	22.8288	±	0.0632	-94.0264	**
34	LPF-DD	34.8714	±	0.0793		
35	LPM-NDD	21.6364	±	0.1717	-20.9625	**
36	LPM-DD	26.9440	±	0.0621		
37	AF-NDD	14.9465	±	0.0780	-49.2423	**
38	AF-DD	27.3160	±	0.2179		
39	AM-NDD	12.8615	±	0.0422	-19.5696	**
40	AM-DD	19.5578	±	0.3098		

n = 5 (Abbreviations used are as per section 3.1.2.)

Table 4.74. Trehalose concentration (mg/ml) in the haemolymph of different sexesn in NDD generations of *A. mylitta*

SN	Age	Mean	±	SE	t Stat	
1	VEF-NDD	7.4848	±	0.0653	18.4530	**
2	VEM-NDD	5.4851	±	0.0474		
3	VMF-NDD	10.2124	±	0.0776	25.4700	**
4	VMM-NDD	7.4760	±	0.0607		
5	VLF-NDD	13.0355	±	0.0727	18.1333	**
6	VLM-NDD	9.8987	±	0.1278		
7	SLF-NDD	15.8074	±	0.0661	104.8937	**
8	SLM-NDD	12.3600	±	0.0627		
9	PPF-NDD	27.9124	±	0.1806	23.3868	**
10	PPM-NDD	23.6620	±	0.0223		
11	MPF-NDD	31.1720	±	0.1385	12.5930	**
12	MPM-NDD	27.9328	±	0.2034		
13	LPF-NDD	22.8288	±	0.0706	5.2380	**
14	LPM-NDD	21.6364	±	0.1920		
15	AF-NDD	14.9465	±	0.0872	27.1378	**
16	AM-NDD	12.8615	±	0.0472		

N = 5 (Abbreviations used are as per section 3.1.2.)

Table 4.75. Trehalose concentration (mg/ml) in the haemolymph of different sexesn in DD generations of *A. mylitta*

SN	Age	Mean	±	SE	t Stat	
1	VEF-DD	3.1512	±	0.0113	-39.2845	**
2	VEM-DD	5.0330	±	0.0378		
3	VMF-DD	7.0630	±	0.0359	-40.2234	**
4	VMM-DD	10.1750	±	0.0515		
5	VLF-DD	13.5211	±	0.0127	-114.2615	**
6	VLM-DD	18.9064	±	0.0396		
7	SLF-DD	22.1808	±	0.0582	36.7387	**
8	SLM-DD	17.9347	±	0.1211		
9	PPF-DD	35.5637	±	0.1098	24.2117	**
10	PPM-DD	30.3947	±	0.1394		
11	MPF-DD	35.6581	±	0.1218	14.6359	**
12	MPM-DD	31.0736	±	0.2165		
13	LPF-DD	34.8714	±	0.0887	64.5623	**
14	LPM-DD	26.9440	±	0.0694		
15	AF-DD	27.3160	±	0.2436	14.2671	**
16	AM-DD	19.5578	±	0.3464		

n = 5 (Abbreviations used are as per section 3.1.2.)

4.6.5. Protein : Quantitative measurements of Total Haemolymph Proteins

Insect utilise protein in a diverse array of physiological and metabolic processes, keeping in view of their importance in insect haemolymph. Ontogenetic availability of haemolymph protein was estimated in NDD and DD generations of A. mylitta. The quantum of protein estimated in terms of mg/ml of haemolymph and the data is presented in Tables 4.76 to 4.88. It was 3.4959 ± 0.0255 in I-NDD, which showed an increasing pattern in subsequent instars viz., 4.1962 ± 0.0983 in II-NDD, 5.1738 ± 0.0471in III-NDD, 6.8038 ± 0.0292 in IV-NDD and finally up to 54.0634 ± 0.4426 in SLF-NDD and 50.2600 ± 0.8123 in SLM-NDD. The protein concentration showed a decreasing trend in pupal stage i.e. 36.7742 ± 0.4291 in LPF-NDD and 24.3060 ± 0.4243 in LPM-NDD as well as in adult stage, 26.3652 ± 0.5494 in AF-NDD and 19.9716 ± 0.7332 in AM-NDD

Similar trends were recorded in DD generations. Total protein was 4.5038 ± 0.0234 in I-DD, 5.8154 ± 0.0330 in II-DD, 5.4952 ± 0.0329 in III-DD, 7.3368 ± 0.0216 in IV-DD. Protein concentration attended it's peak in VLM-DD (58.4076 ± 0.1044) and SLF-DD (65.6628 ± 0.4812). It gradually came down in pupal stage as it was recorded to be 50.7236 ± 0.1243 in LPF-DD and 41.5416 ± 0.2034 LPM-DD and it dipped to further lower value in AF-DD (36.1392 ± 0.0962) and AM-DD (32.0610 ± 0.670).

One-way ANOVA indicated that DD generation had comparatively higher level of protein than that of their respective life stages of NDD generations except in male spinning larvae and male mid pupae of NDD which had comparatively higher level of proteins than their DD generations stages. The difference was found to be significant ($p < 0.01$) in between the I-NDD & I-DD, II-NDD & II-DD, III-NDD & III-DD, IV-NDD & IV-DD, V instar larvae of NDD & DD, different pupal stages of NDD & DD, and AF-NDD & DD. This difference was recorded to be significant ($p < 0.05$) in between PPF-NDD and PPF-DD.

Table 4.76. Total protein concentration (mg/ml) of the haemolymph of different stages in NDD & DD generations of *A. mylitta*

SN	Age	Mean	±	SE	t Stat	
1	I-NDD	3.4959	±	0.0255	-25.2917	**
2	I-DD	4.5038	±	0.0234		
3	II-NDD	4.1962	±	0.0983	-23.8851	**
4	II-DD	5.8154	±	0.0330		
5	III-NDD	5.1738	±	0.0471	-5.3009	**
6	III-DD	5.4952	±	0.0329		
7	IV-NDD	6.8038	±	0.0292	-20.2353	**
8	IV-DD	7.3368	±	0.0216		
9	VEF-NDD	8.7864	±	0.0349	-22.9571	**
10	VEF-DD	10.4970	±	0.0736		
11	VEM-NDD	12.3316	±	0.0673	18.3504	**
12	VEM-DD	14.5804	±	0.1102		
13	VMF-NDD	15.5568	±	0.0794	-24.2087	**
14	VMF-DD	18.4154	±	0.1521		
15	VMM-NDD	29.0872	±	0.4214	-11.2560	**
16	VMM-DD	32.3876	±	0.1263		
17	VLF-NDD	33.7502	±	0.5766	-6.0727	**
18	VLF-DD	37.0778	±	0.0531		
19	VLM-NDD	27.7114	±	0.1430	158.4694	**
21	VLM-DD	58.4076	±	0.1044		
21	SLF-NDD	54.0634	±	0.4426	-16.8017	**
22	SLF-DD	65.6628	±	0.4812		
23	SLM-NDD	50.2600	±	0.8123	10.6230	**
24	SLM-DD	42.5284	±	0.0759		
25	PPF-NDD	48.0492	±	0.4516	-2.9436	*
26	PPF-DD	49.3116	±	0.0837		
27	PPM-NDD	31.7570	±	0.4178	-13.2394	**
28	PPM-DD	40.4086	±	0.3957		
29	MPF-NDD	48.4944	±	0.3018	5.4758	**
30	MPF-DD	50.2182	±	0.0784		
31	MPM-NDD	49.4606	±	0.5217	16.9094	**
32	MPM-DD	40.4814	±	0.0894		
33	LPF-NDD	36.7742	±	0.4291	-41.1181	**
34	LPF-DD	50.7236	±	0.1243		
35	LPM-NDD	24.3060	±	0.4243	-40.7920	**
36	LPM-DD	41.5416	±	0.2034		
37	AF-NDD	26.3652	±	0.5494	-18.3234	**
38	AF-DD	36.1392	±	0.0962		
39	AM-NDD	19.9716	±	0.7332	-17.9755	**
40	AM-DD	32.0610	±	0.0670		

n = 5 (Abbreviations used are as per section 3.1.2.)

t-Test was implied to find out the difference between the mean values of protein concentrations of sexes. In NDD generation initially male V-NDD larvae had high protein but middle and late stage V-NDD, spinning female larvae and pre and mid pupae females had high level of protein than their male counterparts and thereafter in following stages and adult females had significantly higher level of proteins till emergence of adults. The difference was significant ($t < 0.01$) in between VEF-NDD & VEM-NDD, VMF-NDD & VMM-NDD, VLF-NDD & VLF-NDD & VLM-NDD, SLF-NDD & SLM-NDD, PPF-NDD & PPM-NDD, MPF-NDD & MPM-NDD, LPF-NDD & LPM-NDD, and AF-NDD & AM-NDD.

Table 4.77. Total protein concentration (mg/ml) in the haemolymph of different sexesn in NDD generations of *A. mylitta*

SN	Age	Mean	±	SE	t Stat	
1	VEF-NDD	8.7864	±	0.0349	-44.7259	**
2	VEM-NDD	12.3316	±	0.0673		
3	VMF-NDD	15.5568	±	0.0794	-42.6710	**
4	VMM-NDD	29.0872	±	0.4214		
5	VLF-NDD	33.7502	±	0.5766	11.3815	**
6	VLM-NDD	27.7114	±	0.1430		
7	SLF-NDD	54.0634	±	0.4426	4.4163	**
8	SLM-NDD	50.2600	±	0.8123		
9	PPF-NDD	48.0492	±	0.4516	32.6720	**
10	PPM-NDD	31.7570	±	0.4178		
11	MPF-NDD	48.4944	±	0.3018	-1.7676	NS
12	MPM-NDD	49.4606	±	0.5217		
13	LPF-NDD	36.7742	±	0.4291	35.4449	**
14	LPM-NDD	24.3060	±	0.4243		
15	AF-NDD	26.3652	±	0.5494	12.7645	**
16	AM-NDD	19.9716	±	0.7332		

n = 5 (Abbreviations used are as per section 3.1.2.)

In case of DD generation early middle and late fifth instar male larvae had higher protein level than their female counterparts. Thereafter, in spinning larvae, pre, mid and late pupae and adults the level of protein was significantly higher in females. Significant difference ($p < 0.01$) was recorded in between VEF-DD & VEM-DD, VMF-DD & VMM-DD, VLF-DD & VLM-DD, SLF-DD & SLM-DD, PPF-DD & PPM-DD, MPF-DD & MPM-DD, LPF-DD & LPM-DD, AF-DD & AM-DD.

Qualitative measurements of Total Haemolymph Proteins

Qualitative haemolymph protein patterns of A. mylitta were observed through disc polyacrylamide-gel electrophoresis (PAGE) in non-diapausing and diapausing generations, such as fully grown I, II, III, IV instar larvae, early middle and late age V instar larvae and pupae of both sexes and adults females and males and the Rf values and proteinogram is presented in Tables 4.79 & 4.80 and Figures 4.19 and 4.20..

Table 4.78. Total protein concentration (mg/ml) in the haemolymph of different sexes in DD generations of *A. mylitta*

SN	Age	Mean	±	SE	t Stat	
1	VEF-DD	10.4970	±	0.0736	-29.0494	**
2	VEM-DD	14.5804	±	0.1102		
3	VMF-DD	18.4154	±	0.1521	-101.3041	**
4	VMM-DD	32.3876	±	0.1263		
5	VLF-DD	37.0778	±	0.0531	-309.9116	**
6	VLM-DD	58.4076	±	0.1044		
7	SLF-DD	65.6628	±	0.4812	54.2934	**
8	SLM-DD	42.5284	±	0.0759		
9	PPF-DD	49.3116	±	0.0837	26.5110	**
10	PPM-DD	40.4086	±	0.3957		
11	MPF-DD	50.2182	±	0.0784	86.9491	**
12	MPM-DD	40.4814	±	0.0894		
13	LPF-DD	50.7236	±	0.1243	50.8832	**
14	LPM-DD	41.5416	±	0.2034		
15	AF-DD	36.1392	±	0.0962	28.6006	**
16	AM-DD	32.0610	±	0.0670		

n = 5 (Abbreviations used are as per section 3.1.2.)

In non-diapausing generation, as evident from fig. 4.19, the haemolymph of larvae was electrophoretically separated. There were 11 protein bands in I-NDD, 13 bands in each II & III NDD and 14 in IV NDD. The number of bands was recorded to be less, 10 bands in early fifth instar females and 11 in males. The number of bands was observed to be 10 in spinning larvae and pre-pupae of both the sexes, except in pre-pupae females, where only 9 bands were observed. The number of bands increased in mid- and late-pupae of both the sexes, except in late male pupae, where it came down to 9, which further reduced to 8 in adults. In the haemolymph of late female pupae, adult females and adult males, there were 14, 11 and 8 proteins bands, respectively. In general, females had the higher number of protein bands than males. In I NDD, the first band was a slow moving thin band and had very low Rf value (0.01) and the remaining bands had Rf values in between 0.09 to 0.86. The bands 9 and 10 were deeply stained having Rf values of 0.76 and 0.78.

The second and third instar larval haemolymph was represented by 13 bands, slightly higher number of bands than first instar. The first band had slightly changed mobility (0.02) than first instar. The seventh band stained deeply both in second and third instars, having Rf values 0.26 and 0.28, respectively. There were a few slightly faster moving protein bands comprised the major proportion of haemolymph proteins whose Rf values ranged in between 0.53 to 0.54; 0.70 to 0.71 and 0.80 to 0.85 , respectively, in second and third instars.

In fourth instar larvae, total 14 bands of proteins were observed. Deeply stained two bands having Rf of 0.11 and 0.26 appeared which were not present in third instar. Band 6, which stained lightly, also appeared in fourth instar having Rf 0.39. The light stained broader protein bands observed in second and third instar (Rf 0.32 and 0.33) were not present in fourth instar larvae. The protein band having Rf of 0.81 become prominent in fourth instar. A new fast moving protein band also appeared which had the Rf value of 0.94.

The proteinogram of fifth instar early, middle and late female larvae had only 10-11 bands and were lesser than fourth instar. Deeply stained two protein bands having Rf of 0.13 and 0.34 were prominent in early fifth instar stage. The breadth of protein band having Rf of 0.34 increased till spinning larval stage. Similar protein bands also appeared in male fifth instar larvae, but their breadth was lesser than female larvae. In the middle region, five bands having Rf values of 0.34, 0.43, 0.56, 0.64 and 0.74 were more prominent in females than males.

In female prepupal stage, the number of bands reduced to 9 only, which again increased and went up to 14 bands during late female pupal stage. In late female pupae, several deeply stained bands appeared below the deeply stained band having Rf of 0.15. These bands had the Rf of 0.21, 0.32, 0.36, 0.43, 0.59, and 0.68. One fast moving lighter band protein having Rf ranging in between 089 to 0.95 was consistently present starting from fourth larval stage to late female pupal stage. This band was almost consistently present in males too. In late stage male and female pupae, the prominent bands were either disappearing or narrowing in comparison with earlier stages.

Adult females had altogether 11 bands when compared with males, which had only 8 bands. At the origin, 4 heavier bands of proteins were present in adult females having Rf of 0.03, 0.05, 0.10 and 0.13 may be the female specific chromo-proteins or vitellogenic proteins. It was difficult to obtain homology of bands of one sex to other in adults. Band 4 of female (Rf: 0.13) did not had homology with male protein bands. Adult males had more number of fast moving protein bands than females.

No marked differentiation was observed in haemolymph protein bands patterns of the different life stages of diapausing generations when compared with non-diapausing generation. Among the larval haemolymph, there were 11 protein bands in I-DD, 13 bands each in II & III -DD, 15 in IV-DD and the number of protein bands reduced to 11 and 12 in different age groups of fifth instar female and male larvae, respectively.

The number of bands in early and mid pupal stages were recorded to be the same like the fifth instar larvae, however, number of proteins bands were recorded to be 15, 13 and 11 in mid-aged female, late female and male pupae, respectively. Thereafter, in adults of both the sexes, only 10 bands were observed.

There was one deeply stained band, which appeared during fifth instar and continued to remain present till late pupal stage of both the sexes having Rf 0.04 to 0.06 was observed to be thickening and comprised the major protein fraction in both the sexes. This band was gradually increasing up to mid-pupal stage, later it was a thin line in late-pupal stage. This band may represent the diapausing protein, which are prominent in diapausing generations. Two fast moving proteins of Rf 0.97 and 0.98 and the last band of light stained protein also gradually thickened till mid-pupal stage in both the sexes and gradually started to disappear in late pupal stage and no trace of these protein band were present in the adults of both the sexes.

Table 4.79 : Relative Mobility (Rf) values measured from prote in band patterns observed in the haemolymph of different stages of non-diapausing generation of *A. mylitta.*

Table 4.79. Relative Mobility (Rf) values measured from protein band patterns observed in the haemolymph of different stages of non-diapausing generation of *A. mylitta*

Rf	I	II	III	IV	VEF	VEM	VMF	VMM	VLF	VLM	SLF	SLM	PPF	PPM	MPF	MPM	LPF	LPM	AF	AM
(-) 0.00	ORIGIN																			
	0.01	-	-	-	-	-	-	-	-	-	-	-	-	-	-	-	-	-	-	-
	-	0.02	0.02	-	0.02	-	-	-	-	-	-	-	0.02	-	-	-	-	-	-	0.02
	-	-	-	0.03	-	-	-	-	-	-	0.03	-	-	-	0.03	0.03	0.03	-	0.03	-
	-	-	-	-	-	0.04	0.04	0.04	-	-	-	-	-	0.04	-	-	-	-	-	-
	-	-	-	-	0.05	-	-	-	-	0.05	-	-	-	-	-	-	-	-	0.05	-
	-	-	-	-	-	-	-	-	-	-	-	-	0.06	-	0.06	-	-	-	-	-
	-	-	0.07	-	-	-	-	-	0.07	-	-	-	-	-	-	-	-	-	-	-
	-	0.08	-	-	-	0.08	0.08	-	-	-	-	0.08	-	-	-	0.08	-	-	-	-
	0.09	-	-	-	-	-	-	-	0.09	-	-	-	-	-	-	-	0.09	0.09	-	0.09
	-	-	-	-	-	-	-	0.10	-	-	-	-	-	-	-	-	-	-	0.10	-
	-	-	-	0.11	-	-	-	-	-	-	-	-	-	-	-	-	-	-	-	-
	-	-	-	-	-	-	-	-	-	0.12	-	-	-	-	-	-	-	-	-	-
	-	-	-	-	0.13	0.13	0.13	-	-	-	-	-	-	-	-	-	-	-	0.13	-
	-	0.14	-	-	-	-	-	0.14	0.14	-	-	-	-	-	-	-	-	-	-	-
	-	-	0.15	-	-	-	-	-	-	0.15	0.15	-	0.15	0.15	-	-	0.15	-	-	-
	-	-	-	-	-	-	-	-	-	-	-	0.16	-	-	0.16	0.16	-	0.16	-	-
	-	0.18	-	-	-	-	-	-	-	-	-	-	-	-	-	-	-	-	-	-
	0.19	-	-	-	-	-	-	-	-	-	-	-	-	-	-	-	-	-	-	-
	-	-	0.20	-	-	-	-	-	-	-	-	-	-	-	-	-	-	-	-	-
	-	-	-	-	-	-	-	-	-	-	-	-	-	-	-	-	0.21	-	-	0.21
	-	-	-	-	-	-	-	-	-	-	-	0.22	-	-	-	-	-	0.22	0.22	-
	-	0.24	-	-	-	-	-	-	-	-	-	-	-	-	-	0.24	-	-	-	-
	-	0.25	0.25	-	-	-	-	-	-	-	-	-	-	0.25	-	-	-	-	-	-
	-	0.26	-	0.26	-	-	-	-	-	-	-	-	-	-	-	-	-	-	-	0.26
	0.27	0.27	0.27	-	-	-	-	-	-	-	-	-	-	-	-	-	-	-	-	-
	-	-	0.28	0.28	-	-	-	-	-	-	-	-	-	-	-	-	-	-	-	-
	-	-	-	-	-	-	-	0.29	-	-	-	-	-	-	-	-	-	-	-	-
	-	-	0.30	-	-	-	-	-	-	-	-	-	-	-	-	-	-	-	0.30	0.30
	-	-	-	-	-	-	-	-	-	0.31	-	-	-	-	-	-	-	0.31	-	-
	-	0.32	-	-	-	-	-	-	-	-	-	-	-	-	-	-	0.32	-	-	-
	-	-	0.33	-	-	0.33	-	-	-	-	-	-	-	-	-	0.33	-	-	0.33	-
	-	-	-	0.34	0.34	-	-	-	-	-	-	-	-	-	0.34	-	-	-	-	-
	-	-	-	-	-	-	-	-	-	-	-	-	-	0.35	-	-	-	-	-	-
	-	-	-	-	-	0.36	-	-	-	-	0.36	-	-	-	-	-	0.36	-	-	-
	-	-	-	-	-	-	-	0.37	0.37	-	-	0.37	-	-	-	-	-	-	-	-
	-	-	-	-	-	-	-	-	-	-	0.38	-	-	-	-	-	-	-	-	-
	-	-	-	0.39	-	-	-	-	-	-	-	-	-	-	0.39	-	-	0.39	-	-
	-	-	-	-	-	-	0.40	-	-	0.40	-	-	-	-	-	-	-	-	-	0.40
	-	-	-	-	0.43	-	-	-	-	-	-	-	-	-	-	0.43	0.43	-	-	-
	-	-	-	0.44	-	-	-	-	-	-	-	-	-	-	-	-	-	-	0.44	-
	0.45	-	-	0.45	-	-	0.45	-	-	-	0.45	0.45	0.45	-	-	-	-	-	-	-
	-	-	-	-	-	-	-	-	0.47	-	-	-	-	-	-	-	-	-	-	-
	-	-	-	-	-	0.49	-	-	-	-	0.49	-	-	-	-	-	-	-	-	-
	-	-	-	0.50	-	-	-	-	-	0.50	-	-	-	-	-	0.50	-	0.50	-	-
	0.52	-	-	-	-	-	-	0.52	-	-	-	-	0.52	0.52	0.52	-	0.52	-	-	-
	-	0.53	-	-	-	-	-	-	-	-	-	0.53	-	-	-	-	-	-	-	-
	-	-	0.54	-	-	-	-	-	0.54	-	-	-	-	-	-	-	-	-	-	-
	-	-	-	-	-	0.55	0.55	0.55	-	-	-	-	-	-	0.55	-	-	-	-	-
	-	-	-	-	0.56	-	-	-	-	-	-	-	-	-	-	-	-	-	-	-
	0.58	-	-	-	-	-	-	-	-	-	-	-	-	-	-	-	-	-	-	-
	-	-	-	0.59	-	0.59	-	-	-	-	-	-	-	-	0.59	-	0.59	-	-	-
	-	-	-	-	-	-	0.60	-	-	-	0.60	-	-	0.60	-	0.60	-	-	0.60	0.60
	-	-	-	-	-	-	-	-	0.61	0.61	-	-	-	-	-	-	-	-	-	-
	-	-	-	-	-	-	-	-	-	-	-	-	0.62	-	-	-	-	-	-	-
	-	-	-	-	0.64	-	-	-	-	-	-	-	-	-	0.64	-	0.64	-	-	-
	-	-	-	-	-	-	-	0.65	-	-	-	-	-	-	-	-	-	0.65	-	-
	0.66	-	-	-	-	-	-	-	-	-	-	-	-	-	-	0.66	-	-	-	-
	-	-	-	-	-	-	-	-	-	-	-	-	-	-	0.68	-	0.68	-	-	-
	-	-	-	0.69	-	-	-	-	-	0.69	-	-	-	-	-	-	-	-	-	-
	-	0.70	-	-	-	-	-	-	-	-	-	-	0.70	-	-	-	-	-	-	-
	-	-	0.71	-	-	-	-	-	-	-	-	-	-	-	-	-	-	-	-	-
	-	-	-	-	-	-	0.72	-	0.72	-	0.72	0.72	-	-	-	-	0.72	-	-	-
	-	-	-	-	-	0.73	-	-	-	-	-	-	-	0.73	-	-	-	-	-	-
	-	-	-	-	0.74	-	-	0.74	-	-	-	-	-	-	-	-	-	-	-	-
	-	-	-	-	-	-	-	-	-	-	-	-	-	-	-	0.75	-	-	-	-
	0.76	-	-	-	-	-	-	-	-	0.76	-	-	-	-	-	-	-	0.76	-	-
	-	-	-	-	-	-	-	-	-	-	-	0.77	-	-	-	-	-	-	-	-
	0.78	-	-	-	-	-	-	-	-	-	-	-	-	0.78	-	-	-	-	0.78	-
	-	-	-	-	-	-	-	-	-	-	-	-	-	-	0.79	-	-	-	-	-
	-	0.80	-	-	-	-	-	-	-	-	-	-	-	-	-	-	-	-	-	-
	-	-	0.81	0.81	-	0.81	-	-	-	-	-	-	-	-	-	-	0.81	-	-	-
	-	-	-	-	-	-	-	-	-	-	0.83	-	-	-	-	0.83	-	-	-	-
	-	-	-	-	-	-	-	0.84	-	-	-	-	-	-	-	-	-	-	-	-
	-	0.85	0.85	-	-	-	0.85	-	0.85	0.85	-	0.85	0.85	-	-	-	-	-	-	-
	0.86	-	-	-	-	-	-	-	-	-	-	-	-	0.86	-	-	-	-	-	-
	-	-	-	0.87	-	-	-	-	-	-	-	-	-	-	-	-	-	-	-	-
	-	-	-	-	-	-	-	-	-	-	-	-	-	-	-	-	-	-	-	0.88
	-	-	-	-	0.89	-	-	-	-	-	-	-	-	-	-	-	-	-	0.89	-
	-	-	-	-	-	0.93	-	-	-	-	-	-	-	-	-	-	0.93	0.93	-	-
	-	-	-	0.94	0.94	-	0.94	-	-	0.94	0.94	-	0.94	0.94	0.94	0.94	-	-	-	-
	-	-	-	-	-	-	-	0.95	0.95	-	-	0.95	-	-	-	-	-	-	-	-
(+) 1.00	FRONT																			

Table 4.80 : Relative Mobility (Rf) values measured from prote in band paterns observed in the haemolymph of different stages of diapausing generation of *A. mylitta*.

Rf	I	II	III	IV	VEF	VEM	VMF	VMM	VLF	VLM	SLF	SLM	PPF	PPM	MPF	MPM	LPF	LPM	AF	AM
(-) 0.00	ORIGIN																			
	0.01	-	-	-	-	-	-	-	-	-	-	-	-	-	-	-	-	-	-	-
	-	0.02	0.02	-	-	-	-	-	0.02	-	-	-	0.02	0.02	-	-	-	-	-	0.02
	-	-	-	-	0.03	0.03	0.03	-	-	0.03	0.03	0.03	-	-	0.03	0.03	0.03	-	0.03	-
	-	-	-	0.04	-	-	-	0.04	-	-	-	-	-	0.04	-	-	-	-	-	-
	-	-	-	-	-	-	-	-	-	-	0.05	-	-	-	-	-	-	-	0.05	-
	-	-	-	0.06	-	-	-	-	-	-	-	-	0.06	-	0.06	-	-	-	-	-
	-	0.07	0.07	-	0.07	-	-	-	0.07	-	-	-	-	-	-	-	-	-	-	-
	-	-	-	-	-	0.08	0.08	-	-	-	-	0.08	-	-	-	0.08	-	-	-	-
	0.09	-	-	-	-	-	-	-	-	-	-	-	-	-	-	-	0.09	0.09	-	0.09
	-	-	-	-	-	-	-	0.10	-	-	-	-	-	-	-	-	-	-	-	-
	-	-	-	-	-	-	-	-	-	0.12	-	-	-	-	-	-	-	-	-	-
	-	-	-	0.13	-	-	-	-	-	-	-	-	-	-	-	-	-	-	-	-
	-	0.14	0.14	-	-	-	-	-	-	-	-	-	-	-	-	-	-	-	0.14	-
	-	-	-	-	-	-	-	-	-	-	-	-	-	-	-	-	0.15	-	-	-
	-	-	-	-	-	-	-	-	-	-	-	-	-	-	-	-	-	0.16	-	-
	-	0.18	0.18	-	-	-	-	-	-	-	-	-	-	-	-	-	-	-	-	0.18
	0.19	-	-	-	-	-	-	-	-	-	-	-	-	-	-	-	-	-	-	-
	-	-	-	-	0.20	-	-	0.20	-	-	0.20	-	-	-	-	-	-	0.20	-	-
	-	-	-	-	-	-	-	-	0.21	-	-	-	-	-	-	-	0.21	-	-	-
	-	-	-	-	-	-	-	-	-	0.22	-	0.22	-	-	-	-	-	-	0.22	-
	-	-	-	-	-	0.23	0.23	-	-	-	-	-	-	-	-	-	-	-	-	-
	-	0.24	-	-	-	-	-	-	-	-	-	-	0.24	0.24	0.24	-	-	-	-	-
	-	0.25	-	-	-	-	-	-	-	-	-	-	-	-	-	-	-	-	-	-
	-	0.26	-	0.26	-	-	-	-	-	-	-	-	-	-	-	-	-	-	-	0.26
	0.27	0.27	-	-	-	-	-	-	-	-	-	-	-	-	-	0.27	-	-	-	-
	-	-	0.28	0.28	-	-	-	-	-	-	-	-	-	-	-	-	-	-	-	-
	-	-	0.29	-	-	-	-	0.29	-	-	-	-	-	-	-	-	-	-	-	-
	-	-	0.30	-	-	-	-	-	-	-	-	-	-	-	-	-	-	-	0.30	0.30
	-	-	-	-	-	-	-	-	-	0.31	-	-	-	-	-	-	-	0.31	-	-
	-	0.32	0.32	-	-	-	-	-	-	-	-	-	-	-	-	-	0.32	-	-	-
	-	-	-	-	-	0.33	-	-	-	-	-	-	-	-	-	-	-	-	0.33	-
	-	-	-	0.34	0.34	-	-	-	-	-	-	-	-	-	-	-	-	-	-	-
	-	-	-	-	-	-	-	-	-	-	-	-	-	-	-	0.35	-	-	-	-
	-	-	-	-	-	0.36	-	-	-	-	0.36	-	-	0.36	0.36	-	0.36	-	-	-
	-	-	0.37	-	-	-	-	0.37	0.37	-	-	0.37	-	-	-	-	-	-	-	-
	-	-	-	-	-	-	-	-	0.38	-	0.38	-	-	-	-	-	-	-	-	-
	-	-	-	0.39	-	-	-	-	-	-	-	-	-	-	0.39	-	-	0.39	-	-
	-	-	-	-	-	-	0.40	-	-	0.40	-	-	-	-	-	-	-	-	-	-
	-	-	-	-	-	-	-	-	-	-	-	-	0.42	-	-	-	-	-	-	0.42
	-	-	-	-	0.43	-	-	-	-	-	-	-	-	-	-	0.43	0.43	0.43	-	-
	-	-	-	0.44	-	-	-	-	-	-	-	-	-	-	-	-	-	-	0.44	-
	0.45	-	-	0.45	-	-	0.45	-	-	-	0.45	0.45	0.45	-	-	-	-	-	-	-
	-	-	-	-	-	-	-	-	0.47	-	-	-	-	-	-	-	-	-	-	-
	-	-	-	-	-	0.49	-	-	-	-	-	-	-	-	-	-	-	-	-	-
	-	-	-	0.50	-	-	-	-	-	0.50	-	-	-	-	-	0.50	-	0.50	-	-
	0.52	-	-	-	-	-	-	0.52	-	-	-	-	-	0.52	0.52	-	0.52	-	-	-
	-	0.53	-	-	-	-	-	-	0.53	-	-	0.53	0.53	-	-	-	-	-	-	-
	-	-	0.54	-	-	-	-	-	-	-	-	-	-	-	-	-	-	-	-	-
	-	-	-	-	-	0.55	0.55	0.55	-	-	-	-	-	-	0.55	-	-	-	-	-
	-	-	-	-	0.56	-	-	-	-	-	-	-	-	-	-	-	-	-	-	-
	0.58	-	-	-	-	-	-	-	-	-	-	0.58	-	-	-	-	-	-	-	-
	-	-	-	0.59	-	0.59	-	-	-	-	-	-	-	-	0.59	-	0.59	-	-	-
	-	-	-	-	-	-	0.60	-	-	-	0.60	-	-	0.60	-	0.60	-	-	0.60	0.60
	-	-	-	-	-	-	-	-	0.61	0.61	-	-	-	-	-	-	-	-	-	-
	-	-	-	-	-	-	-	-	-	-	-	-	0.62	-	-	-	-	-	-	0.62
	-	-	-	-	0.64	-	-	-	-	-	-	-	-	-	0.64	-	0.64	-	-	-
	-	-	-	-	-	-	-	0.65	-	-	-	-	-	-	-	-	-	0.65	-	-
	0.66	-	-	-	-	-	-	-	-	-	-	-	-	-	-	-	-	-	-	0.66
	-	-	-	-	-	-	-	-	-	-	-	-	-	-	0.68	-	0.68	-	-	-
	-	-	-	0.69	-	-	-	-	-	0.69	-	-	-	-	-	-	-	-	-	-
	-	0.70	0.70	-	-	-	-	-	-	-	-	-	0.70	-	-	-	-	-	-	-
	-	-	-	-	-	-	0.72	-	0.72	-	0.72	0.72	-	-	-	-	0.72	-	-	-
	-	-	-	-	-	0.73	-	-	-	-	-	-	-	0.73	-	-	-	-	-	-
	-	-	-	-	0.74	-	-	0.74	-	-	-	-	-	-	-	-	-	-	-	-
	-	-	-	-	-	-	-	-	-	-	-	-	-	-	-	0.75	-	-	-	-
	0.76	-	-	-	-	-	-	-	-	0.76	-	-	-	-	-	-	-	-	-	-
	-	-	-	-	-	-	-	-	-	-	-	0.77	-	-	-	-	-	0.77	-	-
	0.78	-	-	-	-	-	-	-	-	-	-	-	-	0.78	-	-	-	-	0.78	-
	-	-	-	-	-	-	-	-	-	-	-	-	-	-	0.79	-	-	-	-	-
	-	0.80	-	0.80	-	-	-	-	-	-	-	-	-	-	-	-	-	-	-	-
	-	-	0.81	-	-	0.81	-	-	-	-	-	-	-	-	-	-	0.81	-	-	-
	-	-	-	-	-	-	-	-	-	-	0.83	-	-	-	-	0.83	-	-	-	-
	-	-	-	-	-	-	-	0.84	-	-	-	-	-	-	-	-	-	-	-	-
	-	0.85	0.85	-	-	-	0.85	-	0.85	0.85	-	0.85	0.85	-	-	-	-	-	-	-
	0.86	-	-	-	-	-	-	-	-	-	-	-	-	0.86	-	-	-	-	-	-
	-	-	-	0.87	-	-	-	-	-	-	-	-	-	-	-	-	-	-	-	-
	-	-	-	-	-	-	-	-	-	-	-	-	-	-	-	-	-	-	-	0.88
	-	-	-	-	0.89	-	-	-	-	-	-	-	-	-	-	-	-	-	0.89	-
	-	-	-	-	-	-	-	-	-	-	-	-	-	0.93	-	0.93	0.93	0.93	-	-
	-	-	-	0.94	-	-	-	0.94	-	-	0.94	0.94	0.94	-	0.94	-	-	-	-	-
	-	-	-	-	0.95	0.95	0.95	-	0.95	0.95	-	-	-	-	-	-	-	-	-	-
	-	-	-	-	-	-	-	-	-	-	-	-	-	0.97	-	-	0.97	0.97	-	-
	-	-	-	-	0.98	0.98	0.98	0.98	-	0.98	0.98	0.98	0.98	-	0.98	0.98	-	-	-	-
	-	-	-	-	-	-	-	-	0.99	-	-	-	-	-	-	-	-	-	-	-
(+) 1.00	FRONT																			

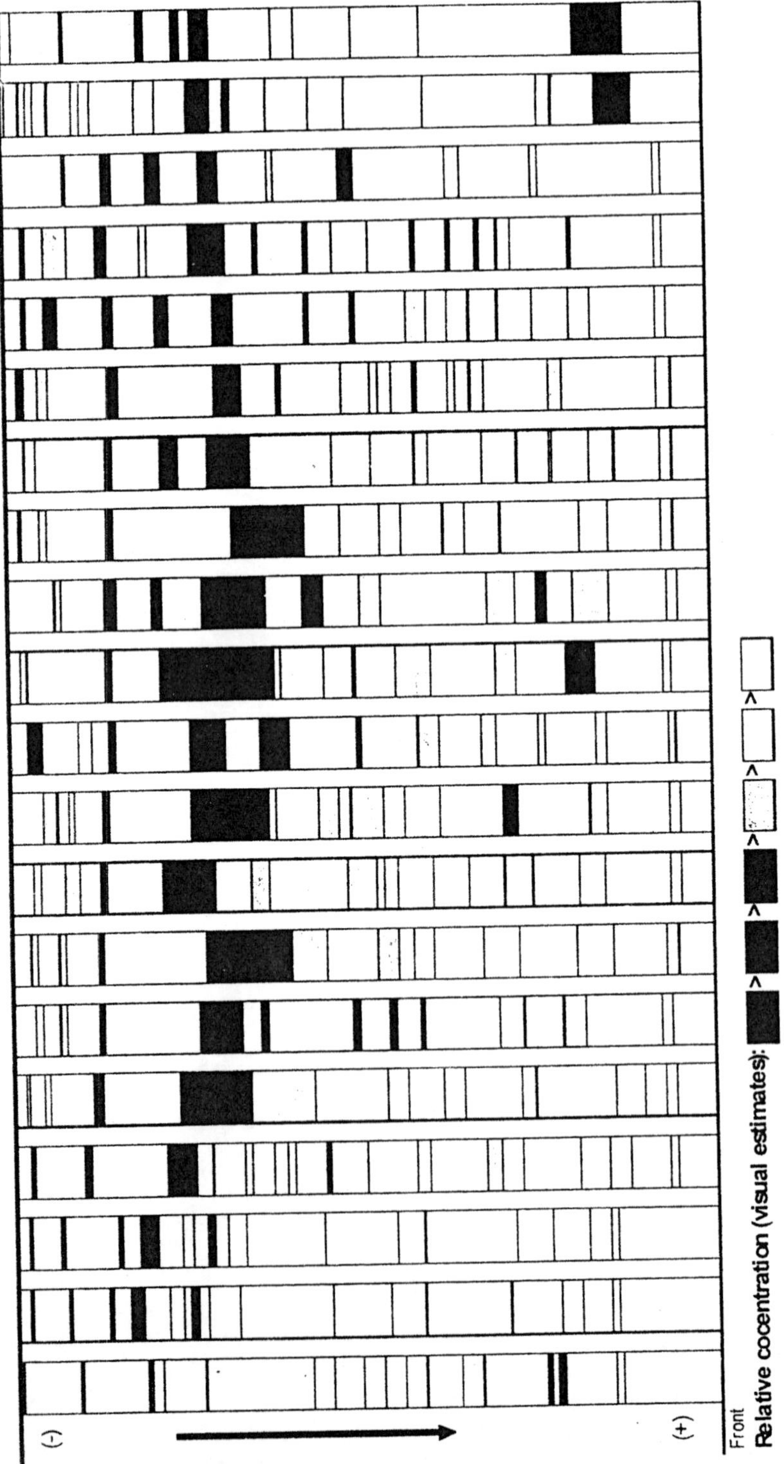
(-)
(+)
Front
Relative cocentration (visual estimates):

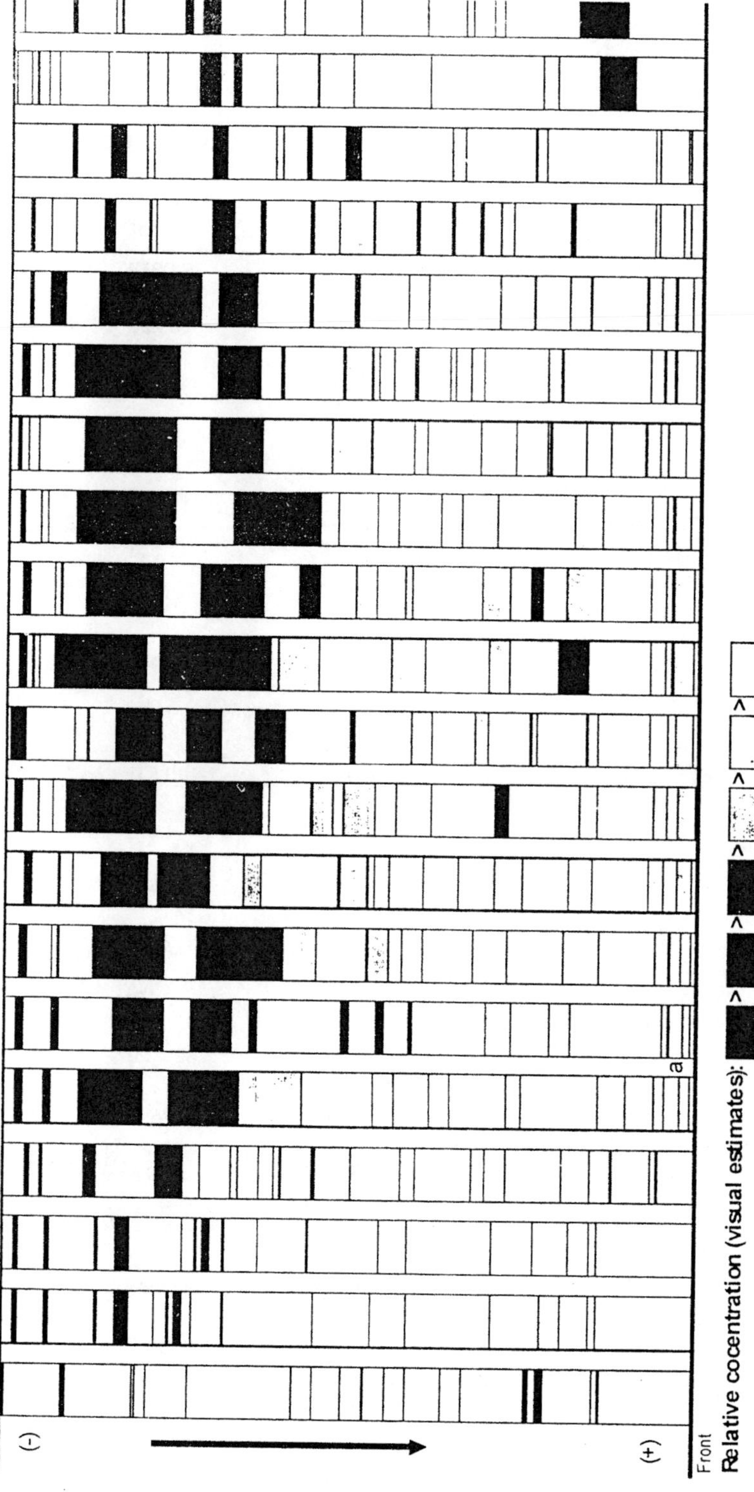
(-)
(+)
a
Front
Relative cocentration (visual estimates):

4.6.6. Total free Amino acids

The ontogenetic availability of quantitative total free amino acids was estimated in mg/ml of haemolymph both in NDD and DD generations of A. mylitta and the data is presented in Table 4.81 to 4.83.

Table 4.81. Total free amino acids concentration (mg/ml) of the haemolymph of different stages in NDD & DD generations of *A. mylitta*

SN	Age	Mean	±	SE	t Stat	
1	I-NDD	4.937	±	0.009	-194.958	**
2	I-DD	7.521	±	0.011		
3	II-NDD	6.713	±	0.004	-83.223	**
4	II-DD	7.509	±	0.008		
5	III-NDD	7.993	±	0.002	-42.079	**
6	III-DD	8.523	±	0.012		
7	IV-NDD	9.382	±	0.014	-40.081	**
8	IV-DD	9.996	±	0.007		
9	VEF-NDD	9.745	±	0.010	-198.497	**
10	VEF-DD	12.003	±	0.006		
11	VEM-NDD	10.247	±	0.003	-217.651	**
12	VEM-DD	12.533	±	0.010		
13	VMF-NDD	11.907	±	0.009	-115.295	**
14	VMF-DD	13.172	±	0.008		
15	VMM-NDD	12.139	±	0.005	-80.451	**
16	VMM-DD	13.616	±	0.019		
17	VLF-NDD	9.624	±	0.003	-547.870	**
18	VLF-DD	12.113	±	0.005		
19	VLM-NDD	9.787	±	0.011	-140.515	**
20	VLM-DD	13.173	±	0.021		
21	SLF-NDD	9.150	±	0.006	4.061	**
22	SLF-DD	9.115	±	0.006		
23	SLM-NDD	9.435	±	0.012	184.492	**
24	SLM-DD	7.346	±	0.008		
25	PPF-NDD	12.407	±	0.006	-317.708	**
26	PPF-DD	15.016	±	0.006		
27	PPM-NDD	13.486	±	0.003	6.501	**
28	PPM-DD	13.461	±	0.002		
29	MPF-NDD	14.801	±	0.005	320.236	**
30	MPF-DD	11.469	±	0.011		
31	MPM-NDD	15.152	±	0.010	204.328	**
32	MPM-DD	12.367	±	0.011		
33	LPF-NDD	11.948	±	0.007	-35.834	**
34	LPF-DD	12.367	±	0.011		
35	LPM-NDD	10.634	±	0.004	-729.013	**
36	LPM-DD	14.192	±	0.003		
37	AF-NDD	18.444	±	0.003	507.609	**
38	AF-DD	15.129	±	0.006		
39	AM-NDD	24.783	±	0.007	375.736	**
40	AM-DD	21.861	±	0.006		

n=5 (Abbreviations used are as per section 3.1.2.)

It was observed that level of free amino acid increased with increase of instar in NDD generation, i.e., from first instar NDD (4.937 ± 0.009) to mid fifth instar female (11.907 ± 0.009) and males (12.139 ± 0.005). Thereafter, the level of free amino acid decreased in both the sexes during spinning larval stage as it was recorded to be 9.150 ± 0.006 in SLF and 9.435 ± 0.012 in SLM. The level of free amino acid observed to be increasing during pre-pupal stages of both the sexes viz. 12.407 ± 0.006 in PPF-NDD and 13.461 ± 0.002 in PPM-NDD and with a fluctuating trend it was recorded to be at its highest level in adults of both the sexes i. e., 18.444 ± 0.003 in AF-NDD and 24.783 ± 0.007 in AM-NDD.

Similar trend of amino acid availability were also recorded in the haemolymph of different stages of DD generation. The level increased from first instar (7.521 ± 0.011) to early fifth instar male (12.533 ± 0.010) and mid fifth instar female larvae (13.172 ± 0.008). Thereafter, it lowered in spinning larval stage (9.115 ± 0.006 in SLF-DD; 7.346 ± 0.008 in SLM-DD). The level of amino acids again increased from pre-pupal stage of both the sexes (15.016 ± 0.006 in PPF-DD & 13.461± 0.002 in PPM-DD) and the highest level was recorded in adults as 15.129 ± 0.006 in AF-DD and 21.861 ± 0.006 in AM-DD.

t-Test was applied to record the differences in between the mean amino acids availability in respective stages of NDD and DD generations. The difference was recorded to be significant ($p<0.01$) in all the larval, pupal and adult stages taken up for study. From first to fourth larval instar the diapausing generation had significantly higher level of amino acids than respective stages of non-diapausing generation. The trend was observed to be some times higher in NDD and sometimes higher in DD generations in the following larval, pupal stages. However in case of adults the AF-NDD (18.444 ± 0.003) had significantly higher amino acids than AF-DD (15.129 ± 0.006), however, reverse was the case in case of males (AM-NDD-24.783 ± 0.007 & 21.861± 0.006-AM-DD).

Table 4.82. Total free amino acids concentration (mg/ml) in the haemolymph of different sexesn in NDD generations of *A. mylitta*

SN	Age	Mean	±	SE	t Stat	
1	VEF-NDD	9.745	±	0.010	-47.626	**
2	VEM-NDD	10.247	±	0.003		
3	VMF-NDD	11.907	±	0.009	-31.520	**
4	VMM-NDD	12.139	±	0.005		
5	VLF-NDD	9.624	±	0.003	-13.472	**
6	VLM-NDD	9.787	±	0.011		
7	SLF-NDD	9.150	±	0.006	-18.827	**
8	SLM-NDD	9.435	±	0.012		
9	PPF-NDD	12.407	±	0.006	-214.917	**
10	PPM-NDD	13.486	±	0.003		
11	MPF-NDD	14.801	±	0.005	-37.073	**
12	MPM-NDD	15.152	±	0.010		
13	LPF-NDD	11.948	±	0.007	148.918	**
14	LPM-NDD	10.634	±	0.004		
15	AF-NDD	18.444	±	0.003	-874.361	**
16	AM-NDD	24.783	±	0.007		

n=5 (Abbreviations used are as per section 3.1.2.)

Between the sexes, the comparisons were also made. In NDD generation in every stage male contained significantly high level of amino acid than females ($p < 0.01$) e.g. in VEF-NDD- 9.745 ± 0.010 & VEM-NDD 10.247 ± 0.003 , in SLF-NDD- 9.150 ± 0.006 & SLM-NDD-9.435 ± 0.012, in adults female 18.44 ± 0.003 & AM-NDD-24.783 ± 0.007. Similar was the case in the larvae and adults male and female of DD generation.

Table 4.83. Total free amino acids concentration (mg/ml) in the haemolymph of different sexesn in DD generations of *A. mylitta*

SN	Age	Mean	±	SE	t Stat	
1	VEF-DD	12.003	±	0.006	-42.590	**
2	VEM-DD	12.533	±	0.010		
3	VMF-DD	13.172	±	0.008	-27.901	**
4	VMM-DD	13.616	±	0.019		
5	VLF-DD	12.113	±	0.005	-51.477	**
6	VLM-DD	13.173	±	0.021		
7	SLF-DD	9.115	±	0.006	209.757	**
8	SLM-DD	7.346	±	0.008		
9	PPF-DD	15.016	±	0.006	231.084	**
10	PPM-DD	13.461	±	0.002		
11	MPF-DD	11.469	±	0.011	-88.350	**
12	MPM-DD	12.367	±	0.011		
13	LPF-DD	12.367	±	0.011	-192.475	**
14	LPM-DD	14.192	±	0.003		
15	AF-DD	15.129	±	0.006	-894.701	**
16	AM-DD	21.861	±	0.006		

n=5 (Abbreviations used are as per section 3.1.2.)

4.6.7. Uric acid (mg/ml in haemolymph)

Uric acid was estimated in NDD and DD generations of A. mylitta in larval, pupal and adult stages. The concentration in mg/ml of haemolymph is presented in tables 4.82 to 4.84. The concentration of uric acid gradually increased in larval instar in NDD generation as it was 0.0098 ± 0.0007 in I-NDD, 0.0110 ± 0.0007 in II-NDD, 0.0150 ± 0.0004 in III-NDD, 0.0194 ± 0.0003 in IV NDD, VMF-NDD (0.0716 ± 0.0010) and VMM-NDD (0.1014 ± 0.0021). A fall was noticed during late V instar followed by an increase in pre-pupal stage (PPF-NDD - 0.0730 ± 0.0020, PPM-NDD-0.0900 ± 0.0006), the level was higher in mid pupal stage too, the level was observed to be decreasing in late pupal stage and it was recorded to be highest in adults (AF-NDD - 1.1126 ± 0.0042, AM-NDD - 1.1812 ± 0.0084).

In DD generation the trend of the concentration of uric acid was similar like that of NDD generation, as it increased from 0.0300 ± 0.0009 in I-DD to 0.0778 ± 0.0009 (VMF) and 0.0960 ± 0.0015 (VMM) . Though it went down and was recorded to be 0.0126 ± 0.0008 in LPF-DD and 0.0192 ± 0.0007 in LPM-DD. The level of uric acid was recorded to be highest in adults (AF-DD - 0.6664 ± 0.0068 and AM-DD - 0.8866 ± 0.0123).

Table 4.82. Uric acid content (mg/ml) of the haemolymph of different stages in NDD & DD generations of *A. mylitta*

SN	Age	Mean	±	SE	t Stat	
1	I-NDD	0.0098	±	0.0007	-18.9190	**
2	I-DD	0.0300	±	0.0009		
3	II-NDD	0.0110	±	0.0007	-21.7568	**
4	II-DD	0.0334	±	0.0006		
5	III-NDD	0.0150	±	0.0004	-34.6418	**
6	III-DD	0.0448	±	0.0012		
7	IV-NDD	0.0194	±	0.0003	-45.4187	**
8	IV-DD	0.0574	±	0.0009		
9	VEF-NDD	0.0830	±	0.0042	5.5355	**
10	VEF-DD	0.0652	±	0.0013		
11	VEM-NDD	0.0508	±	0.0010	-25.8070	**
12	VEM-DD	0.0730	±	0.0008		
13	VMF-NDD	0.0716	±	0.0010	7.7500	**
14	VMF-DD	0.0778	±	0.0009		
15	VMM-NDD	0.1014	±	0.0021	2.1757	*
16	VMM-DD	0.0960	±	0.0015		
17	VLF-NDD	0.0680	±	0.0007	18.5777	**
18	VLF-DD	0.0428	±	0.0011		
19	VLM-NDD	0.0776	±	0.0010	22.1827	**
20	VLM-DD	0.0610	±	0.0007		
21	SLF-NDD	0.0494	±	0.0017	-1.1744	NS
22	SLF-DD	0.0514	±	0.0004		
23	SLM-NDD	0.0556	±	0.0013	-9.8716	**
24	SLM-DD	0.0720	±	0.0006		
25	PPF-NDD	0.0738	±	0.0020	14.7915	**
26	PPF-DD	0.0334	±	0.0014		
27	PPM-NDD	0.0900	±	0.0006	62.8702	**
28	PPM-DD	0.0438	±	0.0004		
29	MPF-NDD	0.0846	±	0.0004	71.8854	**
30	MPF-DD	0.0262	±	0.0009		
31	MPM-NDD	0.0918	±	0.0007	76.5000	**
32	MPM-DD	0.0306	±	0.0009		
33	LPF-NDD	0.0628	±	0.0011	37.0079	**
34	LPF-DD	0.0126	±	0.0008		
35	LPM-NDD	0.0852	±	0.0007	93.3381	**
36	LPM-DD	0.0192	±	0.0007		
37	AF-NDD	1.1126	±	0.0042	48.5574	**
38	AF-DD	0.6664	±	0.0068		
39	AM-NDD	1.1812	±	0.0084	19.4864	**
40	AM-DD	0.8866	±	0.0123		

n=5 (Abbreviations used are as per section 3.1.2.)

t-Test was applied to find out the differences in the mean values of uric acid in different stages of both NDD and DD generations. Uric acid level was significantly higher ($p < 0.01$) in NDD generation than DD generation, particularly in between all the larval stages except in between spinning larvae (SLF-NDD and SLF-DD), where difference was observed to be non-significant. In all other stages of pupae and also in adults, significant ($p < 0.01$) difference was recorded.

t-Test was also applied to find out the difference between the two sexes of NDD generation. The difference was significant ($p < 0.01$) in between VEF-NDD & VEM-NDD, VMF-NDD & VMM-NDD, VLF-NDD & VLM-NDD, PPF-NDD & PPM-NDD, MPF-NDD & MPM-NDD, LPF-NDD & LPM-NDD, AF-NDD & AM-NDD, though the difference was significant ($p < 0.05$) in between SLF-NDD & SLM-NDD.

Similarly, when compared the difference of both the sexes of DD generation, the difference was observed to be significant ($p < 0.01$) in between VEF-DD & VEM-DD, VMF-DD & VMM-DD, VLF-DD & VLM-DD, SLF-DD & SLM-DD, PPF-DD & PPM-DD, LPF-DD & LPM-DD, AF-DD & AM-DD except in between MPF-DD & MPM-DD where the difference was significant at 95 % level.

Table 4.83. Uric acid content (mg/ml) in the haemolymph of different sexesn in NDD generations of *A. mylitta*

SN	Age	Mean	±	SE	t Stat	
1	VEF-NDD	0.0830	±	0.0042	7.1049	**
2	VEM-NDD	0.0508	±	0.0010		
3	VMF-NDD	0.0716	±	0.0010	-12.0263	**
4	VMM-NDD	0.1014	±	0.0021		
5	VLF-NDD	0.0680	±	0.0007	-6.8571	**
6	VLM-NDD	0.0776	±	0.0010		
7	SLF-NDD	0.0494	±	0.0017	-3.7456	*
8	SLM-NDD	0.0556	±	0.0013		
9	PPF-NDD	0.0738	±	0.0020	-7.6030	**
10	PPM-NDD	0.0900	±	0.0006		
11	MPF-NDD	0.0846	±	0.0004	-10.8544	**
12	MPM-NDD	0.0918	±	0.0007		
13	LPF-NDD	0.0628	±	0.0011	-29.9333	**
14	LPM-NDD	0.0852	±	0.0007		
15	AF-NDD	1.1126	±	0.0042	-15.9234	**
16	AM-NDD	1.1812	±	0.0084		

n=5 (Abbreviations used are as per section 3.1.2.)

Table 4.84. Uric acid content (mg/ml) in the haemolymph of different sexesn in DD generations of *A. mylitta*

SN	Age	Mean	±	SE	t Stat	
1	VEF-DD	0.0652	±	0.0013	-4.1457	**
2	VEM-DD	0.0730	±	0.0008		
3	VMF-DD	0.0778	±	0.0009	-13.0668	**
4	VMM-DD	0.0960	±	0.0015		
5	VLF-DD	0.0428	±	0.0011	-16.3441	**
6	VLM-DD	0.0610	±	0.0007		
7	SLF-DD	0.0514	±	0.0004	-23.6298	**
8	SLM-DD	0.0720	±	0.0006		
9	PPF-DD	0.0334	±	0.0014	-6.7698	**
10	PPM-DD	0.0438	±	0.0004		
11	MPF-DD	0.0262	±	0.0009	-3.2262	*
12	MPM-DD	0.0306	±	0.0009		
13	LPF-DD	0.0126	±	0.0008	-7.1170	**
14	LPM-DD	0.0192	±	0.0007		
15	AF-DD	0.6664	±	0.0068	-15.3555	**
16	AM-DD	0.8866	±	0.0123		

n=5 (Abbreviations used are as per section 3.1.2.)

4.6.8. Cholesterol

Cholesterol appears to be transported in haemolymph plasma and may serve as a good energy source. The data pertaining to presence of cholesterol in the haemolymph of different stages of NDD and DD generations are presented in Tables 4.85, 4.86 and 4.87. Among NDD larval instars in I-NDD in its haemolymph cholesterol content and was 1.6674 ± 0.015 mg/ml, in II-NDD it increased (4.6588 ± 0.0124), in III-NDD (3.0680 ± 0.0097), in IV-NDD (2.8622 ± 0.0117). In V instar NDD larvae of different age groups it was comparatively lower. A rise was noticed in mid pupal stage, it was 2.4598 ± 0.0093 in MPF-NDD and 2.0506 ± 0.0225 in MPM-NDD, and it was at its lowest level in adults i.e. 0.9564 ± 0.0086 in AF-NDD and 1.5176 ± 0.0365 in AM-NDD.

In case of DD generation, the cholesterol content was 2.1642 ± 0.0195 in I-DD, 2.8626 ± 0.0226 in II-DD, it went up to 2.7854 ± 0.0152 in III-DD and it lowered to 1.9246 ± 0.0098 in IV-DD. Its concentration was observed in a fluctuating manner in remaining larval instar and a rise was recorded in mid pupal stage i.e. 2.4432 ± 0.0195 in MPF-DD & 2.1592 ± 0.0220 in MPM-DD. It again decreased in late pupal stage and this trend continued till emergence of adults and it was recorded to be lowest in adults as it was 1.1368 ± 0.0072 in AM-DD and 0.3388 ± 0.0127 in AF-DD. The cholesterol content was generally lower in this generation.

Student's t-test was applied to find out the difference in between respective life stages of NDD and DD generation. Significant difference was recorded ($p < 0.01$) in between I-NDD & I-DD to PPF-NDD and PPF-DD, LPF-NDD & LPM-DD, AF-NDD & AF-DD, AM-NDD & AM-DD (Table 4.85). The difference was observed to be significant ($p < 0.05$) in between MPM-NDD & MPM-DD. Rest of the comparisons were observed to be non-significant.

Table 4.85. Cholesterol content (mg/ml) of the haemolymph of different stages in NDD & DD generations of *A. mylitta*

SN	Age	Mean	±	SE	t Stat	
1	I-NDD	1.6674	±	0.0150	-17.5149	**
2	I-DD	2.1642	±	0.0195		
3	II-NDD	4.6588	±	0.0124	69.9202	**
4	II-DD	2.8626	±	0.0226		
5	III-NDD	3.0680	±	0.0097	13.1927	**
6	III-DD	2.7854	±	0.0152		
7	IV-NDD	2.8622	±	0.0117	57.6551	**
8	IV-DD	1.9246	±	0.0098		
9	VEF-NDD	1.8502	±	0.0119	75.1232	**
10	VEF-DD	1.1298	±	0.0061		
11	VEM-NDD	2.0702	±	0.0238	44.5270	**
12	VEM-DD	0.9948	±	0.0053		
13	VMF-NDD	2.5242	±	0.0303	27.6551	**
14	VMF-DD	1.5820	±	0.0093		
15	VMM-NDD	1.7088	±	0.0124	47.0606	**
16	VMM-DD	1.0372	±	0.0077		
17	VLF-NDD	2.1546	±	0.0077	138.0193	**
18	VLF-DD	0.8606	±	0.0093		
19	VLM-NDD	1.4738	±	0.0081	31.5406	**
20	VLM-DD	0.6028	±	0.0301		
21	SLF-NDD	1.8612	±	0.0164	85.2540	**
22	SLF-DD	0.4752	±	0.0153		
23	SLM-NDD	1.2429	±	0.0068	72.3255	**
24	SLM-DD	0.3564	±	0.0082		
25	PPF-NDD	1.7818	±	0.0024	-7.1696	**
26	PPF-DD	1.9034	±	0.0182		
27	PPM-NDD	1.1904	±	0.0030	0.6064	NS
28	PPM-DD	1.1828	±	0.0147		
29	MPF-NDD	2.4598	±	0.0093	0.8945	NS
30	MPF-DD	2.4432	±	0.0195		
31	MPM-NDD	2.0506	±	0.0225	-3.1617	*
32	MPM-DD	2.1592	±	0.0220		
33	LPF-NDD	1.7326	±	0.0097	-23.6818	**
34	LPF-DD	1.9298	±	0.0070		
35	LPM-NDD	0.6252	±	0.0161	-90.2437	**
36	LPM-DD	1.6254	±	0.0171		
37	AF-NDD	0.9564	±	0.0086	35.4475	**
38	AF-DD	0.3388	±	0.0127		
39	AM-NDD	1.5176	±	0.0365	12.1832	**
40	AM-DD	1.1368	±	0.0072		

n = 5 (Abbreviations used are as per section 3.1.2.)

Table 4.86. Cholesterol content (mg/ml) in the haemolymph of different sexesn in NDD generations of *A. mylitta*

SN	Age	Mean	±	SE	t Stat	
1	VEF-NDD	1.8502	±	0.0119	-20.6545	**
2	VEM-NDD	2.0702	±	0.0238		
3	VMF-NDD	2.5242	±	0.0303	27.4737	**
4	VMM-NDD	1.7088	±	0.0124		
5	VLF-NDD	2.1546	±	0.0077	53.5151	**
6	VLM-NDD	1.4738	±	0.0081		
7	SLF-NDD	1.8612	±	0.0164	38.5679	**
8	SLM-NDD	1.2429	±	0.0068		
9	PPF-NDD	1.7818	±	0.0024	153.4164	**
10	PPM-NDD	1.1904	±	0.0030		
11	MPF-NDD	2.4598	±	0.0093	16.3309	**
12	MPM-NDD	2.0506	±	0.0225		
13	LPF-NDD	1.7326	±	0.0097	137.3983	**
14	LPM-NDD	0.6252	±	0.0161		
15	AF-NDD	0.9564	±	0.0086	-17.2457	**
16	AM-NDD	1.5176	±	0.0365		

n = 5 (Abbreviations used are as per section 3.1)

Table 4.87. Cholesterol content (mg/ml) in the haemolymph of different sexesn in DD generations of *A. mylitta*

SN	Age	Mean	±	SE	t Stat	
1	VEF-DD	1.1298	±	0.0061	18.4914	**
2	VEM-DD	0.9948	±	0.0053		
3	VMF-DD	1.5820	±	0.0093	76.7857	**
4	VMM-DD	1.0372	±	0.0077		
5	VLF-DD	0.8606	±	0.0093	13.2749	**
6	VLM-DD	0.6028	±	0.0301		
7	SLF-DD	0.4752	±	0.0153	7.0540	**
8	SLM-DD	0.3564	±	0.0082		
9	PPF-DD	1.9034	±	0.0182	45.4694	**
10	PPM-DD	1.1828	±	0.0147		
11	MPF-DD	2.4432	±	0.0195	7.8165	**
12	MPM-DD	2.1592	±	0.0220		
13	LPF-DD	1.9298	±	0.0070	24.0304	**
14	LPM-DD	1.6254	±	0.0171		
15	AF-DD	0.3388	±	0.0127	-67.8320	**
16	AM-DD	1.1368	±	0.0072		

n = 5 (Abbreviations used are as per section 3.1)

Student's t-test was applied to find out the difference in between the sexes of NDD generation and the difference was significant ($p < 0.01$) in between VEF-NDD & VEM-NDD, VMF-NDD & VMM-NDD, VLF-NDD & VLM-NDD, SLF-NDD & SLM-NDD, PPF-NDD & PPM-NDD, MPF-NDD & MPM-NDD, LPF-NDD & LPM-NDD, AF-NDD & AM-NDD. Adult males had higher cholesterol than females (Table 4.86).

Student's t-test was also applied to observe the difference in between both the sexes of DD generation. There were significant differences ($p < 0.01$) in between VEF-DD & VEM-DD, VMF-DD & VMM-DD, VLF-DD & VLM-DD, SLF-DD & SLM-DD, PPF-DD & PPM-DD, MPF-DD & MPM-DD, LPF-DD & LPM-DD, AF-DD & AM-DD. Like NDD generations adult males had higher cholesterol (1.1386 ± 0.0072) than females (0.3388 ± 0.0127).

5

Discussion

5.1. Introduction

In the present work entitled "Histo-chemical Studies on Diapausing and Non-diapausing Generations of Tropical Tasar Silkworm *Antheraea mylitta* Drury (Lepidoptera: Saturniidae)," general histology of different haemocytes with the help of light microscopy under normal and phase contrast optic and under scanning electron microscopy (SEM); total and differential haemocytes counts in relation to body weight; haemolymph volume of larvae, pupae and adults, physical properties like pH and specific gravity and the presence of osmolytes like Na+, K+, Ca++ and Mg++ of different life stages; presence of general proteins, nucleic acids, PA/S substances, glycogen, bound lipids and alkaline phosphatases in cytoplasm and nucleus of different haemocytes through histochemical tests; haemolymph organic constituents such as sugars in the form of trehalose; nitrogenous constituents in the form of quantitative and qualitative total proteins, total amino acids; uric acids and lipids in form of cholesterol in different developmental stages viz. I. II, III, IV, early, mid and late aged V instar larvae, early, mid and late aged pupae and adults of both the sexes and generations have been considered.

5.2. Types of Haemocytes

The extra cellular body fluid-the haemolymph of insects, like the blood of higher animals has two components: the plasma and the corpuscles or haemocytes (Richards and Davies, 1977). The haemolymph circulates freely within the body bathing the different tissues. Arnold (1979a) and Jones, (1979) have reviewed and reported several types of haemocytes in insects. According to them, in most of the insects, there were three well-defined types of haemocytes viz. Prohaemocytes, Plasmatocytes and Granulocytes and one or more of four other types in some other insects (Coagulocytes, Spherulocytes, Adiphohaemocytes and Oenocytoids). For classification of haemocytes both transmission and phase contrast microscopy were used by various workers. Five types of haemocytes, namely, Prohaemocytes, Plasmatocytes, Granulocytes, Spherulocytes and Oenocytoids have been reported in lepidopteran larvae of *Malcosomia disstria* Hübner by Arnold and Sohi, 1974 and in *Euxoa declarata* Walker by Arnold and Hinks, 1975 and in *Bombyx mori* by Akai and Sato (1971, 1973). Gupta (1979a & b; 1985), while summarizing the types of haemocytes based on their morphological and cytological characteristics, reported seven types of haemocytes in Lepidopteran insects and those were Prohaemocytes, Plasmatocytes, Granulocytes, Spherulocytes, Adipocytes, Coagulocytes and Oenocytoids. Five types of haemocytes were recognised in *Menduca sexta*, namely, Prohaemocytes, Plasmatocytes, Granulocytes, Spherulocytes and Oenocytoids (Horhov and Dunn, 1982). Lea (1986) has reported Prohaemocytes, Plasmatocytes, Adipohaemocytes, Spherulocytes and Oenocytoids in wax moth *Galleria mellonella*. Based on phase contrast and light microscopy, a total of 6 cell types was identified in *Diatraea saccharalis* (Lepidoptera: Pyralidae), Prohaemocyte, Granulocyte, Adipohaemocyte, Spherulocyte, Plasmatocyte and Oenocytoid, all of which had morphological characteristics similar to those described for other Lepidoptera (Barduco, et al., 1988). Four types of haemocytes (Prohaemocytes, Plasmatocytes, Granulocytes and Spherulocytes) and changes in the haemocytes (Prohaemocytes, Plasmatocytes and Granulocytes) during metamorphosis are described in *Lymantria dispar* (Kim, et al., 1990 a & b). The number of Granulocytes and the granules within them increased during the pupal stage. The granules were perhaps composed of polysaccharides, protein and lipid. Lipidous granules were especially rich at this stage (Kim, et al., 1990 a & b). Pathak (1991) reported Prohaemocytes, Plasmatocytes, Granulocytes, Oenocytoids and Adipohaemocytes in *Halys dentata* (Pentotomidae: Heteroptera). Three basic types of haemocytes, namely Plasmatocytes, Cystocytes and Prohaemocytes were identified in *Armigeres subalbatus* (Zahedi, 1993). Ten different types of haemocytes were recorded from the haemolymph of adults of *Blattella germanica* (Pathak and Kulshreshtha, 1993). Five types of haemocytes in *Dysdercus koenigii,* namely the Prohaemocyte, Plasmatocyte, Granular haemocyte, Adipohaemocyte and the Oenocytoid have been reported by Sharma et al., 1998. In haemolymph of larvae of noctuiid moth *Plusia orichalcea*, seven types of haemocytes were reported, namely, Prohaemocytes, Plasmatocytes, Granulocytes, Spherulocytes, Adipohaemocytes, Oenocytoids and Coagulocytes (Pathak and Saxena, 1994). Seven types of haemocytes were also reported in *Spodoptera litura* (Kurihara et al., 1992a & b) and in *Lymantria dispar* (Butt and Shield, 1996). The most common types of haemocytes have been described in species of diverse orders, namely, *Polistes hebraeus,* (Ahmad, 1988); six triatomune species (Azambuza, et al., 1991); *Aulacophora foveicollis* and *Mylabris pustulata* (Ahmad, 1992); in *Leucophaea maderae* (Fenoglio et al., 1993); *Cimex rotundatus* (Sonwane and More, 1993); in *Lymantria dispar* (Butt and Shield, 1996); *Dactylopius confusus* (Joshi, and Lambdin, 1996); *Anopheles albimanus* (Hernandez et al., 1999) and in *Acheta domesticus* (De Silva et al., 2000).

Several workers have reported type of haemocytes in silk producing insects. For example, Raichoudhury and Sengupta (1959) named seven types of haemocytes in *Bombyx mori*, but Nittono (1960) reported only five types in same species. Five types of haemocytes (Prohaemocytes, Plasmatocytes, Granulocytes, Cytocytes or Spherulocytes and Oenocytoids) have been reported in *Hyalophora cecropia* (L.), *Antheraea polyphamous* (Cram.) and *Samia cynthaea* (Drury) by Lea and Gilbert (1966). Again in *Philosamia ricini* (*Samia cynthia ricini*), five types of haemocytes, namely, Prohaemocytes, Plasmatocytes, Granulocytes, Cytocytes and Oenocytoids were reported (Begum et al. 1992a & b) and five types (Prohaemocytes, Plasmatocytes, Granulocytes, Cytocytes and Oenocytoids) in *Bombyx mori* (Balavenkatsubbaiah et al., 2001. Similarly, in different species of genus *Antheraea* five types (Prohaemocytes, Plasmatocytes, Granulocytes, Spherulocytes and Oenocytoids) of haemocytes, were reported such as in *Antheraea pernyi* by Beaulaton and Monpeyssin (1976) and in *Antheraea assama* by Bardoloi and Hazarika (1992 and 1995). The type of haemocytes observed in the present study in haemolymph of tropical tasar silkworm *Antheraea mylitta* was similar to the type of haemocytes *A. pernyi* and *A. assama*. Thus, the results are in conformity with the findings of above authors.

5.3. Structure of Haemocytes

The classes of insect haemocytes are also characterised by various biochemical and other cytological parameters, e.g., appearance of mature cells, amount of DNA, volume, numbers, types of cytoplasmic inclusions and the relative size of nucleus and cell body. These parameters, separately, or in combination, relate the structure of cells (Arnold, 1972; Gupta, 1985) and these characteristics have been helpful to characterise the different orders of insects (Wigglesworth, 1959; Jones, 1962).

Arnold (1974, 1979a), while classifying the Prohaemocytes based on their ultra structure, described them to be small, usually round or ellipsoidal cells with a relatively larger nucleus. The chromatin of the nucleus was visibly granular and evenly distributed.

Plasmatocytes were highly pleomorphic cells of variable size and were characterised by their large, round to ovoid nucleus enveloped in an equal or larger volume of finely granular cytoplasm nucleus.

The Oenocytoids were observed to be very distinctive cells of widely variable size and shape characterized by the small round, eccentric nucleus and large volume of superficially homogeneous, basophilic or neutrophilic cytoplasm.

Spherulocytes were very distinctive haemocytes, characterized mainly by the large oval or spherical inclusions that often filled the cytoplasm and obscured the nucleus. The spherules were PAS positive in number of insects (Lea and Gilbert, 1966; Gupta and Sutherland, 1967; Arnold and Salkeld, 1967).

Granulocytes were reported to be the compact cells of variable size, usually round or disk shaped, with a relatively small nucleus enveloped in a large volume of cytoplasm which characteristically contained many prominent PA/S positive granules. The granular content of the cytoplasm were quite variable. However, several other types of haemocytes were also described by Arnold (1974, 1979a) and these were cytocytes, podocytes and vermiform cells.

In the present study two types of Plasmatocytes, namely, elliptical and fusiform were observed in *A. mylitta*. Arnold (1961) observed that Plasmatocytes as motile cells, which change their form readily under some conditions. They tend to maintain a slightly twisted, fusiform shape while circulating in haemolymph, presumably to some physical or functional advantage (Arnold and Hinks, 1975). Probably, such cells were fixed during course of study in *A. mylitta* and they were observed to be fusiform in shape. The fusiform shape of Plasmatocytes has been determined by the distribution of microtubules, which are located in the cell periphery (Crossly, 1975). Variable shapes of Plasmatocytes have also been reported with special emphasis on the fusiform type in *Euxoa annir* and

E. lutulents (Arnold, 1979b). Presumably, fusiform Plasmatocytes have been observed due to peripheral distribution of microtubules in *A. mylitta.*

Akai and Sato (1973) while working on the ultra structure of the larval haemocytes of the silkworm, *Bombyx mori* L. stated the Prohaemocytes to be characteristically contained few cytoplasmic organelles, fusiform Plasmatocytes had elongated nucleus and cytoplasm with well-developed organelles; Granulocytes were polymorphic in size and shape characterised by secreting vesicles; Spherulocytes were characterised by cytoplasmic spherules, which contained fine homogeneous granular material; Oenocytoids were large cells characterised by large cytoplasmic inclusions, which were composed of clusters of fibrous proteins arranged in concentric circle. Similar features may be true with the haemocytes of *A. mylitta* as they were also five previously mentioned types.

Neuwirth (1973) observed four haemocyte types in the late last larval instar of *Galleria mellonella*, namely, Plasmatocytes, Granulocytes, Spherulocytes and Oenocytoids. Their structures were similar to other lepidopteran haemocytes and like that in *A. mylitta* as observed in the present study. Based on light microscopic studies, the structure of the five types of the larval haemocytes in pink bollworm, Pectinophora gossypiella Sanders were described. With the help of phase contrast microscope Prohaemocytes were observed to be small with simple cellular organisation; Plasmatocytes were of two types: spindle shaped and ovoid cells, having large pseudopodia-like structure; Spherulocytes contained many spherules; Oenocytoids possessed bundles of microtubule-like structure (Raina, 1976). In the present study also two types of Plasmatocytes, namely, elliptical and fusiform were observed in *A. mylitta*. Arnold (1961) observed that Plasmatocytes as motile cells, which change their form readily under some conditions. They tend to maintain a slightly twisted, fusiform shape while circulating in haemolymph, presumably to some physical or functional advantage (Arnold and Hinks, 1975). Probably, such cells were fixed during course of this study in *A. mylitta* and they were observed to be fusiform in shape. The fusiform shape of Plasmatocytes has been determined by the distribution of microtubules, which are located in the cell periphery (Crossly, 1975). Variable shapes of Plasmatocytes have also been reported with special emphasis on the fusiform type in *Euxoa annir* and *E. lutulents* (Arnold, 1979 b). Presumably, fusiform Plasmatocytes have been observed due to peripheral distribution of microtubules in *A. mylitta*.

Akai and Sato (1973), while working on the ultra structure of the larval haemocytes of the silkworm, *Bombyx mori* L. observed that Prohaemocytes characteristically contained few cytoplasmic organelles, fusiform Plasmatocytes had elongated nucleus and cytoplasm with well-developed organelles; Granulocytes were polymorphic in size and shape characterised by secreting vesicles; Spherulocytes were characterised by cytoplasmic spherules, which contained fine homogeneous granular material; Oenocytoids were large cells characterised by large cytoplasmic inclusions, which were composed of clusters of fibrous proteins arranged in concentric circle. Similar features may be true with the haemocytes of *A. mylitta*.

Bartninkaite, (1995) has reported the comparative morphology of the haemocytes of larvae of four species of ermine moths viz., *Yponomeuta padellus, Y. cognatellus* [Y. *cagnagellus*], *Y. evonymellus* and *Y. malinellus* of the family Yponomeutidae on the basis of cytochemical techniques. Ribeiro, et al., (1996) have reported the haemocytes of the armyworm *Mythimna unipuncta* (Lepidoptera: Noctuidae). Sharma et al., (1998) reported that Adipohaemocytes of the bug *Dysdercus koenigii* were unique in possessing nucleolar ribosomes and a few of the Adipohaemocytes and the Oenocytoids had two nuclei.

Gupta (1985) while reviewing the size of haemocytes of different insect orders reported that generally, Prohaemocytes measured 6-10 µm in width and 10-14 µm in length; Plasmatocytes measured 3.3-5 µm in width and 3.3 to 4.0 µm in length, Granulocytes measured 4-32 µm in width and 10-45 µm long, Spherulocytes 9.25 µm long and 5-10 µm wide and Oenocytoids 16-54 µm or more of variable size.

In the present study in *Antheraea mylitta* Prohaemocytes measured 8.500 to 18.063 µm (mean 11.794 ± 1.138 µm) in diameter, cell area ranged in between 56.768 to 256.342 µm2 and volume ranged in between 321.685 to 3,086.362 µm3. Oenocytoids' length measured in between 14.88 to 25.50 µm and width ranged in between 14.88 to 25.50 µm, volume ranged in between 1,724.03 to 8,685.48 µm3; Plasmatocytes were seen in two shapes, elliptical (length range - 12.75 to 23.38 µm; width range 10.63 to 19.13 µm; cell volume range 753.95 to 4,071.32 µm3); and fusiform (length range - 17.00 to 34.00 µm; width range - 8.50 to 17.00 µm; cell volume range 643.37 to 39.64 µm3); Spherulocytes (length range - 12.75 to 29.75; width range-4.25 to 12.75 µm volume range - 120.63 to 1809.48 µm3); Granulocytes (length range-12.75 to 25.50µm; width range- 8.50 to 23.38µm; volume range-482.53 to 7298.22µm3). - (Figs. 5.1. - 5.4., Tables 4.1 & 4.2).

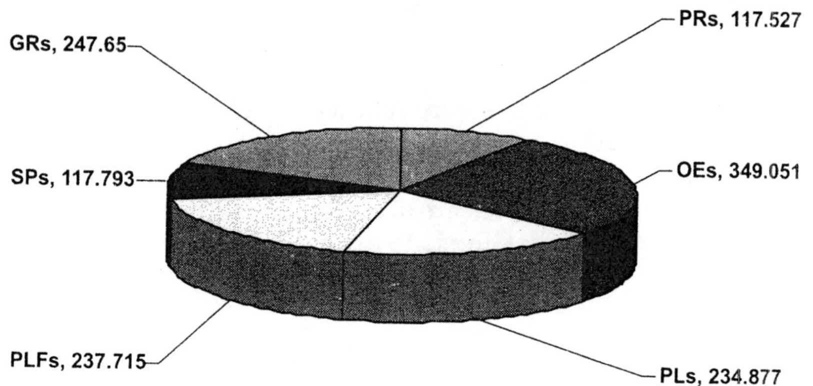

Fig. 5.1. Comparison of cell area (µm2) of different haemocytes of *A. mylitta*.

Nuclear area of Prohaemocytes ranged in between 31.932 to 173.852 µm2 (mean 86.053 ± 17.290 µm2), nucleus vs. cell area ranged in between 0.563 to 0.810 (mean 0.732 ± 0.027); of Oenocytoids in between 14.192 to 31.932 µm2 (mean 20.224 ± 2.790 µm2), nucleus vs. cell area 0.028 to 0.100 (mean 0.058 ± 0.007), of Plasmatocytes of ellipsoidal type 56.768 to 106.440 µm2 (mean 78.865 ± 5.513 µm2), nucleus vs. cell ratio 0.200 to 0.625 (mean 0.335 ± 0.049) and of Plasmatocytes of fusiform type 56.768 to 283.839 µm2 (mean 150.009 ± 24.772 µm2) nucleus vs. cell area ratio 0.333 to 1.125 (mean 0.631 ± 0.076). Spherulocytes and Granulocytes had irregular shaped nucleus hence it was not

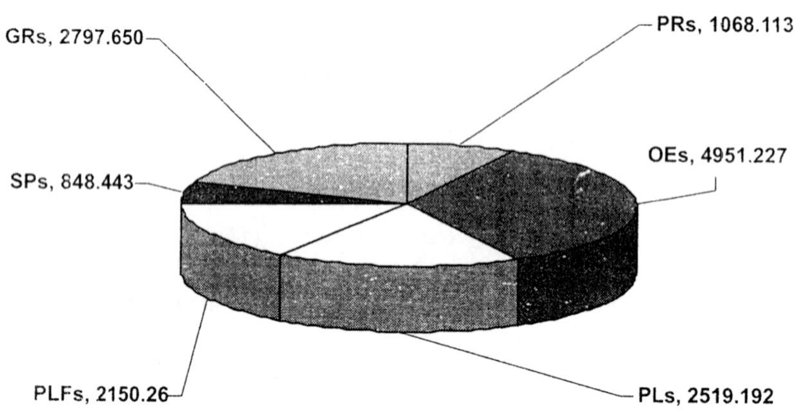

Fig. 5.2. Comparison of cell volume (µm3) of different haemocytes of *A. mylitta*.

possible to work out such ratio. The nucleus to cell area ratio (n/c) was in order of Prohaemocytes > Plasmatocytes > Oenocytoids, whereas n/c could not be worked out in other two cell types, i.e., Spherulocytes and Granulocytes, because of irregular shape of the nucleus (Table 4.1). Similar trend of ratio were also observed in other lepidopteran species, *Malcosoma disstria* (Arnold and Sohi, 1974).

The size of the different haemocytes were relatively higher in

A. mylitta as reported in other lepidopteran by Gupta (1985) though their structure were similar to those of earlier reported after Arnold, 1974; 1979a and Gupta, 1985. Therefore, the structures of the haemocytes are in conformity with the structure as reported after Jones, 1979; Arnold, 1979a & b; Gupta, 1979a & b, 1983, 1985.

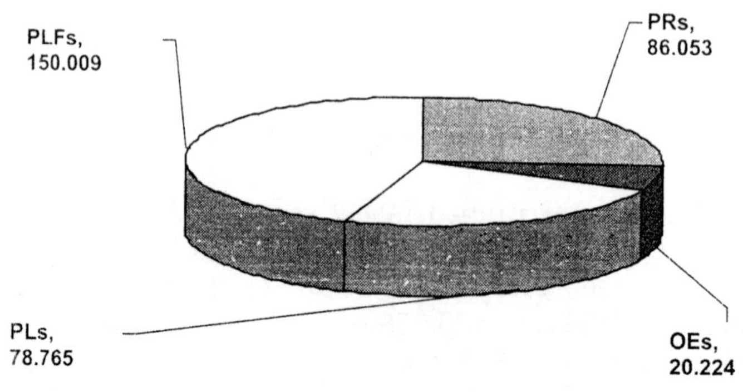

Fig. 5.3. Comparison of nuclear area (µm2) of different haemocytes of *A. mylitta*.

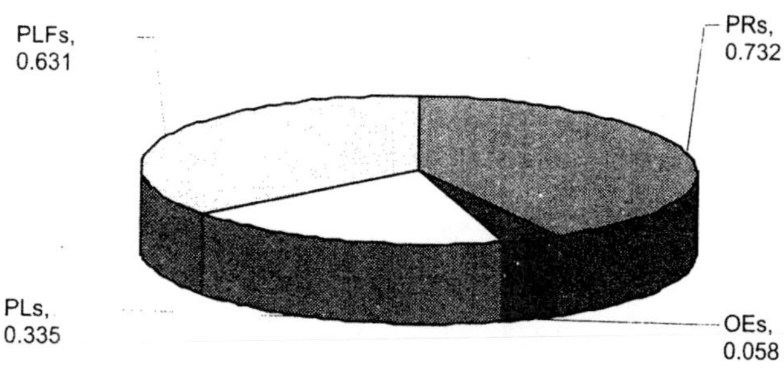

Fig 5.4. Comparison of ratio of nuclear area vs. cell area of different haemocytes of *A. mylitta*.

Based on similar morphological features, the haemocytes were classified in *A. mylitta* (Figs. 4.1. to 4.17). Five types of structurally different haemocytes have also been reported in *Antheraea pernyi*, of these, the three types of haemocytes were reported from one source, i.e., Prohaemocytes, which appear to develop by transitional stages into Plasmatocytes and Granulocytes (Beaulaton and Monpeyssin, 1976). The same case may be true in case of *A. mylitta*.

Relationship between haemocytes classes have been explained in detail by Arnold, 1952. He explained that the haemocytes, with the possible exception of the Oenocytoids, apparently develop originally from a common source, the Prohaemocytes and the development follows two diverged lines, on the one hand it results in the formation of the phagocytic cells, the Plasmatocytes and on the other hand, it produces the Spherulocytes, which are potentially but not actively phagocytic.

The Spherulocytes appear to develop from Prohaemocytes that were smaller and more compact than normal. Later, Hinks and Arnold (1977), while working on the haemocytopoiesis in Lepidoptera, suggested that haemocytes fall into two categories: (i) those derived from circulating haemocytes, mainly granular and spherule cells, and (ii) those derived from haemopoetic organs (Prohaemocytes, Plasmatocytes and Oenocytoids).

This concept may also be true in case of present study because above five types of cells were observed in the haemolymph of *A. mylitta* also. While reviewing the works on the evolution of haemocyte types, Gupta (1985) stated that one haemocyte can and does differentiate into another type, it may be possible that in taxa that are reported to possess other types besides Prohaemocytes, Plasmatocytes, and Granulocytes that the last perhaps, further differentiated into Spherulocytes, Adipocytoids, Coagulocytes, and Oenocytoids, although not necessarily in that order. Hence, nothing can be said in certainty about the origin of five types of haemocytes as observed in *A. mylitta* in present study.

In Lepidoptera, haemocytes are reported to be more extensively diversified than in most other insect orders and are strangely varied between and within the family (Arnold, 1972; Arnold and Sohi, 1974; Arnold and Hinks, 1975). The presence of haemocytes and their types differed in *A. mylitta* in present study when compared with other species of Lepidoptera as described under review of literature and in the preceding paragraphs.

Insect haemocytes are known to be capable of rapidly reducing the numbers of circulating foreign material/particles by either phagocytosis, nodule formation or encapsulation (Salt, 1970). This way they are of great help in wound repair, coagulation, melanization and immobilisation of invading organisms by way of encapsulation or phagocytosis (Ratcliffe and Rowley, 1979; Ratcliffe, 1993; Pech et al., 1995; Butt and shield, 1996; Lavine and Strand, 2002). Plasmatocytes and Granulocytes are the main haemocytes, which are involved in phagocytosis of foreign micro-organisms (Jones, 1962; Salt, 1970; Arnold, 1974; Whitcomb et al., 1974; Rowley, 1977). Gupta (1985, 1986, 1991a & b) has also explained that the haemocytic immunity in insects is accomplished by phagocytosis, encapsulation, nodule formation, immunologic factors, coagulation, and poison detoxification mechanism. He has further explained that phagocytosis is accomplished in three stages: (i) recognition for foreign body, (ii) its ingestion, and (iii) final disposal or clearance from the body. Encapsulation isolates the foreign micro-organism, if it is too large to be phagocytised. Here also the Plasmatocytes and Granulocytes are primarily involved though Oenocytoids may also act (Zachary, et al., 1975; Nappi, 1974& 1975; Pionar, 1974; Ratcliffe and Rowley, 1979).The process of encapsulation is accomplished in two stages (i) recognition of foreign body and (ii) capsule formation and melanization (Gupta, 1985). In case of nodule formation, a nodule is formed around a melanized core of coagulum, which is produced by the Granulocytes, later this contained by three layers of Plasmatocytes (Ratcliffe and Rowley, 1979). Similar functions of Plasmatocytes, Granulocytes, and Oenocytoids may also be true by the haemocytes of *A. mylitta* as it is also reared outdoor and is directly exposed to the pathogens. Comparable action of haemocytes are also reported in case of parasitisation of *Pseudaletia separata* by *Apanteles kariyai* (Tanaka, 1987); detoxification of deltamethrin in *Philosamia ricini* (Begum et al., 1998); encapsulation by haemocytes in *Pseudoplusia includens* (Pech and Strand, 1996; Loret and Strand, 1998). Recent workers have described that insect immune system is further subdivided into humoral and cellular defence responses. Humoral defences include the production of antimicrobial peptides (Meister et al., 2000; Lowenberger, 2001), reactive intermediates of oxygen or nitrogen (Bogdan et al., 2000; Vass and Nappi, 2001), and the complex enzymes cascade that regulate coagulation or melanization of haemolymph (Muta and Iwanga, 1996; Gillespie et al., 1997). Similar function of haemocytes can not be ruled out in case of A. mylitta though these may be considered as future research strategies.

5.4. Haemogram (Total and Differential Haemocyte Counts) in *Antheraea mylitta*

The haemogram is a statement of the haemocyte picture in an insect at a given time. It includes the total number of haemocytes (total haemocyte count or THC) in a standard quality of haemolymph (Jones, 1962), together with an estimate of the relative number of haemocytes in the different categories (differential haemocyte count or DHC). Differential haemocyte count is obtained by identifying suitable number of cells in a random sample (Jones, 1962). Ideally, the haemogram includes the haemolymph volume (Arnold, 1974) and the number can be expressed in a more meaningful manner by way of expressing these figures in a absolute number of each category, which may represent an estimate of the total populations as derived from total and differential counts in relation to haemolymph volume. Shapiro (1979b) has indicated that Tauber and Yeager (1934) was first to make such study in some insects; later he studied THC in some orders.

5.5. Total haemocyte count (THC)

Total haemocyte counts (THC) recorded in different life stages/ages of non-diapausing generations (NDD) of A. mylitta were 43, 620.36/mm3 in I instar (I-NDD), 46.498 in II instar (II-NDD), 57,996.10 in III instar (III-NDD), a little lower (55,955.10mm3) in IV instar, a rise in V-instar early females (69,167.96/mm3), V-instar early males (60,824.26/mm3), V-instar mid-age females (75,234.66/mm3), V instar mid-age males (77,526.36mm3), followed by a decrease in V instar late-age females (60,904.80/mm3), V instar late age males (62,259.10/mm3), spinning female larvae (56,627.20/mm3) and spinning male larvae (57,054.12/mm3). A steep rise in THC was recorded in pre-pupal stages of both the sexes: 73,290.82/mm3 in female pre-pupa and 78,265.44/mm3 in male pre-pupa. After that the THC values went down sharply to 11,543.86/mm3 in late female pupae, 11,528.32/mm3 in late male pupae and lowest in adults (female: 8,756.52/mm3 and males: 8,533.80/mm3) (Table 4.3; Figs. 5.5 & 5.6).

Total haemocyte counts (THC) during different life-stages/ages of diapausing generation (DD) of *A. mylitta* showed similar increasing trend as mean THC values were 43,370.30/mm3 in I-instar (I-DD), 45,542.40/mm3 in II-instar (II-DD), 55,536.56/mm3 in III-instar (III-DD), a slightly lowered THC values (54,990.32/mm3) in IV-instar (IV-DD) and a rise again in female (VEF-DD) to 67,363.22/mm3 and in male V instar (VEM-DD) to 62,876.82/mm3. The count rose further in V instar mid-aged female (VMF-DD) to 70,967.98/mm3 and in V instar mid-aged male (VMM-DD) to 75,097.96/mm3. In spinning stages the THC declined; in spinning female larvae (SLF-DD) the count was 55,456.72/mm3 and in male (SLM-DD) it was 56,073.16/mm3, a steep rise in THC was recorded in a pre-pupa

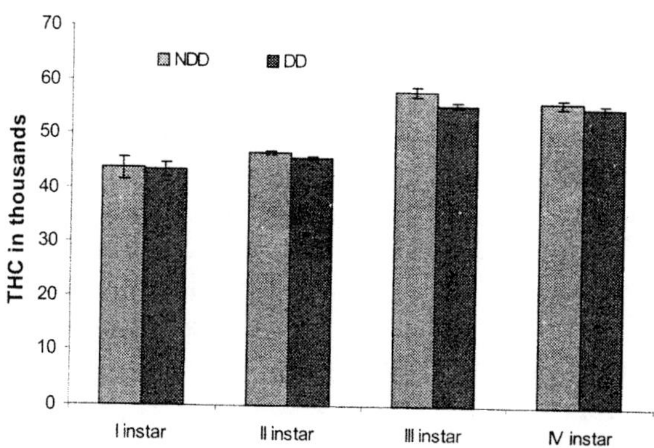

Fig. 5.5. Comparison between non-diapausing and diapausing generations for total haemocyte counts (THC) in early instars of *A. mylitta*.

female (PPF-DD) and male (PPM-DD), in former the count was 75,846.08/mm3 and in later the count was 75,678.84/mm3. Thereafter, THC values declined progressively and in adults; in female (AF-DD), it counted 8,740.84/mm3 and in males (AM-DD) their number was 8,536.56/mm3 - (Table 4.4; Figs. 5.5 & 5.6). Matrix showing significant means differences among different life-stages of both non-diapausing and diapausing generations of *A. mylitta* have been presented in Table 4.5.

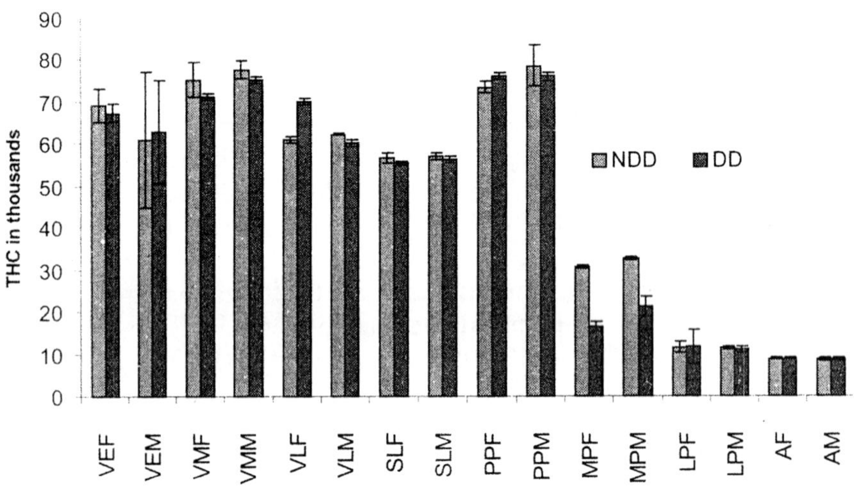

Fig. 5.6. Comparison between non-diapausing and diapausing generations of *A. mylitta* for total haemocyte counts (THC) of V instars larvae, pupae and adults.

These THC values were recorded in heat-fixed condition and heat-fixed method has been reported to give consistent results as observed in *Prodenia eridania* (Rosenberger and Jones, 1960). The THC may normally vary greatly with amount of available haemolymph, sex, stages of development and other physiological states (Arvy et al., 1949; Arnold, 1952). There may be variations in means of THC of various insects of the same stage and physiological status, which may be from 1.8 (Wheeler, 1961) to 5.3 fold (Jones and Tauber, 1951). In the present study, THC values were recorded to be 7.8 times in larvae (V-instar females), 8.37 times in pre-pupal females when compared with adult females in non-diapausing generation. Such many-fold variations were also recorded in diapausing female and male THC. Similar results were also obtained by Arnold, 1952 in the different life stages of Mediterranean flour moth, *Ephestia kuhniella*. He observed that the cells increased in numbers at a relatively constant rate during larval life and reached a peak at pupation. In case of *A. mylitta* too, during larval stage, number of haemocytes increases with the increase in larval instars in both non-diapausing and diapausing generations. He also observed a slight increase in number of cells at the time of emergence of adults. This type of increase was not found in the present study with *A.*

mylitta. Apparently, the haemocytes of *A. mylitta* are most active during larval stages. Probably they are also required during pupation for removal of histolysed tissues as their population was higher during those stages (PPF and PPM). Their small numbers in adults implied a lesser degree of activity. This is in agreement with Arnold, 1952. Wheeler (1963) further elaborated that THC significantly increased prior to ecdysis, abruptly fell at ecdysis, and remained about the same for 24 hours and these changes were related to the haemolymph volume in *Periplaneta americana*. A variation in THC counts was also reported in large milkweed bug, *Oncopeltus fasciatus* (Feir, 1964a), with gradual increase in days after ecdysis in O. fasciatus (Feir, 1979). The THC was reported to be higher in pre-adult stages of cockroaches than adults (Arnold, 1972) and similar results were found in the present study carried out in *A. mylitta*.

Previous workers also suggested that changes in THC with respect to age and development occur and it was generally higher in pre-adult stages (Tauber and Yeager, 1936), in *Sarcophaga bullata* (Jones, 1956) and in *Bombyx mori* (Nittono, 1960). Such cyclic increase and decrease in the THC was also most conspicuous at the last ecdysis in *Periplaneta americana*, too (Wheeler, 1963).

The embryonic origin of insect haemocytes is by proliferation into the epineural sinus from the ventrally located median strand of mesoderm shortly after retraction of yolk (Bronskill, 1959; Ivanova-Kasas, 1959). Post embryonically, haemocyte may arise from division of circulating cells in the haemocoel (Tauber, 1936; Yeager, 1945; Arnold, 1952; Nittono, 1960; Shapiro, 1968; Arnold and Hinks, 1976) or they may arise from their release from haemocytopoietic organs within which they multiply and differentiate (Arvy, 1956). It has been reported that in Lepidoptera and Diptera, the haemocytes are principally released just before or in early stages of pupation (Arvy, 1953, 1956). However, Shrivastava and Richards (1965) believed that there was a direct transition from germinal Prohaemocytes through Plasmatocytes to Adipohaemocytes (granular haemocytes) and other haemocytes.

The sudden rise in THC of *A. mylitta* at its pre-pupal stages in both the non-diapausing and diapausing generations may probably be due to release of haemocytes from the haemopoetic organs.

There are of course other interpretations of haemocytopoiesis in Lepidoptera and it suggests that the haemocytopoietic tissue produces haemocyte population, initially and to a large extent in the form of Prohaemocytes and Plasmatocytes (Monpeyssin and Beaulaton, 1978; Beaulaton, 1979) and these basic haemocytes are pluripotent, and the main source of other types. Similar may be true in case of *A. mylitta*.

Usually, the haemocyte are most numerous during the stage of active growth, say during larval life and their number decline just prior to pupation (Bhadur and Pathak, 1971; Arnold, 1974; Shapiro, 1979a). The increase in number of haemocytes with increase in age of pre-adult stages in *Periplaneta americana* (Wheeler, 1963); *Bombyx mori* (Nittono, 1960; Wago and Ishikawa, 1979; Balvenkatsubbaiah et al., 2001); Hyalophora cercropia L. (Lea, 1964); *Halys dentata* (Bhadur and Pathak, 1971); *Galleria mellonella* (Shapiro, 1979 a & b) and *Blattella germanica* (Hazarika and Gupta, 1989) have been reported. An increase of THC in pupal stage was also observed in *Lymantria dispar* (Butt and Shield, 1996). The results in present study are in conformity with the findings of above authors. Similar increase in THC have also been reported in *Antheraea assama* (Bardoloi and Hazarika, 1995), in *Pieris brassicae* (Bauer et al., 1998), in Rhyncoris *marginatus* (Ambrose et al., 1999), in *Philosamia ricini [Samia cynthia ricini]* (Begum et al., 1998) and progressive increase in larval instars of *A. mylitta* (Sharan et al., 2002).

The difference between the sexes of non-diapausing generation was observed to be significant ($p < 0.01$), though V instar early female had higher THC than males, trend was reversed as V instar mid-aged males had higher THC than V instar mid-aged females, males had significantly higher THC than females up to mid-pupal stage and in adult stage, adult females had higher ($p < 0.01$) THC than males (Table 4.6; Fig. 5.7). When compared the THC between the sexes of diapausing generation, the trend was like that of non-diapausing generation (Table 4.7; Fig. 5.8).

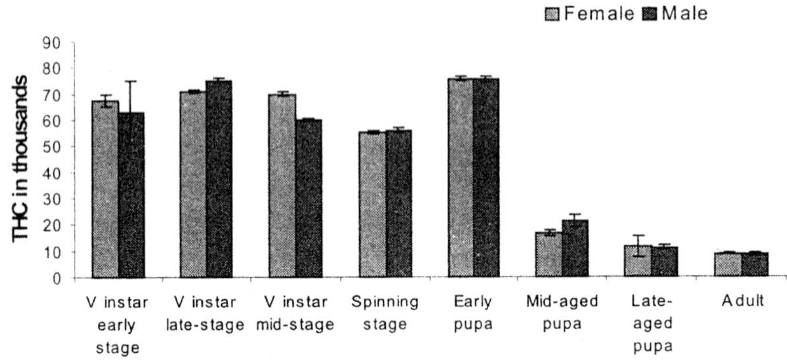

Fig. 5.7. Comparison between sexes of non-diapausing generation of *A. mylitta* for total haemocyte count (THC).

In general, variations in THC due to sexes alone are not significant (Ogel, 1955; Clark and Chadbourne, 1960; Wheeler, 1961). However, relatively minor differences have been reported in two sexes of insects (Tauber and Yeager, 1934, 1936; Lea, 1964; Hoffmann, 1967a & b; Bhadur and Pathak, 1971). In the present study, V instar early females and adult females had higher THC than males in both the generations which is in agreement with the work of Arvy et al.(1949) in adult females of *Mantis* and Webley (1951) in African migratory locust. Further the mid-aged V instar, late V instar, spinning fifth instar larvae, early pupal stage and mid-pupal stages of NDD generation and mid-aged V instar, spinning fifth instar and mid-pupal stage of DD generation had higher mean THC in males than in females which is in conformity with the observations of Smith (1938) on *Periplaneta americana*.

A comparison was also made to observe the difference of level of THC in non-diapausing and diapausing generations of *A. mylitta*. Significantly higher number of THC was recorded in non-diapausing *A. mylitta* from I to IV instar and early aged V instar female larvae.

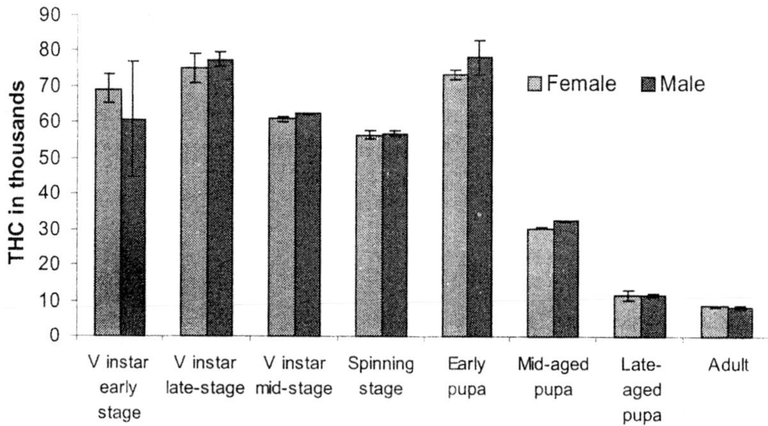

Fig. 5.8. Comparison between sexes of diapausing generation of *A. mylitta* for total haemocyte count (THC).

The difference was observed to be non-significant in mid aged V instar larvae of both the generations. The level of THC was observed to be sometimes high in non- diapausing generation during late V instar and spinning larvae, different aged pupae and sometimes THC was recorded to be higher in above stages in diapausing generation in both the sexes. No marked differentiation was observed in adults of both the sexes of non-diapausing and diapausing generations.

The decrease in THC during diapause have been reported in *Sesamia cretica* which comes to normal level like non-diapausing individuals in adult stage after termination of diapause (EL Mandarawy, 1997). Raina and Bell (1974) have also observed a significant reduction in the level of THC in diapause state, which comes to normal level upon termination of diapause in *Pectinophera gossypiella* Saunders. Similar results were not obtained in case of *A. mylitta,* whereas in mid-aged diapausing pupae representing the true diapause state had higher THC level once compared with respective stage of non-diapausing pupae (Table 4.5), though after termination of diapause, there was no significant difference in THC value of adult females and males of both the sexes and generations. Contrary to this, no significant differences were observed between counts in non-diapausing and diapausing larvae and pupae, or between the sexes within the each group in the pink bollworm, *Pectinophora gossypiella* Saunders (Clark and Chadbourne, 1960), even though there was a change in the number of categories.

Changes in population of haemocytes are also reported with increase in instar or further advanced life stages in insects, such as in *Dysdercus koenigii* (Saxena and Tikku, 1990); in *Lymantria dispar* (Kim et al., 1990 a & b); in Diatraea saccharalis (Strand and Noda, 1991); in bacteria infected larvae of *Mythimia separata* (Sha and Xie, 1992); in

ß-ecdysone treated Dysdercus cingulatus in relation to eclosion, sex and mating; in *Spilostethus hospes* (Sanjayan et al., 1996), in microsporidian infected *Galleria mellonella* (Tonka, 1997); in different instars and further advance stages of *Cotesia glomerata* (Bauer et al., 1998) and in parasitized *Spodoptera littoralis* (Hegazi et al., 1998).

5.6. Absolute number of haemocytes in relation with body weight and haemolymph volume

In the present study mean body weight of *A. mylitta* larvae, pupae and early-emerged adults were taken out in both of its non-diapausing and diapausing generations. The weight of the larvae gradually increased with advancing instars till V instar larval stage, decease in weight was observed in spinning larvae of both the sexes in both the generations. Weight showed a decreasing trend till emergence of adults. Female pupae and adults had higher weight than males (Figs. 5.9 & 5.10).

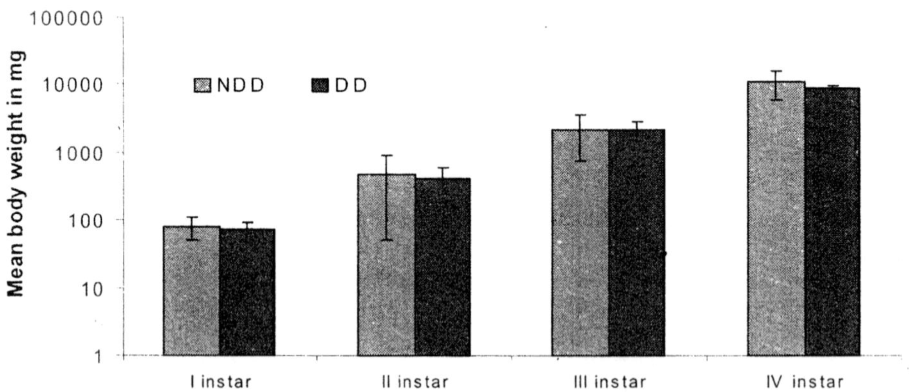

Fig 5.9. Comparison between non-diapausing and diapausing generations of *A. mylitta* for mean body weights in early instars.

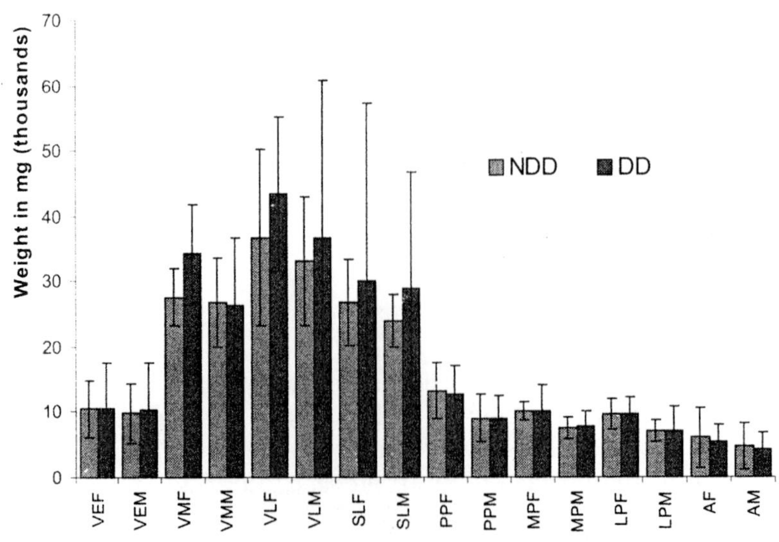

Fig 5.10. Comparison between non-diapausing and diapausing generations of *A. mylitta* for mean body weights of V instar, pupae and adults.

The volume of the haemolymph varies widely according to age and developmental stages in insects (Florkin and Jeuniaux, 1974). In the present study in A. mylitta, the volume in I instar larvae of non-diapausing generation was higher than I instar larvae of diapausing generation. This trend was observed up to IV instar (fig. 5.11). Mid-aged V instar female larvae and late-aged V instar larvae in diapausing generation had higher volume than their non-diapausing counter parts. Spinning larvae of both non-diapausing and diapausing generations had almost equal volume in both the sexes. The difference in volume was non-significant in pupal stages of both the sexes of both the generations. However, non-diapausing female had slightly higher haemolymph volume than diapausing females (Tables 4.8. & 4.9; Fig 5.12).

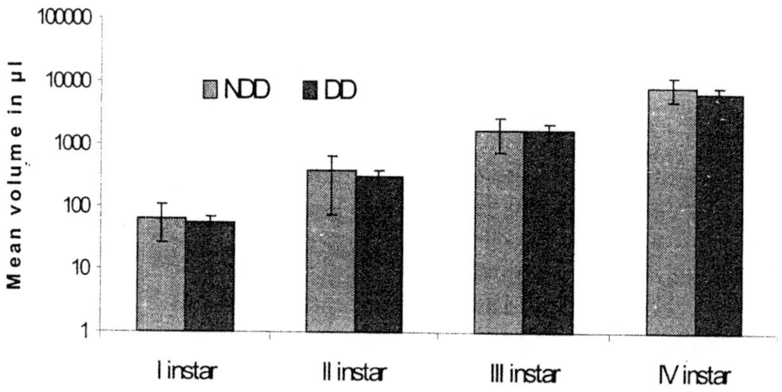

Fig. 5.11. Comparison between non-diapausing and diapausing generations of *A. mylitta* for mean haemolymph volume of early instars.

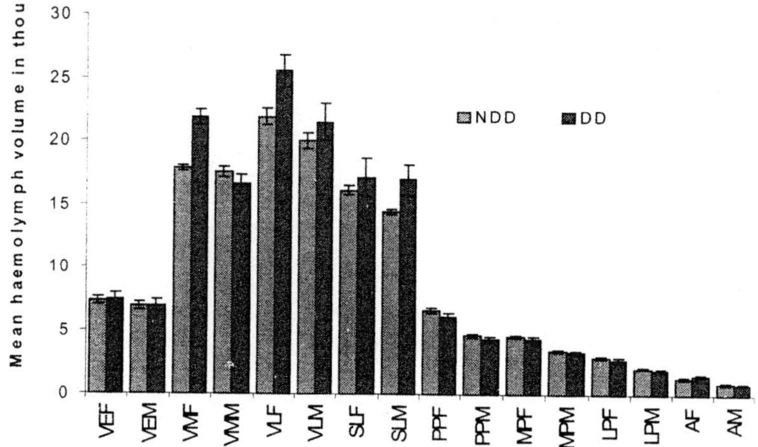

Fig. 5.12. Comparison between non-diapausing and diapausing generations of A. mylitta for mean haemolymph volume of V instar larvae, pupae and adults.

Within the sexes, male larvae, pupae and adults had significantly ($p < 0.01$) less volume than females. This trend was observed to be true in case of life-stages of diapausing generations too (Tables 4.8. & 4.9; Figs. 5.13. & 5.14). The volume of haemolymph increased with increase of larval instars in *A. mylitta* and this trend is in conformity with the observations made after Florkin and Jeuniaux (1974). In Lepidoptera, the abundant haemolymph of pupa is utilised and almost none remains in adults (Heller, 1932; Florkin, 1937a & b). Florkin (1937a & b) investigated changes in haemolymph volume of *Bombyx mori* from the start of spinning to eclosion. Up to eleventh day after spinning, the haemolymph volume was static, thereafter it declined and remained static and after eclosion further reduction of haemolymph volume occurred. Florkin and Jeuniaux, (1974) observed that the haemolymph volume increases with respect to the dry weight of the body; during the end of every larval instar. The resulting increased hydrostatic pressures help in rupturing of old cuticle and expansion of new integument prior to its hardening. A short time after ecdysis, the haemolymph volume decreases (Loughton and Tobe, 1969). The present study supports the above view because early V instar larvae had less volume than IV instar larvae, pupae also had lesser volume than spinning larvae, adults had further less volume than pupae of both non-diapausing and diapausing generations. During the development of *Sarcophaga,* the haemolymph volume was also found to be decreasing (Jones, 1956, 1967b). The haemolymph volume was expressed as a percent of body weight (Jones, 1956) or as volume (in microliter, µl) (Jones, 1967b). The haemolymph volume increased in *Euxoa* during larval development (Arnold and Hinks, 1976), age-dependent changes in *Schistocerca gregaria* (Lee, 1961). The present study in *A. mylitta* also supports the view of above authors.

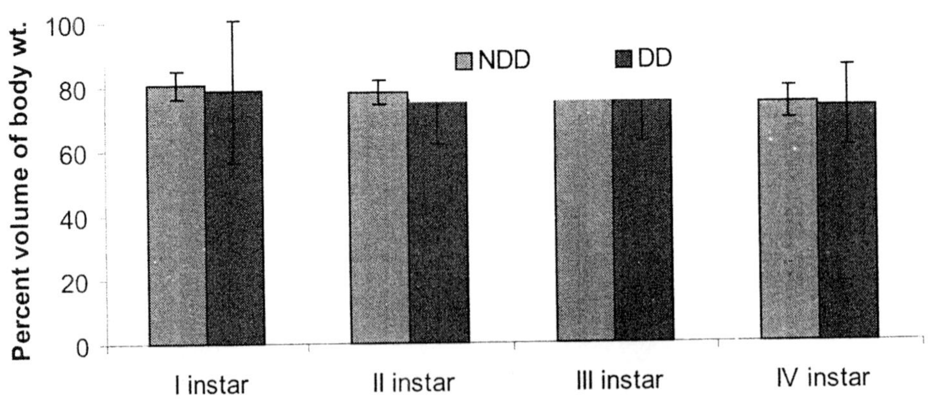

Fig. 5.13. Comparison between non-diapausing and diapausing generations of *A. mylitta* for mean haemolymph volume as percent of body weight of early instars.

In case of *A. mylitta*, the percent volume against body weight was higher from larval to late pupal stage in non-diapausing generation when compared with the body weight to volume ratio in diapausing generation. The percent volume was higher in adult females and males of diapausing generation than non-diapausing males and females of *A. mylitta.*

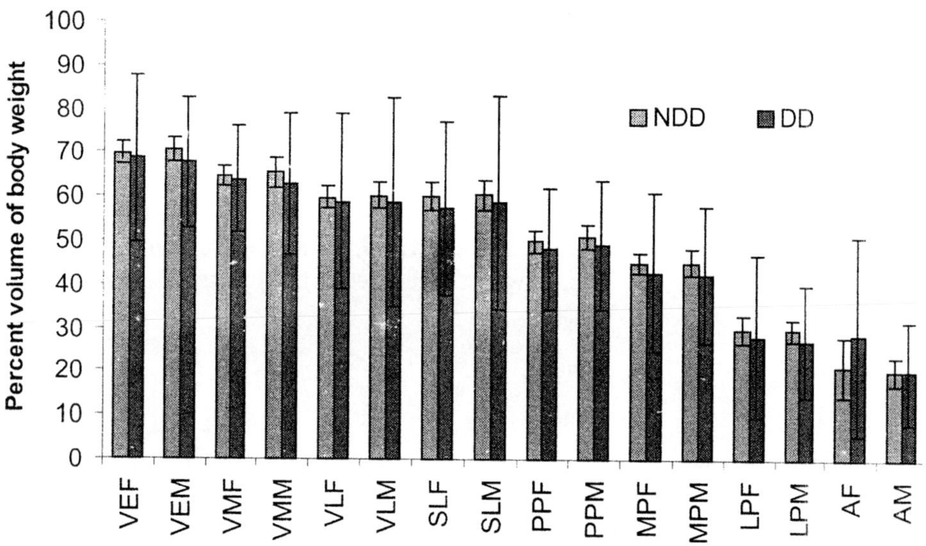

Fig. 5.14. Comparison between non-diapausing and diapausing generations of *A. mylitta* for mean haemolymph volume as percent of body weight of V instar larvae, pupae and adults.

Wheeler (1963) observed that inter moult larvae had higher percent volume in *Periplaneta americana* when compared with the adults. Although, in *Galleria*, the haemolymph volume (as percent body weight) did not change significantly in VII instar larvae, but the amount of haemolymph increased as they gained weight (Shapiro, 1968). Jones (1967b) found that haemolymph volume (as percent body weight) remained unchanged (34 %) during feeding period and gradually decreased in newly formed pupae (16 %). The same trend was also observed in case of *A. mylitta* in the present study. The variation in haemolymph volume percentage in similar age life-stages in insects are also reported (Richardson et al., 1931; Zielinska and Wroniszewska, 1957). The percent volume range in the similar stage *A. mylitta* life stages also varied considerably, the range has been shown in (Tables 4.8. & 4.9). These results are in conformity with the findings of the above authors.

Results are also there that haemolymph volume percentage decreases as a result of starvation (Wharton et al., 1965) and prolonged withholding of food and water from larvae of American roaches failed to change haemolymph volume percent (Buxton, 1930); and in other beetles (Fraenkel and Blewett, 1944). Insect haemolymph volume vary from 1 to 20,000 µl and it is a major water storage compartment (Mullin, 1985). It was observed that in normally hydrated adult Locusta, haemolymph contains about 20 % of body weight (Burton et al., 1972). Haemolymph volumes may fluctuate in response to many activities or conditions, such as due to feeding (Edney, 1977; Bernays and Chapman, 1974), flight (Gringorten and Friend, 1979), starvation (Wharton, 1965); moulting (Nicolson, 1980) and desiccation (Nicolson et al., 1974). In certain studies, the endocrine extracts effect the haemolymph volume (Pathak, 1991) or hormones directly or indirectly influence the haemolymph volume, thereafter, may affect the haemocyte count (Feir, 1979; Pathak, 1983, 1986). Effect of feeding on different host plants is also reported to affect haemolymph volume in *Antheraea assama* (Hazarika et al., 1994).

The changes in haemocyte populations and haemocyte types combining the total haemocyte count (THC), differential haemocyte count (DHC) and haemolymph volume, the number of haemocytes can be approximated in different developmental stages of insects (Shapiro, 1968; Arnold and Hinks, 1976). The absolute number of haemocytes can be calculated based on THC and haemolymph volume (Smith, 1938; Wheeler, 1963; Jones, 1967a & b; Shapiro, 1968; Arnold and Hinks, 1976). The absolute number of haemocytes in different larval instars of *A. mylitta* was calculated and is presented in Tables 4.8. & 4.9 and Figs 5.15 & 5.16.

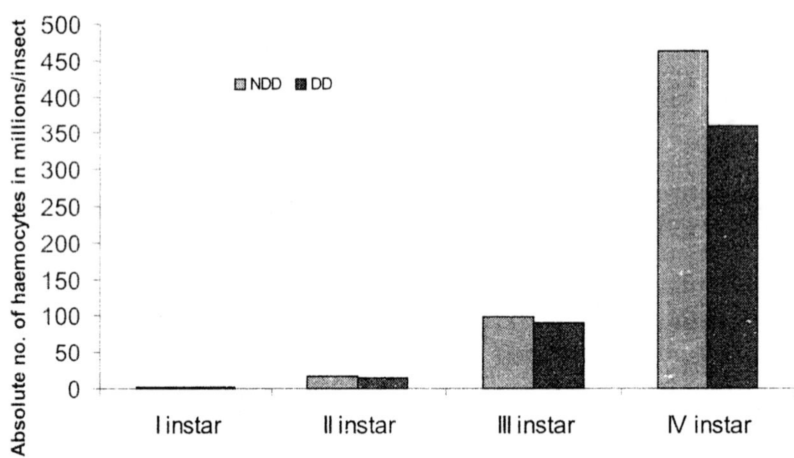

Fig. 5.15. Comparison between non-diapausing and diapausing generations for absolute haemocyte count of early instars of *A. mylitta*.

It may be seen from the table that up to fourth instar, the absolute number of haemocytes were higher in non-diapausing than their respective stages of diapausing generation. Thereafter, absolute number of haemocytes increased in early fifth instar male larvae of diapausing generation when compared with non-diapausing generation. Though, a definite trend of higher or lower absolute number of haemocytes was not observed between the two generations during fifth instar larval stages of the sexes and generations. However, the absolute number was higher in adult females of diapausing generation and reverse was the trend in case of adult males. Absolute number of haemocytes nearly halved in newly eclosed adult females and males when compared with pupal stages. This study supports the view of other workers that the bulk amount of haemolymph last after emergence/ eclosion of adult (Heller, 1932; Florkin, 1937a; Webley, 1951; Lee, 1961; Wheeler, 1963; Florkin and Jeuniaux, 1974). Thus, it appears that in different life-stages of *A. mylitta*, the level of THC, DHC and the absolute number of haemocytes in haemolymph does not remain constant in spite of changes in haemolymph volume. The haemocytes/haemogram of *A. mylitta* changes quantitatively and qualitatively during growth, development, and a new vista is open to further analyse the factors involved in these changes.

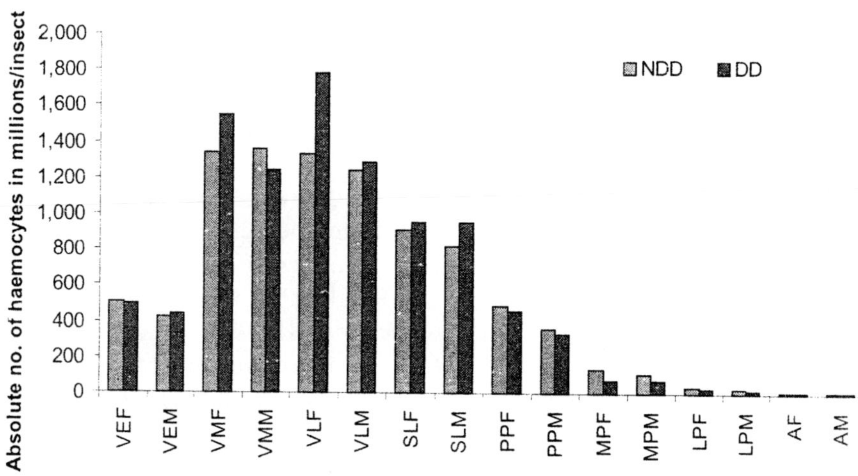

Fig. 5.16. Comparison between non-diapausing and diapausing generations for absolute haemocyte count of V instar larvae, pupae and adults of *A. mylitta*.

5.7. Differential haemocyte count (DHC)

Differential haemocyte count (DHC) or the proportion of different haemocyte types was also determined in the larval, pupal and adult stages of *A. mylitta* in both its non-diapausing and diapausing generations. The detail mean data are shown in Tables 4.10 to 4.33.

In non-diapausing generation, I instar larva showed highest percentage of Granulocytes (51.066) followed by Plasmatocytes (30.766), Prohaemocytes (8.166), Spherulocytes (7.300) and Oenocytoids (0.566). In II instar the order of different types of haemocytes were: Plasmatocytes (46.233) > Granulocytes (28.466) > Spherulocytes (16.633) > Prohaemocytes (5.833) > Oenocytoids (1.233). In III instar, the level of Granulocytes (34.724) and Plasmatocytes (35.631) was almost equal, Oenocytoids (0.933), Prohaemocytes (8.532) and Spherulocytes were slightly higher in percentage (19.898) than Prohaemocytes. In IV instar, haemocytes mainly comprised of Granulocytes (39.665), Plasmatocytes (30.027) and Spherulocytes (20.063). In V instar females in their early stages among three major groups of haemocytes the levels of order were Granulocytes - 57.600, Plasmatocytes - 22.333, Spherulocytes - 14.866, Prohaemocytes - 3.367 and Oenocytoids - 1.633 and this order continued till spinning larval stage SLF: GR - 51.966; PL - 27.833; SP - 15.433. Same trend was found from V instar early males till spinning larvae. Although, Prohaemocyte was at their low level in spinning stage (PR in SLF - 0.733; SLM - 0.766), Oenocytoids were at slightly higher level (SLF - 3.800; SLM - 3.266). Up to late pupal stage, major proportion of haemocytes was of Plasmatocytes, Granulocytes and Spherulocytes (LPF: SP - 12.600;

GR - 34.533; PL - 26.433 & LPM: SP - 9.166, lower than female; GR - 41.233, higher than female; PL - 34.500, higher than female). However, in adults Granulocytes were highest followed by Plasmatocytes and Spherulocytes (AF-NDD: GR - 53.966; PL - 36.966; SP - 5.333 & AM-NDD: GR - 49.333; PL - 46.633; SP - 3.333. In mid-pupal stage of non-diapausing generation, 1/3rd proportion of haemocytes were of degenerated cells (MPF - 30.200; MPM - 32.600) and in late pupal stage, it ranged in between 32.600 (MPM - NDD) to 30.200 (MPF - NDD) - (Tables 4.10. to 4.33; Figs. 5.17. & 5.18).

Differential haemocyte counts (DHC) of different larval instars, pupal and adult stages of diapausing (DD) generation also indicated that first instar larvae contained Granulocytes, Plasmatocytes and Spherulocytes (GR - 50.633, PL - 30.533 & SP - 8.800) though next higher were Prohaemocytes (7.566). In II instar, Granulocytes came down with increase in Spherulocytes and Plasmatocytes. In III instar Prohaemocytes percent went up to 10.433, this was highest among all the life stages. The number of Granulocytes again rose up (51.333) followed by Plasmatocytes (25.900), Spherulocytes (15.466) in VEF-DD. Males (VEM-DD) had Granulocytes (58.100), Plasmatocytes (19.667), Spherulocytes (14.200), Oenocytoids (2.500) and Prohaemocytes (2.300). In mid-aged larval stage, male had higher Granulocytes (50.300) than females (47.366) and higher percentage of Granulocytes was recorded in pre-pupal stage (PPF - 52.300; PPM - 52.833). In adult stages, major proportion of haemocytes was contained by Granulocytes (AF - 50.100; AM - 52.867) and Plasmatocytes (AF - 41.367; AM - 39.300). Mitotic cells were at their peak in I instar diapausing generation (1.633) and in prepupal (PPF-DD -3.333 and PPM-DD-3.033) and late-pupal stage (LPF-DD: 2.033; LPM-DD: 1.000) - (Tables 4.10. to 4.33; Figs. 5.17. & 5.18).

Around one third proportion of haemocytes was degenerated cells which were observed in mid-pupal stage (MPF-DD: 30.333; MPM-DD: 29.366) and one tenth in late pupal stage (LPF - 9.666; LPM - 9.866) - (Tables 4.10. to 4.33).

In *Forficula aricularia*, the Prohaemocytes were higher in early stage of larval development and thereafter a rise was noticed (Arvy and Lhoste, 1946). During larval-life, the Prohaemocytes steadily decrease, while Plasmatocytes steadily increase, Spherulocytes attained a maximum in the third larval stage in *Leptinotarsa decemlineata* (Arvy et al., 1948). In *Sarcophaga* larva, Plasmatocytes pre-dominate (Jones, 1956). As the larvae develop, the Plasmatocytes differentially decrease and the Granulocytes increase. In last stage larvae, the spherule cells increase and they were reported to filled with tyrosinase, thereafter, a decrease in their number was reported in *Sarcophaga bullata* (Jones, 1956). This increase may be due to the sudden release of differentiated cells from compact haemocytopoietic organs or to the re-entrance of previously adhering haemocytes into circulation; it was not due to mitosis of circulating haemocytes (Jones, 1956). Eventually, both granular haemocytes and spherule cells disappear and apparently only Plasmatocytes were found in the newly emerged adults (Yeager, 1945). In Mediterranean floor moth, *Ephestia kuhniella*, the more than 90 % of the total types were accounted for Plasmatocytes and Spherulocytes. Prohaemocytes comprised relatively fraction in all larval stages except the first larval stage in *Ephestia kuhniella* (Arnold, 1952). The same was true in case of *A. mylitta* in the present study, though their percentage had two peaks: first in I instar and second in III instar, in both non-diapausing and diapausing generations. As reported by

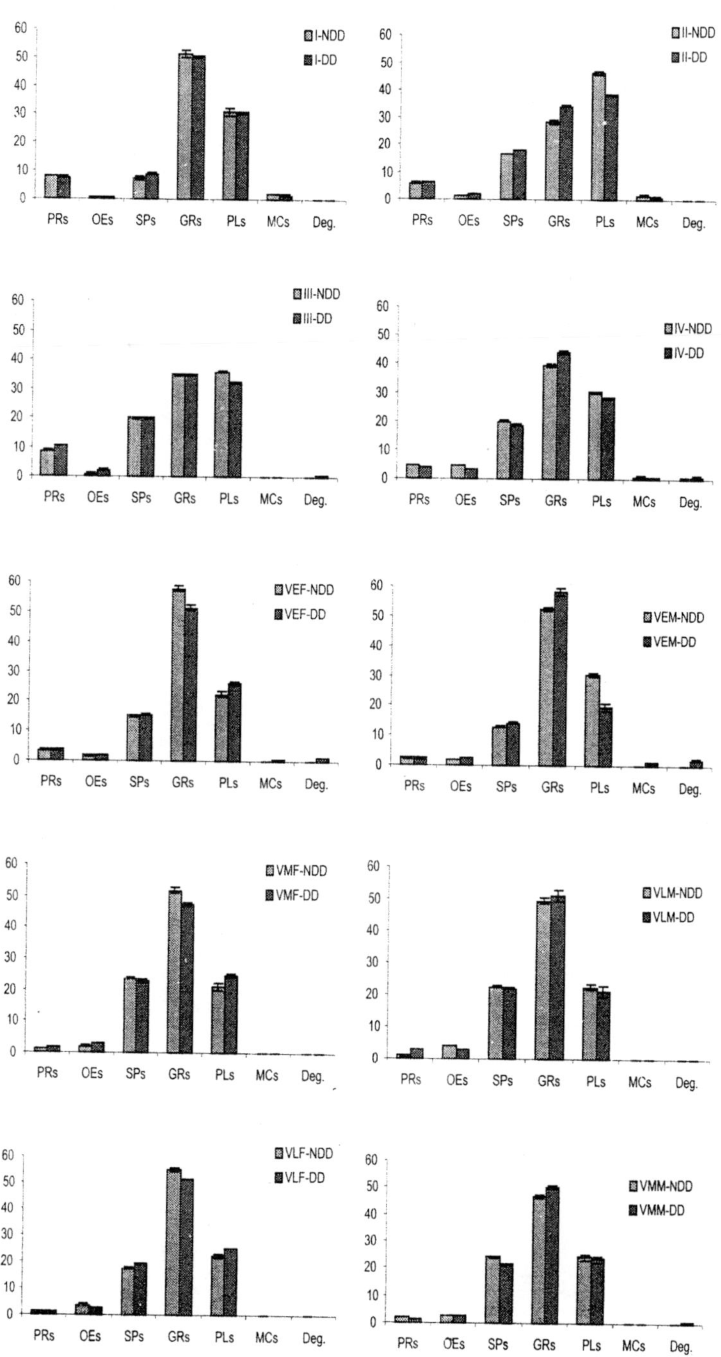

Fig. 5. 17. Comparison between percentage (y-axis) of different haemocyte all larval stages (except spinning stage) of non-diapausing and diapausing generations of *A. mylitta*.

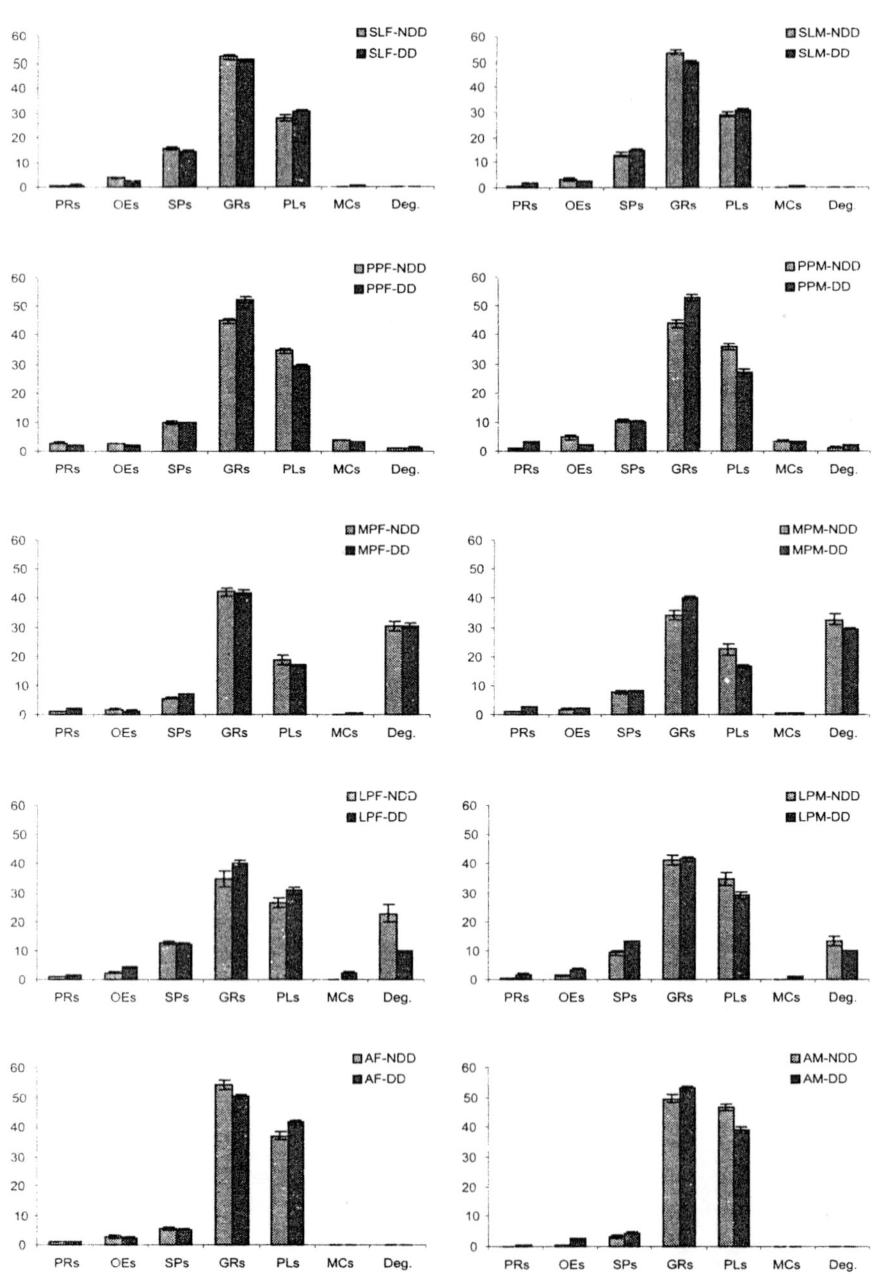

Fig. 5.18. Comparison between percentage (y-axis) of different haemocytes of spinning larvae, pupae and adult stages non-diapausing and diapausing generations of A. mylitta.

above authors, the major proportion of haemolymph cells was comprised of Spherulocytes and Plasmatocytes in all the stages of *A. mylitta* also. The findings in *A. mylitta* support, to some extent, the view of above authors. The Spherulocytes formed approximately the numerical reciprocal of the Plasmatocytes (Arnold, 1952). The same was true in case of *A. mylitta* too. With decrease in number of Granulocytes, the number of Plasmatocytes increased or vice-versa case was observed in case of *A. mylitta*. The number of degenerated cells were high in stages prior to emergence of adults (Arnold, 1952) and it was equally true in case of *A. mylitta* in the present study, since highest percentage of degenerated cells were observed in mid- and late-pupal stages of *A. mylitta* in both non-diapausing and diapausing generations. Nittono (1960), while working on DHC of *Bombyx mori* L. indicated that - (a) Prohaemocytes and Spherulocytes were highest in percentage on the 1st day after ecdysis in both larval and pupal stages; (b) Plasmatocytes are higher in percentage near later half of the each larval and pupal stages; (c) Granulocytes were highest in proportion in the later half of each stadium, decreasing at the end of pupal life; (d) Oenocytoids were highest at the beginning of the fourth and fifth larval stages, decreasing before pupation. In the present study too, the Granulocytes were highest in the later half of the fifth instar larvae, though this trend was not observed in pupal stages. The trend of Oenocytoids population percent was also found to be decreasing in A. mylitta at the time of pupation, particularly in mid-pupal stage. A comparative study of the haemocytes (haemolymph cells) of cockroaches was done by Arnold (1972). He observed that Granulocytes dominate the picture in both nymphal and adult haemolymph. In case of *A. mylitta* too, highest proportion of Spherulocytes was observed in adults of both sexes in both non-diapausing and diapausing generations (Tables 4.14 to 4.16). Granulocytes were also observed to be highest in *Malcosoma disstria* (Arnold and Sohi, 1974). Arnold and Hinks (1975) inferred that Granulocytes and Spherulocytes are the only classes of haemocytes that divide frequently enough to account for their continued increase in number in *Euxoa decalarata*. This may be true in case of *A. mylitta* too, because their percentage dominated in both non-diapausing and diapausing generations.

Mitotic division of circulating Prohaemocytes and Granulocytes has been reported for a number of different species in Lepidoptera, Arnold (1952) for *Anagasta* (*Ephestia kühniella*), Lea and Gilbert (1966) for *Hyalophora cercropia*, Shapiro (1968) for *G. mellonella*, Arnold and Hinks (1976) for *E. decalarata*. Arnold & Hinks (1976) and Hinks & Arnold (1977) have demonstrated that dual origin of haemocytes takes place in insect as discussed in previous pages. Among all these haemocytes, only Prohaemocytes divide in circulation to become Plasmatocytes and side by side, liberation of haemocytes from haemocytopoietic organs are responsible for increase in number of different classes of haemocytes. The mitotic cells well observed in the haemolymph of *A. mylitta* and the same were recorded to be highest in pre-pupal stages indicating thereby the release of differentiated haemocytes and further increase in their numbers due to mitotic division. Major proportion of Granulocytes was also reported in haemolymph of *Manduca sexta* (Horhov and Dunn, 1982), Adipohaemocytes in *Galleria mellonella* (Lea, 1986), Plasmatocytes and Granulocytes in *Halys dentata* (Pathak, 1991), Plasmatocytes and Granulocytes in *Lymantria dispar* (Butt and Shield, 1996).

In larval stage Lepidoptera Granulocytes and Plasmatocytes are the only haemocytes capable of adhering to foreign surfaces, and together only comprise more than 50 % of haemocytes in circulation (Lackie, 1988 a & b; Ratcliffe, 1993; Strand and Pech, 1995). Plasmatocytes and Granulocytes were also observed to be in major proportion in larval, pre-pupal and pupal stages of *Lymantria dispar* (Kim et al., 1990b; Butt and Shield, 1996); in adult *Dysdercus koenigii* (Sharma et al., 1998); increasing number of Granulocytes and fluctuating Plasmatocytes in *Rhynocoris marginatus* (Ambrose et al., 1999); in *Philosamia* ricini (Begum et al., 1994; 1998); in *Spodoptera litura* (Kurihara et al., 1992a & b); in muga silkworm *Antheraea assama* (Bardoloi and Hazarika, 1995); in Spilostethus hospes (Sanjayan et al., 1996); only Granulocytes in *Rhynocoris marginatus* (Ambrose et al., 1996b) and in *Perigallia ricini* (Jeyakumar et al., 1995). The results observed in the present study in *A. mylitta* are in conformity with the findings of above workers. Though, Jones, 1956 have reported that in newly emerged adults of *Sarcophaga* bullata Spherulocytes disappeared and apparently only Plasmatocytes were found. This was not true in case of A. mylitta as both Granulocytes and Plasmatocytes were recorded to be highest in numbers in adults. As far as the function of Granulocytes and Plasmatocytes are concerned, they help the process of encapsulation in lepidopterans (Wiegand et al., 2000) and in others Granulocytes are the first cells to bind to the target and Plasmatocytes attaching thereafter (Sato et al., 1976; Schmit and Ratcliffe, 1977, 1978; Lackie et al., 1985; Pech and Strand, 1996).

When compared the DHC of non-diapausing and diapausing generations of *A. mylitta,* it was observed that trend was not very fixed as sometimes in pre-adult stages level of Prohaemocytes was higher in many larval stages in diapausing generation than non-diapausing. In diapausing generation females had higher Plasmatocytes than males in pre-pupae, late pupae and adults (Tables 4.32 & 4.33).

Oenocytoids were generally higher in percentage in mid-aged V instar of diapausing generation, and lower in non-diapausing generation and in late pupae females were higher than males in diapausing and non-diapausing generation both and adults had higher Oenocytoids in females of NDD (Tables 4.11 to 4.13).

Granulocytes were higher in percentages in diapausing generation I, II & IV instars than their respective stages of non-diapausing generation. Early V instar larval stage of NDD females had higher percentage than females of DD. Mid-pupae had comparatively less percentage of Spherulocytes of both the sexes of non-diapausing and diapausing generations. Late female pupae of non-diapausing generation had higher percentage of Spherulocytes than diapausing generation, and reverse was the case in males though in adults: AF-NDD and AM-DD had higher percentage than AM-NDD and AF-NDD (Tables 4.15 to 4.17).

Plasmatocytes were more in non-diapausing generation than diapausing generation up to IV larval stage of A. mylitta. Further, up to late pupal stage, Plasmatocytes were higher in non-diapausing generation than diapausing generation, except in VEF-NDD, VMF-NDD and VLF-NDD. The adult females of non-diapausing generation had lower Plasmatocytes than AF-DD and the result was just opposite in case of adult males (Tables 4.18 to 4.20).

Similarly, when compared the presence of mitotic cells in non-diapausing and diapausing generation, the trend was fluctuating as sometimes, it was higher in non-diapausing generation and sometimes in diapausing generation. In some cases non-significant differences were observed (Tables 4.21 to 4.23).

Degenerated cells were generally higher in non-diapausing generation than diapausing and were recorded to be highest in mid pupal and late pupal stage (Tables 4.22 to 4.24).

When compared the presence of different haemocytes in between the sexes of non-diapausing and diapausing generations separately, the trend was quite fluctuating. However, in non-diapausing generation, the Prohaemocytes, Oenocytoids and Spherulocytes were higher in late-pupae and adult females than male pupae and females. In case of diapausing generation Prohaemocytes, Oenocytoids and Spherulocytes were higher in male late pupae and adults than female late pupae and adults (Tables 4.25 to 4.30; Figs 5.19. & 5.20).

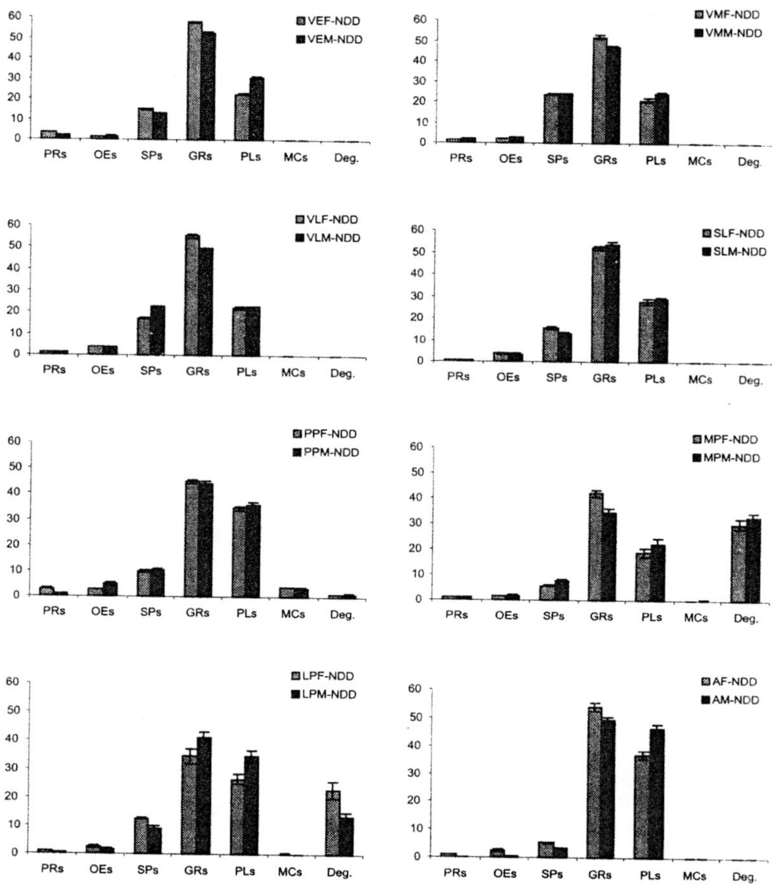

Fig. 5.19. Comparison between sexes for percentage (y-axis) of different haemocytes of V instar larvae, pupae and adults of non-diapausing generation of *A. mylitta*.

Granulocytes were higher in late male pupae than females and in adults; the trend was opposite to it in non-diapausing generation. In case of diapausing generation, late pupae and adult males had higher Granulocytes than female pupae and adults (Table 4.28).

Plasmatocytes were higher in late pupae and adult males of non-diapausing generation, and in diapausing generation, adult females had higher Plasmatocytes than males (Table 4.29).

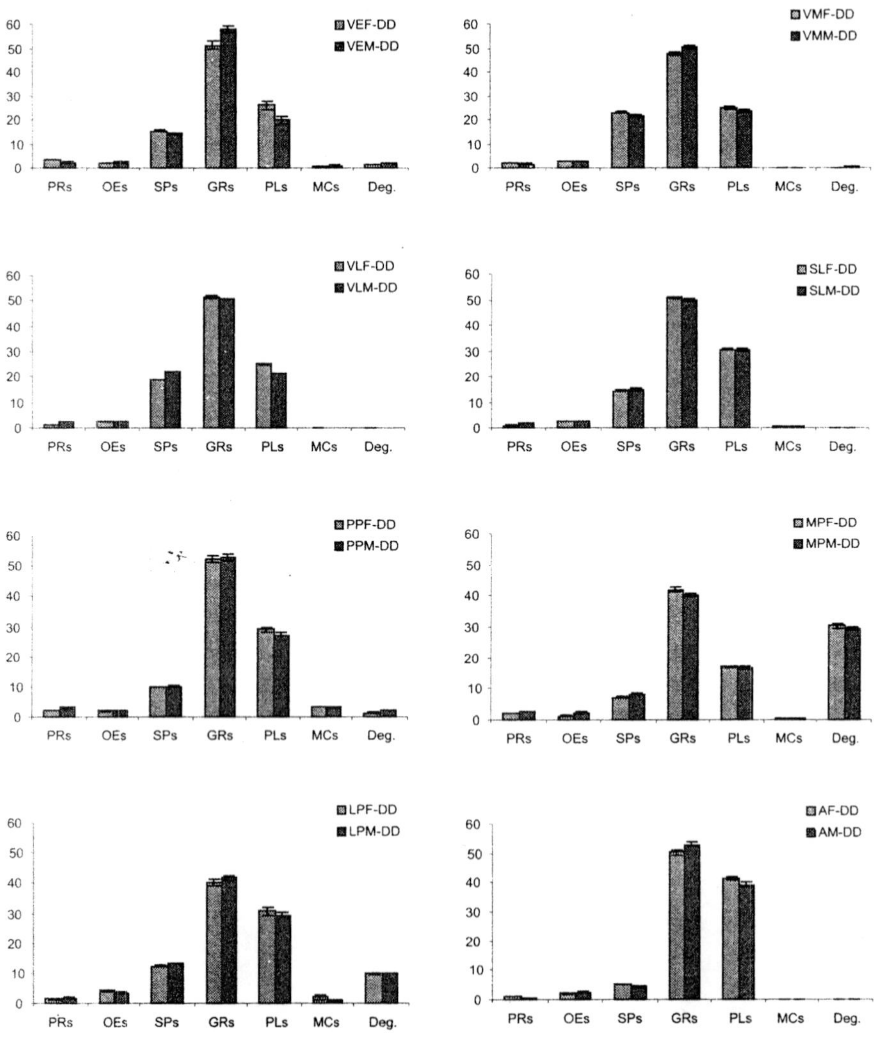

Fig. 5.20. Comparison between sexes for percentage (y-axis) of different haemocytes of V instar larvae, pupae and adults of diapausing generation of *A. mylitta*.

In mitotic cells percentage, no marked difference was recorded in between sexes of non-diapausing generation. The same was true in diapausing generation too, except in early, mid fifth instar larvae and mid-pupal stage (Table 4.30).

The differences in the mean percentage of degenerated cells also were non-significant in non-diapausing generation between sexes except in late-pupal stage, where in females, it was higher than males. In diapausing generation too, the difference of mean percentage was non-significant in advance stages of development and even in adult females and males (Table 4.31).

Haemogram changes markedly during the developmental stages (Jones. 1965; Hoffmann, 1970; Bhadur and Pathak, 1971). In holometabolous insects, the Plasmatocytes account for most increases during the larval growth and granular haemocytes seem to account for the peak in number prior to pupation (Arnold, 1974). Present observations in A. mylitta are in conformity with the observation of Arnold, 1974. When compared between the sexes, the counts for females may be higher than those for males during the insect's development and tend to remain higher during adult life (Arnold, 1974). Relatively, minor differences have been reported in the two sexes of some insects (Tauber and Yeager, 1936; Lea, 1964; Hoffmann, 1967a & b; Bhadur and Pathak, 1971). No significant differences were observed in DHC of non-diapausing and diapausing Pectinophora gossypiella (Clark and Chadbourne, 1960; Raina and Bell, 1974).

In case of A. mylitta the difference were significant in all the classes of haemocytes of non-diapausing and diapausing generations as explained in preceding paragraphs. This may be attributed to the demand and functional activities of different haemocytes in various life-stages and their level of development. In case of Sesamia cretica, during diapause, it was observed to be a significant reduction in the numbers of all the haemocytes, upon termination of diapause, the haemocytes level increased (El Mandarawy, 1997). However, in the present study, this reduction was observed only in the case of A. mylitta diapausing pupae i.e., mid-pupal stage, though it came to normal level like non-diapausing generation after termination of diapause.

Various fluctuation in DHC of haemocytes are also reported because of formation of basement membrane during wound healing in diapausing Hyalophora cecropia pupae (Clark and Harvey, 1964) and after ecdysis (Wigglesworth, 1955); Granulocytes are chief phagocytic cells in Pieris rapae crucivora Boisduval (Takada and Kitano, 1971), they increase in pharate pupal stage when reorganisation of tissues occurs (Raina and Bell, 1974). Spherulocytes show a positive correlation with silk production (Nittono, 1960; Lea, 1964).

Oenocytoids were relatively low in all the stages of non-diapausing and diapausing generations, than other haemocytes in a.mylitta which may be due to their greater adhesiveness which supports the view of Neuwirth (1973).

The decrease in the level of THC and Plasmatocytes at the on-set of diapause and a subsequent increase at the time of diapause termination (late pupal and adult stages) essentially follows the pattern of oxygen consumption that was shown to occur in various holometabolous species during diapause and development (Slama, 1960; Harvey, 1962; Raina and Bell, 1974). Spherulocytes have been suggested to transport cuticular components (Sass et al., 1994), while Oenocytoids contain cytoplasmic phenoloxidases precursors that likely play a role in melanization of haemolymph (Ashida and Dhoke, 1980; Iwama and Rashida, 1986; Jiang et al., 1997). In Lepidoptera, the granular cells and Plasmatocytes are the only haemocyte types reported to be phagocytic (Lavine and Strand, 2001, 2002). Similar roles may perhaps have been played by Granulocytes and Plasmatocytes in A. mylitta.

5.8. Histochemistry of *A. mylitta* haemocytes

Haemolymph performs several functions in insects. It serves as the transport milieu for the exchange of essential materials. The present study was restricted to localisation of general protein, bound lipids, nucleic acids, glycogens (carbohydrates), alkaline phosphatases activity in the haemocytes of larvae, pupae and adults of *A. mylitta*. The detail results of these tests are presented in Tables 4.34 to 4.54).

General proteins were observed to be localised more in the nucleus of Prohaemocytes, Spherulocytes, Granulocytes and Plasmatocytes of *A. mylitta*. Cytoplasm contained little lower concentration of general proteins of these haemocytes than their nucleus. The concentration was almost alike in nucleus and cytoplasm of Granulocytes, Spherulocytes and Plasmatocytes from spinning larvae till emergence of adults. Prohaemocytes cytoplasm had lower protein reaction when compared with cytoplasm of Spherulocytes and Granulocytes. Oenocytoids nuclei and cytoplasm showed lower presence of proteins in comparison with nuclei and cytoplasm of Prohaemocytes, Spherulocytes, Granulocytes and Plasmatocytes. In Oenocytoids nuclei, the proteins were available in traces. The same was the case in both non-diapausing and diapausing generations (Table 4.34).

Nucleic acid (DNA) were also localised with the help of Feulgen test. Nuclei contained highest level of DNA in Prohaemocytes, Spherulocytes, Granulocytes and Plasmatocytes up to fifth instar larvae. Oenocytoids contained little lower DNA when compared with remaining haemocytes. Prohaemocytes, Spherulocytes, Granulocytes and Plasmatocytes of pupae and adults contained lower proteins, when compared with larval stages though it was available in traces in Oenocytoids of pupae and adults. The trend was alike in both non-diapausing and diapausing generations (Table 4.35). No DNA was localised in cytoplasm of these cells.

Histochemical tests performed for localisation of both DNA and RNA in the cytoplasm and nucleus of *A. mylitta* haemocytes showed very strong presence of these materials in nuclei of Prohaemocytes, Spherulocytes, Granulocytes and Plasmatocytes up to late fifth instar larvae. Thereafter, the concentration of these materials came down in pupae and adults. Cytoplasm of these cells contained nucleic acids in strong (++) proportion in Plasmatocytes of all the stages, weak (+) in Prohaemocytes of spinning, late pupae and adult stages of non-diapausing generation and in Prohaemocytes of pupae and adults of diapausing generation. The Oenocytoids contained little lower level of these materials in cytoplasm weak (+) and nucleus (up to VLM). The nuclei of Oenocytoids of pupae and adults indicated the presence of these materials in traces (±). This was also true in case of diapausing generation (Table 4.36).

PA/S substances were absent in the nuclei of these cells though they were present in cytoplasm. The reaction was very strong (+++) in non-diapausing larval stages of *A. mylitta* up to fifth instar Prohaemocytes, Spherulocytes, Granulocytes and Plasmatocytes and thereafter up to adult stage the reaction was strong (++). The Oenocytoids contained PA/S substances in strong (++) intensity in larval stages and in traces (±) from spinning larvae to adults of non-diapausing generation (Table 4.37).

In case of diapausing generation PA/S substances were comparatively less. Prohaemocytes, Plasmatocytes and Spherulocytes of larval stages, except in Granulocytes, female spinning larvae showed weak presence of PA/S substances in cytoplasm (Table 4.37). There were no changes in status of PA/S substances between non-diapausing and diapausing generation. Under Acridine orange test, intense fluorescence ranging in colour from yellow to red were seen. The colour was different for different classes of haemocytes (Fig 4.18a & b).

Bound lipids were absent in nuclei but were present in cytoplasm of all the haemocytes in both the generations and with Acetone Sudan Black B showed very strong (+++) reaction in Spherulocytes up to VLM, strong (++) in Prohaemocytes up to spinning female larval stage (SLF), in Granulocytes up to prepupal stage of males (PPM), weak (+) in Plasmatocytes up to late male fifth instar (VLM) , in Oenocytoids up to VLM in NDD generation and strong (++) in Prohaemocytes, Oenocytoids, Spherulocytes up to late fifth instar of male (VLM), in granulocytes up to spinning larval stage of male (SLM) and weak (+) in Plasmatocytes up to late fifth instar larvae of male (VLM) in DD generation; in traces (±) in Plasmatocytes from female spinning larvae (SLF) to adult females (AF) in both the generations; in Oenocytoids from female spinning larvae (SLF) to adult females(AF) in NDD (Table 4.38).

Very strong (+++) reaction of glycogen was recorded in Prohaemocytes up to spinning larval stage and in Granulocytes up to fifth instars, strong (++) in Oenocytoids up to fifth instar of non-diapausing generation. In haemocytes of latter stages, the reaction was weak (+). In cytoplasm of Spherulocytes, glycogen was either very weakly present or was available in traces (±). The same was the case in Plasmatocytes. Little change was observed in diapausing generation, such as Prohaemocytes of spinning larvae and LPM-DD had weak (+) glycogen. Glycogen was absent in the nuclei of all the haemocytes (Table 4.39).

Weak (+) presence of alkaline phosphatases activity in the cytoplasm was observed in Spherulocytes and Granulocytes till mid-pupal age both in non-diapausing and diapausing generation. In Prohaemocytes, it was observed in traces (±) (Table 4.40).

With the sexes general protein, nucleic acids, PA/S substances in Prohaemocytes showed very strong (+++) to strong (++) reaction in the cytoplasm of spinning male larvae, late-age male pupae and female adults of non-diapausing generation; glycogen in female and male spinning larvae and late-age pupae of non-diapausing generation, bound lipids in spinning female and male larvae of same generation. Bound lipids were higher in Spherulocytes of pre-pupae of non-diapausing generation, and it was strong (++) in Granulocytes of male pre-pupae. When tested for the presence of alkaline phosphatases in the cytoplasm of nuclei of different types of haemocytes in its NDD and DD generations, its presence was observed only in the cytoplasm of the PRs, SPs and GRs. The reactions showed the presence of alkaline phosphatases either week or in traces. In rest of the haemocytes and sexes, no difference was recorded (Tables 4.41 to 4.54).

Histochemical techniques have also been used in studying the intra-cellular lipids and neutral fats in *Ephestia kühniella*. Spherulocytes were reported to be loaded with fat globules at the time of pupation (Arnold, 1952). Arnold and Salked (1967) also applied such histochemical tests in the haemocytes of *Blaberus giganteus*. They observed that only cytoplasm of each haemocyte category showed some positive reaction to the histochemical tests. A few, small PA/S-positive granules in Prohaemocytes, numerous in Plasmatocytes and highest in Granulocytes and Spherulocytes were present. No histochemical distinction was evident at the level of these tests between granules of Granulocytes and Spherulocytes but they observed differences in the size of the inclusions and possibly in the concentration of materials in them. These studies support the observations made in present study in *A. mylitta* where cytoplasm of all the haemocytes showed strong PA/S reaction in almost all the life stages. Bound lipids were generally high in Spherulocytes of A. mylitta and support the view of above authors. DNA, like other insects, was confined to the nucleus of these haemocytes. There are also other reports wherein insects haemocytes are reported to contain glycogen (Vaney and Maigon, 1905; Yeager and Munson, 1941); neutral mucopolysaccharides (Wigglesworth, 1955, 1956a & b; Bronskill, 1960; Nittono, 1960); fat or phospholipids (Munson and Yeager, 1944; Arnold, 1952; Vercauteren and Aerts,, 1958); although cholesterol, acetycholinesterase, choline acetylase are absent in haemocytes (Arnold, 1952; Vercauteren and Aerts, 1958; Colhoun, 1959; Yushkevich, 1960). The classes of insect haemocytes are also reported to be characterised by various biochemical and cytological parameters e.g., appearance of mature cells, amount of DNA, types of cytoplasmic inclusions, etc. (Arnold, 1972). In case of *A. mylitta* too, the differential presence of DNA was observed in nuclei of different haemocytes. Generally, it was lower in Oenocytoids in larval forms and was available in traces in pupae and adults. Histochemical tests conducted in the haemocyte cytoplasm of *Galleria mellonella* have also shown strong presence of general proteins, PA/S substances in Plasmatocytes, Granulocytes and Spherulocytes and weak presence of these substances in Oenocytoids (Neuwirth, 1973). Presence of general protein, PA/S substances and bound lipids in Prohaemocytes and Spherulocytes of *Calpodes ethlius* is reported (Lai Fook, 1973). Granular haemocytes have often been called Coagulocytes, Amoebocytes, Phagocytes, Adipohaemocytes, hyaline cells, etc. (Lai Fook, 1973) and are reported to be structural analogues of human haemolymph platelets (Baerwald and Boush, 1970). A progressive accumulation of lipid droplets are reported in granular haemocytes (Gupta and Sutherland, 1967), which are storage sites (Lai Fook, 1973).

Spherulocytes are reported to be in greater number in diapausing pink ball worm (Clark and Chadbourne, 1960). In case of *A. mylitta* Spherulocytes of non-diapausing generation showed more bound lipids than diapausing generation in larval, pupal and adult stages. Akai and Sato (1973) found mucopolysaccharides and mucoproteins in Spherulocytes and suggested that they are the source of haemolymph proteins. Nittono (1960) said that there was a positive correlation between the presence of spherules in Spherulocytes and silk production.

Akai and Sato (1973) have suggested that Oenocytoids are perhaps involved in protein synthesis and they may be in some way involved in storage and transport of body metabolites. In the present study, cytoplasm and nuclei of Oenocytoids showed positive reaction for the presence of general protein, PA/S substances, bound lipids (only in larval forms) and glycogen in Oenocytoids of *A. mylitta* supports the above view. High PA/S was also reported in

Granulocytes and Spherulocytes of Manduca sexta in comparison with Prohaemocytes and Plasmatocytes (Horhov and Dunn, 1982).

Babers (1941) have shown presence of glycogen in haemocytes of *Prodenia* larva. It is also reported in haemocytes of *Galleria* (Ashhurst and Richards, 1964), in Granulocytes of Locusta (Brehlin et al., 1975); in haemocytes of *Calliphora* (Crossley, 1968); in *Blaberus* (Moran, 1971); and in *Antheraea pernyi* (Beaulaton and Monpeyssin, 1976). In A. mylitta the presence of glycogen was high in larval Prohaemocytes and Granulocytes cytoplasm. It was also present in Oenocytoids and Plasmatocytes with relatively low intensity. Prohaemocytes of diapausing generations contained lower amount of glycogen than non-diapausing generation in *A. mylitta*.

Electron micrographs have shown the presence of lipids in haemocytes, generally, lipid content is high before pupation, indicating that some haemocytes parallel the lipid metabolism of fat body cells, both in *Ephestia* (Arnold, 1952) and in *Galleria mellonella* (Ashhurst and Richards, 1964). The concentration/presence of lipids was recorded to be higher in non-diapausing and diapausing Prohaemocytes and Spherulocytes of *A. mylitta* prior to pupation and supports the view of above authors, though, in diapausing generation stages, it was observed to be a bit lower. Such interpretation of presence of lipids have also been done by other authors, e.g., in Ephestia (Smith, 1938; Grimstone et al., 1967); in *Calliphora* (Crossley, 1968, 1975; Zachary and Hoffmann, 1973) and in *Pectinophora* (Raina, 1976).

Histochemical tests have also shown that most of the granules of Granulocytes contain glycoprotein or mucopoplysaccharides (Francois, 1974, 1975 a & b; Costin, 1975). Occasionally they may contain lipid droplets (Gupta, 1979 a & b). Spherulocytes contain neutral or acid mucopolysaccharides and glycomucoproteins (Vereauteren and Aerts, 1958; Nittono, 1960; Ashhurst and Richards, 1964; Gupta and Sutherland, 1967; Costin, 1975; Beaulaton and Monpeyssin, 1976 & 1977). Histochemically, Oenocytoids are reported to contain tyrosinase (Dennell, 1947), protein (Akai and Sato, 1973) and PA/S positive granules indicating the presence of glycoprotein, mucopolysaccharides and sulphonated, periodate-reactive sialomucin (Costin, 1975).

In Oenocytoids of *A. mylitta* in their cytoplasm in larvae of non-diapausing and diapausing generations PA/S substances were present; in nucleus of Oenocytoids the presence of general protein was more in larvae, mid-pupae and adults of non-diapausing and diapausing generations than their cytoplasm. The presence of afore discussed substances in the present study supports the findings of above authors.

Chippandale (1970a & b) was of the opinion that in *Diatrea*, haemocytes are of minor importance in synthesis of bulk protein. Wigglesworth (1979a) considered haemocytes to be active sites for the synthesis of proteins and they are reported to take part in labelling of amino acids and synthesizing vitellogenins and other proteins (Buhlmann, 1974); in production of injury proteins in *Hyalophora cecropia* (Berry et al., 1964); in production of pro-activators and hemagglutinins in *Blaberus craniifer* (Anderson et al., 1972); in production of lysozymes in *Galleria* and *Calliphora erythrocephala* (Crossley, 1979). Histochemical tests have revealed the presence of glycoprotein in certain haemocytes: Spherulocytes in *Bombyx mori* (Akai and Sato, 1973); Granulocytes in *Thermobia domestica* (Francois, J., 1974, 1975a & b); Plasmatocytes, Spherulocytes and Oenocytoids in *Locusta migratoria*

(Costin, 1975); and Granulocytes in *Antheraea pernyi* (Beaulaton and Monpeyssin, 1976). Histochemicals were applied to identify the types of cells in *Dactylopius confusus* by Joshi and Lambdin (1996) and Granulocytes contained granules showing positive result with periodic acid-Schiff (PAS) test. The Granulocytes and the granules within them increased during pupal stage in *Lymantria dispar*. The granules were perhaps composed of polysaccharides, proteins and lipids. Granules rich of lipids were reported at this stage (Kim et al., 1990a, b). Though, this was not true in the case of *A. mylitta*. Histochemical tests confirmed the higher presence of these granules up to late larval stage in the present study. Reaction showed lower presence of these materials in pupae and adults. Such histochemical tests were also applied with the five types of haemocytes of last instar larvae and pupae of *Agrotis segetum* (Cebesoy and Ayvali, 1996), which indicated that Prohaemocytes gave positive with pMAb and PAS; Plasmatocytes, Granulocytes and Oenocytoids gave positive reaction with pMaG and PAS test and Sudan Black B, but the protein test gave a negative reaction. The pMaG and PAS tests were positive for spherule cells, whereas negative reaction was observed with Sudan Black B. Only Plasmatocytes and Granulocytes gave positive reactions with Toluidine blue and neutral red. In the present study in cytoplasm of Plasmatocytes, Spherulocytes and Prohaemocytes of larvae showed very strong PA/S positive reaction though it was only strong in Oenocytoids up to their larval stages of non-diapausing generation, their presence were weaker in diapausing generations.

Insect haemocytes are concerned with a number of functions such as synthesis of certain haemolymph proteins, transfer and supply of the nutrients and reserves to different tissues, phagocytosis, wound healing, initiation of hormone production and transformation of connective tissue etc. Therefore, presences of enzymes like alkaline phosphatase were tested in the haemocytes of *A. mylitta*. Its presence was observed to be weak in the cytoplasm of Spherulocytes and Granulocytes till mid-pupal stage and in traces up to spinning larvae Prohaemocytes. Some workers have reported its negative or doubtful presence in insect haemolymph (Vercauteran and Aerts, 1958; Colhoun and Yushkevich cited in Jones (1962), Bandopadhaya (1970). Though, a few workers have demonstrated it's presence in insect haemolymph such as Vasuki and Dikshit, 1968 in *Poecilocerus picta*; Bhoumik, 1972 in *Periplaneta americana*; Kukra and Weiser, 1973 in *Barathra lorassica* and Nath and Butler, 1973 in black carpet beetle. From the forgone account, it appears that haemocytes in insects may or may not show alkaline phosphatases activity. Denielli (1958) stated that histochemical tests and localisation of an enzyme depends upon the initial fixation, substrate used, product of hydrolysis and end product of dye pattern in the site of enzyme activity. The histochemical tests performed in *A. mylitta* indicated different type of reactions.

Gupta (1985), while reviewing the types of haemocytes in insects of different orders, indicated that ultra structurally, only seven types of cells have so far been identified, namely, Prohaemocytes, Plasmatocytes, Granulocytes, Spherulocytes, Adiphohaemocytes, Oenocytoids and coagulocytes. Though only five types of haemocytes were identified in the present study, and they were Prohaemocytes, Plasmatocytes, Spherulocytes, Granulocytes and Oenocytoids in *A. mylitta*.

Haemocytes perform many functions: Plasmatocytes and Granulocytes actively participate in phagocytosis, encapsulation and nodule formation. Coagulation is caused by

Granulocytes. In addition lysozymes, located in the lysosomes of haemocytes, play an important role in insect immunity. Haemocytes probably carry hormones to target organs (Gupta, 1985). Haemocytes contain polysaccharides, proteins and lipids in *Lymantria dispar* (Kim et al., 1990a), acid phosphatases (Kim et al., 1990 b), granulocytes releasing nutrients for egg development in *Simulium vittatum* (Luckhart, 1992) and high molecular weight proteins in Manduca sexta (Beck et al., 1996). Plasmatocytes and Granulocytes also participate in injury repair. From the present study, it is indicated that haemocytes synthesise protein, contain glycogen, lipids etc. in haemolymph of *A. mylitta.*

5.9. Physical properties and chemical composition of haemolymph of *Antheraea mylitta*

Physical properties such as pH, specific gravity and haemolymph biochemicals such as protein, amino acid, trehalose, cholesterol, uric acid and cations like Na+, K+, Ca++ and Mg++ were measured in the haemolymph of important life stages of *A. mylitta* because haemolymph acts as storage-reservoir essential for many materials for a variety of insects' life-processes (Mullin, 1985).

5.9.1. Specific gravity

The range of specific gravity was 1.0100 to 1.0550 in non-diapausing generation and 1.0100 to 1.0560 in diapausing generation. In pre and mid-aged pupal stages, the mean difference was non-significant between the two generations, though, in non-diapausing females of mid-pupal stage, the specific gravity was lower (1.0217), than diapausing females (1.0380). In remaining stages, the specific gravity was significantly higher in diapausing generation ($p < 0.01$) - (Table 4.55; Fig. 5.21).

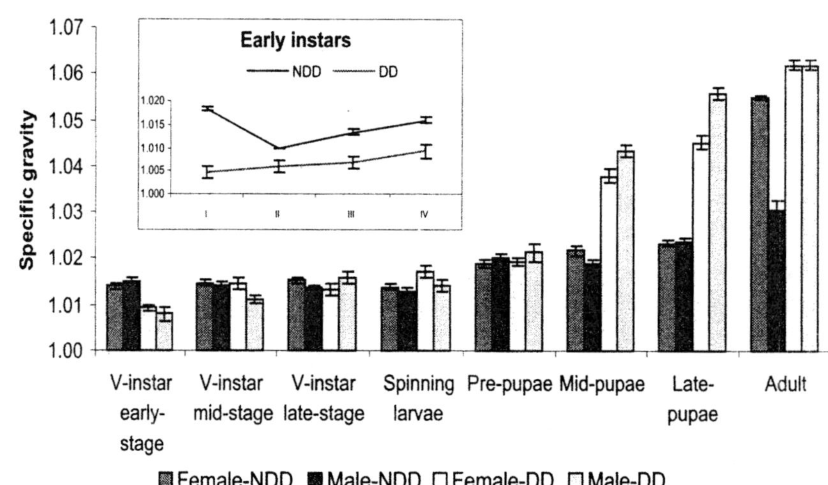

Fig. 5.21. Comparison between mean values of specific gravity (I = ±SE) of non-diapausing & diapausing generations and females & males of *A. mylitta*.

Within the sexes, there was no difference in means of most of the stages, except in between female mid-pupa (MPF) and male mid-pupa (MPM) and adults (AF & AM). Females had higher values than males. In diapausing generation, the mean difference was significantly high in male mid- (MPM) and late-pupae (LPM) than females, though the reverse was the case in spinning larval stage (Table 4.56 & 4.57; Fig. 5.21).

Specific gravity of different orders of insects are reported to range from 1.012 to 1.062, for example: in *Dysticus* - 1.026; in *Hydroplulus* - 1.012 (Barrat and Arnold, 1910); in *Calliphora* - 1.021 (Hopf, 1940); in *Gastrophilus* - 1.062 (Levenbook, 1950); in *Phormia* - 1.018 (Hopf, 1940); in *Apis* - 1.045 (Ducceschi, 1902); in *Bombyx mori* larvae - 1.032 to 1.041 (Ducceschi, 1902; Fredericq, in Ducceschi, 1902); in *Diolephila* larvae - 1.0307 (Heller and Moklowska, 1930); in *Prodenia* larvae (Babers, 1938) and in *Periplaneta* - 1.0162 to 1.0182 (Yeager and Fay, 1935).

Further, Buck (1953); Terra et al., (1974); Dahlman (1974); Mack and Vanderberg (1978) have also reported that specific gravity ranges in between 1.012 to 1.070 in insects. The level of specific gravity in the present study in *A. mylitta* has been observed to fall within the range as observed by above authors.

In other insects, osmolarity values have been reported to range between 215 and 593 mOsm, corresponding to freezing point depressions of 0.40 - 1.10°C (Buck, 1953; Joshua et al., 1973). Insects appear to be quite capable of strongly regulating osmotic pressure under a wide range of fluctuation of volume (Nicolson, 1980; Mullin, 1985). Ontogenetic differences in osmolarity have also been observed (Joshua et al., 1973; Cohen and Patana, 1982). However, the specific gravity reflects in rough way the amount of dissolved material in haemolymph (Buck, 1953).

5.9.2. pH value (Hydrogen ion concentration).

Reports on hydrogen ion concentration (pH) in insect haemolymph show values ranging in between 6.0 to 8.2 (Mullin, 1985; Chapman, 1980; Wigglesworth, 1972). In the present study, mean range of hydrogen ion concentration in *A. mylitta* was observed to range in between 6.87 to 6.93 in non-diapausing generation and 6.71 to 6.91 in diapausing generation. No significant difference was observed in the concentration of hydrogen ion in between the sexes of both the generations, except in between non-diapausing mid-aged female (MPF-NDD) and male (MPM-NDD) pupae, where MPF showed significantly higher ($p < 0.05$) pH than MPM (fig. 5.22).

Buck (1953), while reviewing the hydrogen ion concentration in haemolymph of different insect species stated that it appears to vary inter-specifically in between about pH 6.0 and pH 7.5, and intra-specifically up to 0.7 pH unit. Most of the values in insects were observed to fall slightly on the acid side of neutrality, but in some, such as *Chironomus*, it is distinctly alkaline, pH 7.2 to 7.7. the pH have been reported to range in Lepidoptera larvae well in the acidic side of neutrality, such as in *Bombyx mori*, larvae and pupae ranged within pH 6.4 to 6.7 (Glaser, 1925; Demajanowski et al., 1932; Epstein, 1930); in *Pieris brassicae* pupae - pH 6.50 to 6.77 (Brecher, 1925); in *Galleria mellonella* - pH 5.8 to 6.4 (Taylor et al., 1934); in *Pieris rapae* pupae - pH 5.9 to 6.4 (Fink, 1925); in P. rapae larvae - pH 6.75 to 7.56 (Craig and Clark, 1938), in *Prodenia eridania* larvae - pH 6.40 to

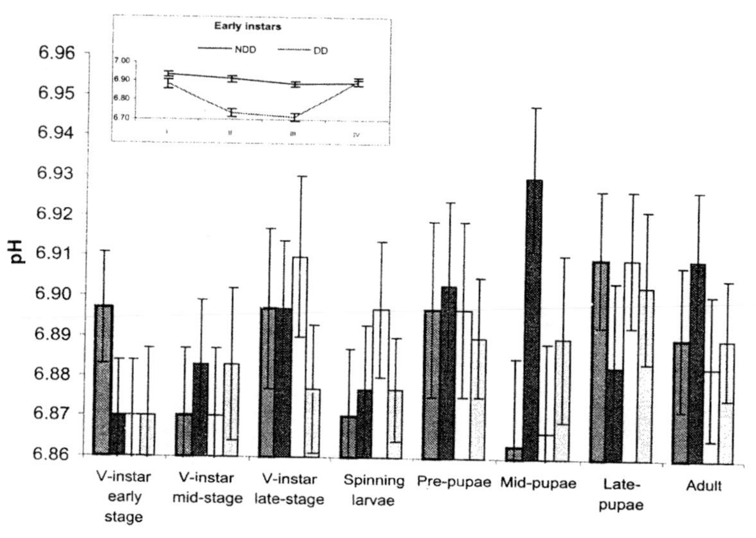

Fig. 5.22. Comparison between mean values of pH (I = ±SE) of non-diapausing & diapausing generations and females & males of *A. mylitta*.

6.70 (Babers, 1938). The pH values are further reported to be in a narrow range between pH 6.4 to 6.8 (Wyatt, 1961; Joshua et al., 1973; Burton et al., 1972). The buffering capacity is also reported to be minimal in this physiological range (Buck, 1953; Wyatt, 1961). Srivastava and Mathur (1966) reported the pH values of 21 species of Lepidopteran larvae to vary between pH 6.2 to 6.9. The hydrogen ion concentration in between pH 6.68 to 6.74 in larvae of *Hyalophora cecropia* (Jungries, 1979); 6.54 to 6.59 in larvae of *Manduca sexta* (Jungries, 1978). pH 6.7 in *Manduca sexta* (Racioppi and Dahlman, 1980), and in *P. brassicae* (Natochin et al., 1988) ; 6.5 in *Heliothis virescens* (Bindokas and Adams, 1988) ; 6.6 in *L. dispar* (Pannabecker et al., 1992) have also been reported.

In the present insect, the haemolymph pH also normally varies in between pH 6.87 to 6.93 in non-diapausing generation and pH 6.71 to 6.91 in diapausing generation which supports the view of above authors (Table 4.58 to 4.60).

Although, the haemolymph pH are reported to be alkaline, in *Drosophilla* (pH 7.1); *Sciara* (pH 7.2); *Chironomus* (pH 7.48) (Beadle, 1939) in the last instar larva of *Pieris* (pH 7.17) and *Heliothis* (pH 7.23) (Craig and Clark, 1938). A transient rise in pH during moulting has been reported (Jungries, 1978; 1979). The pH towards acid side may be due to the availability of metabolic products in haemolymph in the form of acid (Wigglesworth, 1972). The buffering power of the haemolymph is probably due to bicarbonates, inorganic phosphates, prcteins, amino acids and to a small extent urates and when a slightly acidic pH buffering capacity plotted against pH values, always gives a U-shaped curve in insects (Babers, 1938; Wigglesworth, 1972). Thus, the pH in the present study as observed in *A. mylitta* fall within the range of buffering capacity range of other insects.

5.9.3. Haemolymph Cations

5.9.3.1. Sodium (Na+)

The Na+ concentration in both non-diapausing and diapausing generation in the present study ranged in between 1.1500 to 1.8600 and 1.1900- 5.9900 mEq/l, respectively (Table 4.61; Fig. 5.23).

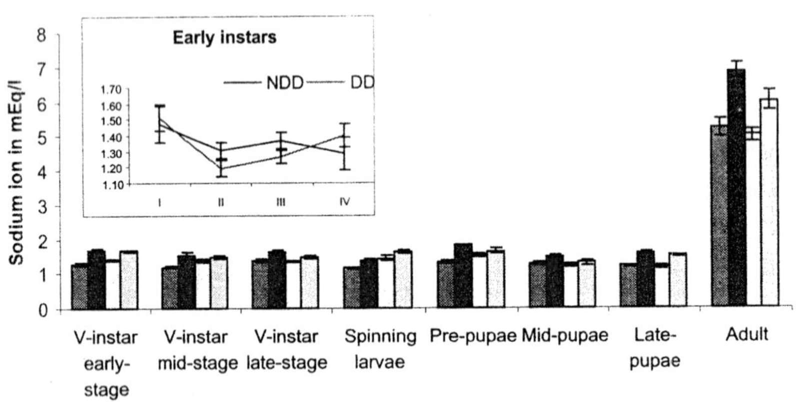

Fig. 5.23. Comparison between mean values of Na+ (I = ±SE) of non-diapausing & diapausing generations and females & males of *A. mylitta*.

Between the sexes of non-diapausing generation and diapausing generation both males had generally higher Na+ than females in diapausing generation (Table 4.62, 4.63; Fig. 5.23).

5.9.3.2. Potassium (K+)

K+ ion concentration ranged from 45.8200-64.6400 mEq/l in non-diapausing and 47.6900 - 60.8200 mEq/l in diapausing generations-(Table 4.64; Fig 5.24).

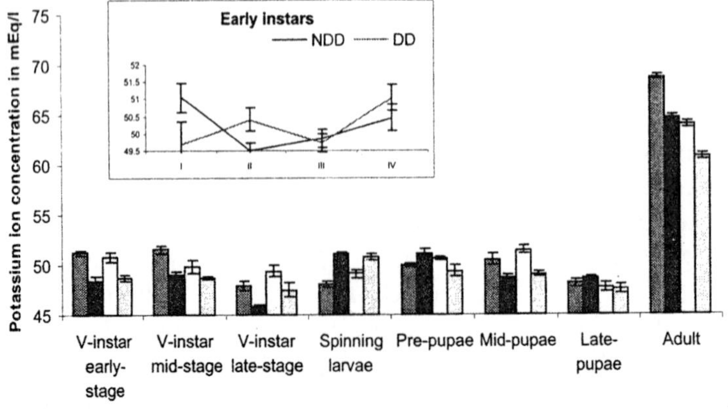

Fig. 5.24. Comparison between mean values of K+ (I = ±SE) of non-diapausing & diapausing generations and females & males of *A. mylitta*.

Within the sexes of non-diapausing generation, females of early, middle and late fifth instar larvae had higher K+ than males; the trend was reversed in spinning larvae and pre-pupae. Again in mid-pupal and adult stage, females had higher K+ than their male counterparts (Table 4.65; Fig 5.24).

In diapausing generation too, similar trend of K+ ion in haemolymph was recorded. Significant differences in mean values of K+ in haemolymph of late pupal stages of the sexes in both the generations was not observed (Table 4.66).

5.9.3.3. Calcium (Ca++)

Concentration of Ca++ ranged from 18.7700 to 30.6900 mEq/l in non-diapausing and 20.3800 to 26.6700 mEq/l diapausing generations. In II and IV instar larvae, early and middle fifth instar larvae of both the sexes and late male fifth instar larvae, the mean difference was non-significant in between non-diapausing and diapausing generations. However, late female of fifth instar diapausing generation had higher Ca++ (23.9000 mEq/l) than females of fifth instar non-diapausing generation (19.9700 mEq/l). Spinning larvae, pre-pupae and mid-pupal stages of diapausing generation had significantly ($p < 0.05$, $p < 0.01$) different Ca++ than in non-diapausing generation (Table 4.67). This difference was non-significant in adult females of non-diapausing and adult males of diapausing generations. Adult males of non-diapausing generation had higher Ca++ than adult males of diapausing generation.

In between the sexes, the mean quantity of Ca++ was significantly higher ($p < 0.05$ and $p < 0.01$) in non-diapausing generation males than females (Table 4.68), except in early fifth instar. In diapausing generation, the difference was non-significant between the sexes except between late-pupae female & late pupae male, early female fifth instar larvae & early male fifth instar, mid-aged female fifth instar & male mid-aged fifth instar larvae of diapausing generation, where males had higher Ca++ than females (Table 4.69; Fig. 5.25).

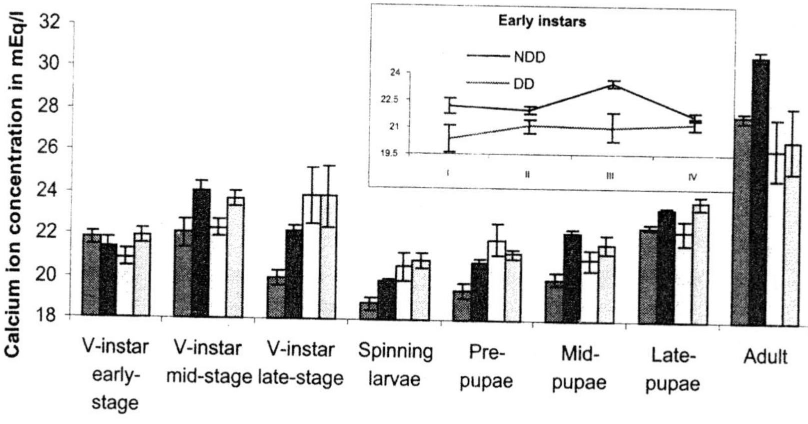

Fig. 5.25. Comparison between mean values of Ca++ (I = ±SE) of non-diapausing & diapausing generations and females & males of *A. mylitta*.

5.9.3.4. Magnesium (Mg++)

Concentration of Mg++ ranged between 29.8390 to 69.0470 mEq/l in non-diapausing and 32.2280 to 64.3100 mEq/l in diapausing generations. The concentration of Mg++ was higher in late larval stages of both the generations (VLF-NDD - 69.0470 mEq/l, VLF-DD - 64.3100 mEq/l; VLM-NDD - 65.3380 mEq/l; VLM-DD - 61.1540 mEq/l). The trend was rather fluctuating in later stages. In pre-pupal stages of diapausing generation, significantly higher Mg++ was recorded in PPM-DD, MPF-DD, LPM-DD and AF-DD (Table 4.70) than non-diapausing generations.

In between the sexes, the mean quantity of Mg++ was significantly higher ($p < 0.05$ and $p < 0.01$) in non-diapausing generation females than males (Table 4.71), particularly in early, middle and late fifth instars. In case of spinning larvae, males had higher Mg++ than females. In remaining life stages, females had higher Mg++ than males (Table 4.71). In diapausing generation, the difference was non-significant between the sexes in spinning larval stage, though males had higher Mg++ than females. In rest of the life stages, females had higher Mg++ than males of diapausing generation (Table 4.72; Fig. 5.26).

The concentration of the quantitatively most important inorganic ions of the haemolymph of about 50 insects were reviewed and reported after Buck, 1953. These ions were Na+, K+, Ca++, Mg++, Cl^-, etc. Among Lepidopteran larvae, the concentration of Na+ was 10 mMq/l in *Attacus polyphemus*, 11 mM/l in *Attacus pernyi*, 3 mM/l in *Saturnia parvonia* pupae, 7 mM/l in *Cecropia sp*. pupae, 4 mM/l in *Saturnia pyri* pupae (Drilhon, 1934), 14 mM/l in *Bombyx mori* larvae, 9 mM/l in *B. mori* pupae (Akao, 1935; Bone, 1945) and 3 mM/l in *Samia walkori* pupae (Gese, 1950); though it's concentration varied from 3 to 174 mM/l in different insect species (Buck, 1953).

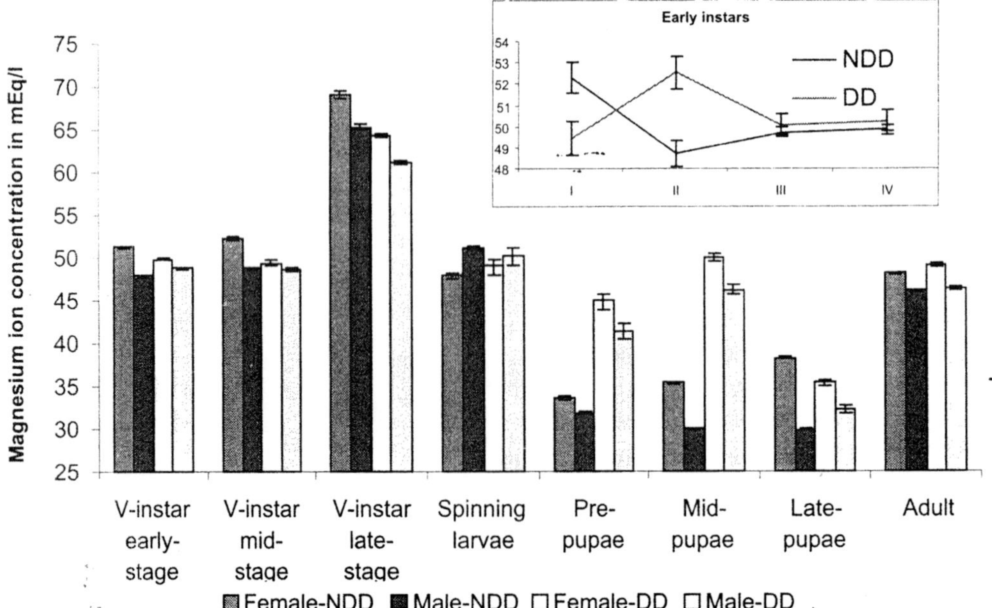

Fig. 5.26. Comparison between mean values of Mg++ (concentration of Mg++ is at y-axis; I = ±SE) of non-diapausing & diapausing generations and females & males of *A. mylitta*.

The concentration of K+ was 43 mM/l in pupae of *Attacus pernyi*, 35 mM/l in pupae of Attacus polyphemus, 50 mM/l in pupae of *Cecropia sp.*, 39 mM/l in pupae of *Saturnia parvonia*, 41 mM/l in pupae of *Saturnia pyri* (Drilhon, 1934); 33 to 40 mM/l in larvae of *Bombyx mori* (Bialaszewicz and Landan, 1938; Bone, 1945; Tobias, 1948b); 43 mM/l in pupae of *Bombyx mori* (Akao, 1935); and 42 mM/l in *Samia walkori* (Gese, 1950); though K+ concentration ranged in between 8 to 62 mM/l in different insect species (Buck, 1953).

Florkin and Jeuniaux (1974) further summarised the presence of haemocations in insects of different orders, particularly in larvae, pupae and adults of Lepidoptera. In *Antheraea mylitta* larvae, the Na+ was 1.3, K+ 49.7, Ca++ 21.9 and Mg++ 37.7 (in Duchateau et al., 1953); in *Bombyx mori* 3rd instar larvae were: Na+ - 3.4, K+ - 41.8, Ca++ - 80.8, Mg++ - 150.5; in 4th instar, Na+ - 6.0, K+ - 39.4, Ca++ -15.0, Mg++ - 148.4, in 5th instar Na+ - 14.6, K+ - 16.1, Ca++ - 24.5, Mg++ - 101.0, in pre-pupa, K+ - 8.2, K+ - 59.2, Ca++ - 26.5, Mg++ 92.5, in adults Na+ - 14.3, K+ - 36.1, Ca++ - 14.5, Mg++ - 44.6 (Duchateau et al., 1953; Bialaszewicz and Landan, 1938).

From the above references, it is amply clear that detailed ontogenetic availability has not been reported in a particular insect species. In case of *A. mylitta* adults, in the present study, the concentration of Na+, K+ and Mg++ was recorded to be higher, though a fluctuating trend was noticed in different stages of non-diapausing and diapausing generations of *A. mylitta*. However, the presences of cations in the larvae of *A. mylitta* are similar to that reported by Duchateau et al., 1953. Some generalisation has been made regarding the relative presence of cations (inorganic constituents) and their importance as osmolar effectors (Florkin and Jeuniaux, 1974). The sum of Na+, K+, Ca++ and Mg++ contributes almost half of the osmotic pressure in more primitive Pterygotes (Odonata, Orthoptera, Dermoptera and Hetroptera), here Na+ and Cl^- are the major osmoeffectors (Mullin, 1985). In higher Endopterygotes (Lepidoptera, Hymenoptera and Coleoptera), the inorganic molecules become much more important as osmoeffectors (Mullin, 1985). The levels of Na+ was lowest in *A. mylitta*, whereas the concentration of K+, Mg++ and to some extent Ca++ was high in all the life-stages. This higher presence of K+ and Mg++ in

A. mylitta is similar to that reported in other Lepidoptera (Florkin and Jeuniaux, 1974). It is possible that the higher level of Mg++ and K+ found in higher Endopterygotes may be attributed to evolution (Mullin, 1985) along the Angiosperms. Progressive adaptations to plant diets (high K+ and Mg++), regulation of steady state haemolymph cationic species developed having low Na+, high K+ and Mg++ in Lepidoptera (Wyatt, 1961; Florkin and Jeuniaux, 1974). Similar may be the case with *A. mylitta* as its haemolymph contained high level of K+ and Mg++, as observed in the present study. Similar trends of presence of these cations have also been reported in other Lepidoptera e.g., in *Hyalophora cecropia*, *Manduca sexta* and *Danaus plexippus* (Jungries et al., 1973); in *Galleria mellonella* (Florkin and Jeuniaux, 1974); in *Manduca sexta* (Racioppi and Dahlman, 1980); in *Spodoptera exigua* (Cohen and Patana, 1982); in *Pieris brassicae* (Natochin et al., 1988); in *H. virescens* (Binodokas and Adams, 1988);and in *Morimus funereus* (Jankovic Hladni et al., 1992). Summary of osmolyte concentrations in haemolymph of fifth instar gypsy moth *(L. dispar)* larvae were also measured and cations were reported to be 67.0 mM/l.

Haemolymph composition tend to vary in response to various conditions or activities such as ontogenetic effects (Jungreis et al., 1973; Schin and Moore, 1977), diet (Jungreis et al., 1973; Shimizu, 1982), distribution pattern in various haemocoel areas (Pichon, 1970) and temperature stress (Cohen and Patana, 1982). Similar studies in future may essentially be required in case of *A. mylitta*.

5.9.4. Trehalose

Carbohydrates are important in insects, since they function as major energy source and utilised as the major exoskeleton component. Insect haemolymph contains trehalose, a non-reducing dimer of a-glucose, in high concentration (Wyatt and Kalf, 1956, 1957) and fluctuation in the level of trehalose reflects upon the physiological state of insects.

The level of trehalose was generally low in non-diapausing generation than that of diapausing generation as it was 0.8501 mg/ml in I instar non-diapausing generation larvae, 1.0502 in I instar diapausing generation larvae; 2.3226mg/ml IV instar non-diapausing generation larvae, 2.9045mg/ml in IV instar diapausing generation larvae; in V instar early aged female larvae of non-diapausing generation, it was 7.4848 mg/ml; in same stage of diapausing generation in same sex, it was 3.1512 mg/ml. Gradually, the level of trehalose was recorded to be higher in diapausing generation after mid-aged V instar female and male stages, till emergence of adult (Table 4.73; Fig. 5.27).

The comparison of trehalose availability was also made in between the two sexes of non-diapausing and diapausing generations. The females in general had significantly higher mean level of trehalose in all the stages (Table 4.74; Fig. 5.27).

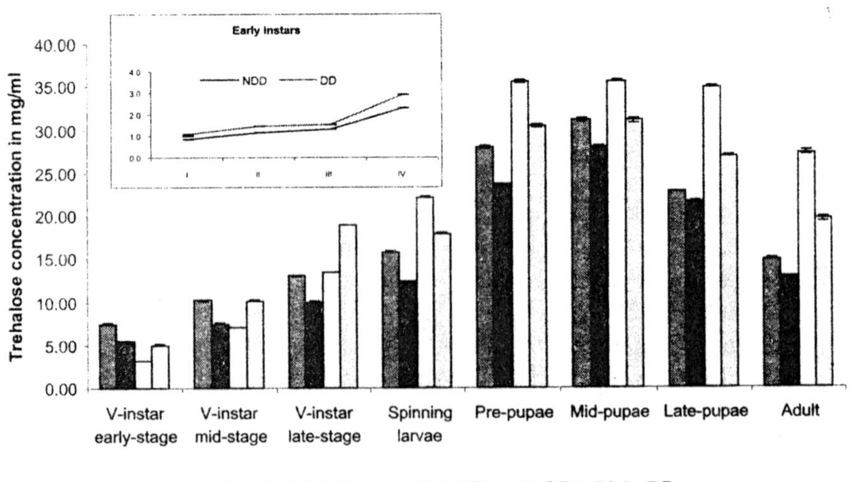

Fig. 5.27. Comparison between mean values of Trehalose (I = ±SE) of non-diapausing & diapausing generations and females & males of *A. mylitta*.

The case was not so in diapausing generation as till late larval stage, males had generally higher level of trehalose, thereafter, the females had higher level of trehalose than males till emergence of adults (Table 4.75; Fig. 5.27).

Wyatt (1967), while reviewing the level of different sugars in the haemolymph of insects stated that when compared with vertebrates, the total carbohydrates content of insect haemolymph was observed to be high. This level is generally greater (0.5 % higher) but may reach as high as by 8.1 % (Wyatt, 1967). Among Lepidopteran insects, the level of trehalose reported to be 1 to 3 mg/ml in *Bombyx mori* larvae (Duchateu Bosson et al., 1963); 4.5 - 6.0 mg/ml in larvae and 2.0 mg/ml in pupae of B. mori (Saito, 1963; Wyatt and Kalf, 1956 & 1957); in V instar larvae of *Antheraea pernyi* it ranged in between 3.2 to 8.8 mg/ml (Smolin, 1960; Egorova, 1963); in *Antheraea polyphemus* larvae and pupae it ranged in between 10-14 mg/ml (Wyatt and Kalf, 1957); in *Hyalophora cecropia* larvae the range was 12.0 to 19.0 mg/ml (Wyatt and Kalf, 1957); 10.0-15.0 mg/ml in developing pupae to adults of *Hyalophora cecropia* (Wyatt, 1967); in *Samia ricini* pupae in diapause the range was from 2.0 to 5.0 mg/ml (Wyatt, 1967); in *Samia cynthia* larvae - 5.0 - 15.0 mg/ml (maximum just before spinning) (Chang et al., 1964; Liu and Feng, 1965); in *Manduca sexta* larva 30.1 to 71.8 μml/ml (Kramer et al., 1978a & b; Jungries, 1978); in *Manduca sexta* and *Manduca quinquemaculata* larvae - 25.0 μml/ml (Kramer et al., 1978 a & b); in *Sphinx chersis* - 133 μml/ml (Kramer et al., 1978 a & b). Carbohydrates in an increasing trend in first to fifth instar larvae of *A. mylitta* from 0.325 to 1.425 mg/ml (Poonia and Mishra, 1975); increased trend in different season/generation of *A. mylitta* larvae (Sinha et al., 1998); trehalose - 3.0 to 4.0 mg/ml in male pupae and 8.0 to 12.0 mg/ml in female pupae in a fluctuating manner during diapause termination of *A. mylitta* (Chaudhury et al., 1993); increased level of trehalose in diapausing females and males and it's gradual decline during diapause termination or during emergence of adults of *A. mylitta* (Ojha et al., 1997).

The findings of the present study are in conformity with findings of the above authors because the level of trehalose increased with increased number of instars. Higher levels of trehalose in diapausing generation, especially, in male pre-pupae and females mid-pupae of *A. mylitta* were observed in the present study. Females generally had higher level of trehalose than males in advance stages of development.

The level of trehalose are reported to vary in larval forms of pure and cross breeds of *Bombyx mori* and an increase is also noticed with the advancement in age of larvae (Sowri and Sarangi, 2002). The level of trehalose are also reported to decrease on the final day of fifth instar larvae may be due to greater utilisation to furnish fuel for active synthesis of silk and other physiological process like movement involved in spinning (Clegg and Evans, 1961; Sowri and Sarangi, 2002). Trehalose levels are also related to moulting, metamorphosis and diapause (Wyatt, 1967; Jungries and Wyatt, 1972; Sakamoto and Horie, 1979; Hirano and Yamashita, 1980). Though it's level are reported to increase towards larval maturation in *Bombyx mori*, decrease at the larval-pupal transformation, and then again increasing during adult development (Hirano and Yamashita, 1983), carbohydrates in pupal moult of *Manduca sexta* (Siegert, 1995) and in A. mylitta (Sinha et al., 1998).

The role of sugar and sugar alcohols are reported in diapause as Hayakowa and Chino (1982a & b) maintained that diapausing insects can be categorised into two types in terms of glycogen metabolism, one is sugar alcohol accumulating type, while the other is trehalose accumulating type. Changes in carbohydrates are also related to cryprotection in insects (Pullin and Bale, 1989). Further, Hayakawa and Chino (1982 a & b) demonstrated a temperature dependent interconversion between glycogen and trehalose in diapausing pupae of the saturniid moth Philosamia cynthia and this inter conversion was mainly due to activation of glycogen phosphorylase and reverse (trehalose to glycogen) due to activation of glycogen synthatase.

Similar studies may form the future course of action in the field of haemolymph biochemical parameters of *A. mylitta.*

5.9.5. Haemolymph proteins

5.9.5.1. Quantitative haemolymph proteins

Protein concentration in insect haemolymph is similar to that of man and other vertebrates, and generally higher than that of the internal fluid of other invertebrates (Florkin and Jeuniaux, 1974). Proteins are among the most complex of all known chemical compounds and the most characteristic of living organisms. The haemolymph proteins of insects have been investigated from various viewpoints. These include (i) mapping of protein patterns in various species, (ii) the identification of protein composition at successive developmental stages by both electrophoretic and immunological techniques, (iii) proposal of possible fluctuations of the protein components on the basis of enzymological and histochemical tests, and (iv) the analysis of both site and mechanism of synthetic processes by isotopic labelling (Chen and Levenbook, 1966 a & b and revised after Buck, 1953; Wyatt, 1961; Chen, 1966). Under the present study, the ontogenetic availability of haemolymph protein was estimated in non-diapausing and diapausing generations of A. mylitta. In non-diapausing generation, the haemolymph protein level increased with increase in number of instars (I-NDD - 3.4959; IV-NDD - 6.8038), many-fold increase in SLF-NDD (54.0634) & SLM-NDD (50.2600), a downward trend in pre-pupal stage - PPF-NDD (48.0492), PPM-NDD (31.7570), which fell gradually in late pupal stages till emergence of adults (AF-NDD - 26.3652; AM-NDD - 19.9716) - (Table 4.76 to 4.78; Fig. 5.28).

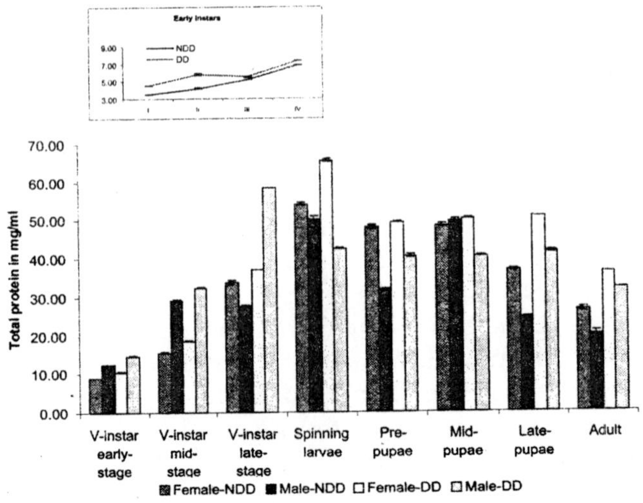

Fig. 5.28. Comparison between mean values of Total Protein (I = ±SE) of non-diapausing & diapausing generations and females & males of A. mylitta.

Similar was the trend recorded in diapausing generation from I to spinning larval stage having comparatively higher protein than NDD generation, except in SLM-NDD, which was higher than respective stage of DD generation (Table 4.76; Fig. 5.28).

In between the sexes, comparisons were made in both the generations. Females had higher level of haemolymph in VLF-NDD, SLF-NDD, PPF-NDD, LPF-NDD and AF-NDD and SLF-DD, PPF-DD, MPF-DD, LPF-DD and AF-DD (Tables 4.77 & 4.78; Fig. 5.28).

A number of studies have shown that haemolymph proteins change during development of silk moths e.g., in *Cecropia* silkworm (Telfer and Williams, 1953 a & b; Telfer, 1954); in giant silk moths (Laufer, 1959) and in other insects (Chen, 1966; Wyatt, 1980).

The change in each antigenic reported to be independent of the others during the course of pupae and adults (Telfer, 1954). Protein concentration was reported to be lower in diapausing pupae and minimum level of protein was recorded in adults *Cynthia* (Laufer, 1960). During development, a progressive genesis of new haemolymph protein fractions occurred in *Malcosoma americanum* (Whittaker and West, 1962); decreased proteins in late pupae and adults of *Malcosoma* and *Rothschildia* (Loughton, 1965); in Lepidoptera or Diptera (Wyatt and Pan, 1978).

Various studies agree in showing that the total content of haemolymph proteins increase during larval development, this is most rapid during the time approaching pupation (Chen, 1956; Chen, 1966). A similar increase is also reported in Culex (Chen, 1959); in *Bombyx* (Florkin, 1937; Wyatt, 1956); in *Popillia japonica* (Ludwig, 1954); in wax moth *Galleria mellonella* (Denuce, 1958); in *Antheraea proylei* Jolly protein changes during development (Sinha and Sinha, 1994); in *Phormia regina* (Chen and Levenbook, 1966a & b); it changes during metamorphosis in *Formica rufa* (Schmidt and Schwanski, 1975). Such trend of increase in haemolymph protein during course of larval development followed by decrease/sharp decline in pupal stages and further decrease in adults are also reported in *Phormia regina* (Chen and Levenbook, 1966 a & b); in *Samia cynthia ricini* (Karaki, 1969); in *Pieris brassicae* (Chippandale and Kilby, 1969); in *Ephestia kuhinella* (Yoo and Lee, 1974); in *Dendrolimus spectrabilis* (Yoo and Lee, 1975); in *Rhynchsciara spectrabilis* (Bianchi and Terra, 1976). Changes in protein pattern are also reported in *Ostrinia nubilalis* during pre-pupal differentiation (Chippendale and Beck, 1966); in larvae and pupae of *Chironomus tentans* (Firling, 1977); higher level of protein in haemolymph of females than males in *Leptinotarsa decemlineata* (Dortland and de Korte, 1978). Chippendale (1970 a & b) attributed the decrease in haemolymph protein content of *Diatraea grandiosella* in pharate pupal stage to the uptake of certain specific proteins by the fat body and this was also true in *Malcosoma grandiosella* (Loughton and West, 1965). Haemolymph proteins also contribute in cuticle formation and as they are taken up by epidermal cells, converted to peptides and utilised in epidermis (Koeppe et al., 1972; Koeppe and Gilbert, 1973).

Similar trends of protein concentration were also observed in the present study in the haemolymph of *A. mylitta*. During larval growth, holometabolous insects synthesise and accumulate specific proteins termed as larval storage proteins (Roberts and Brock, 1981a & b; Levenbook, 1985; Levenbook and Bauer, 1984). The findings of changes in protein patterns during larval, pupal and adult stages in *A. mylitta* (Sinha et al., 1985) also support the trends of protein concentration observed in the present study during non-diapausing and diapausing generations. This is also supported by the works on protein estimation in larvae and pupae of *Spodoptera litura* (Prasad and Nath, 1985); and in haemolymph of Bombyx mori (Nagata and Yoshitake, 1988). Higher protein level was also reported in females of *A. mylitta* ecoraces than males (Kar et al., 1994). Presence of diapause specific proteins is also reported in insects (Brown, 1980; Adedokun and Denlinger, 1985; Joplin et al., 1990). Concentration of proteins in the haemolymph of *P. gossypiella* (Salama et al., 1992); in *Hyphantria cunea* (Keun et al., 1995); in *L. dispar* (Masler and Kovaleva, 1997); male specific proteins in *Galleria mellonella* (Lee et al., 1998) have also been reported. However, the present study, more precisely, reflects upon the ontogenetic availability of protein in haemolymph of *A. mylitta*, which may be further helpful to work out on storage proteins, diapause associated proteins etc. in *A. mylitta*.

5.9.5.2. Qualitative haemolymph proteins

Insect haemolymph, the coelomic fluid, circulates freely around the various tissues in haemolymph and is involved in chemical exchanges, which occur between them (Buck, 1953). Haemolymph of insects serves as an excellent barometer in determining the biochemical development in insects. A number of studies have shown that haemolymph proteins change during development of silk moth (Telfer and Williams, 1953; Telfer, 1954; Laufer, 1959). Earlier works (Loughton and West, 1965; Nakasone and Kobayashi, 1965 and Doira, 1968) revealed that as the larvae proceed gradually through programmed morphogenesis, the proteinogram becomes complex with the increasing number of bands. It was also true in case of *A. mylitta* as the number of bands increased from first to fourth instar in both non-diapausing and diapausing generations (Tables 4.79 & 4.80; Figs. 4.19 & 4.20). Levenbook and Bauer (1984) described larval haemolymph proteins as insect storage proteins because haemolymph proteins are synthesised and secreted by fat body in larval haemolymph during larval growth and feeding (Miller and Silhack, 1982). No single description applies equally to the several storage proteins described (Levenbook and Bauer, 1984), but they may be characterised by the different criteria (Roberts and Brock, 1981 a & b; Levenbook, 1985). First, these proteins are few in number, generally two or three, for any particular species. Second, the site of synthesis is predominantly, if not exclusively, the fat body, which secrets them into the haemolymph. Third, the concentration of storage proteins in the haemolymph increases enormously with the increase in number of instars, especially during last instars and spinning larval stages. The deeply stained bands comprising larger proportion of haemolymph proteins in late fifth instars and spinning larvae were also observed in both non-diapausing and diapausing generations in

A. mylitta in present study. These proteins are reported to account for most, up to 80% of the total haemolymph proteins in insects. Remnants of storage proteins may carry over into adult life, but soon, after adult eclosion, only traces of these proteins can be detected. After cessation of feeding, the haemolymph proteins to a greater or lesser extent are taken by the fat body in the form of protein storage granules. These storage proteins are highly aromatic, containing an exceptionally higher proportion of tyrosine and phenylalanine residues (Levenbook and Bauer, 1984). The number of protein band reduced in later stage of development in *A. mylitta* and this may be attributed either to resorption of haemolymph proteins by the fat body followed by deposition of the same in the form of potentially crystalline protein storage at the time of larval pupal development (Thomson, 1975; Tojo et al., 1978). Disappearance of certain bands at certain stage of insect and their reappearance in subsequent stages is not clearly understood, but Thomson (1975) stated that such a situation can arise during ultimate fate of fat body cell membranes and cytolysis followed by the release of the proteins into the haemolymph.

Specific diapause associated proteins have been reported in insects (Brown, 1980). Three such haemolymph proteins are reported in Lymantria decemlineata (De Loof and De Wilde, 1970). It is suggested that a hormone originating from corpora allatta regulates the concentration of diapause associated proteins and diapause development (De Wilde, 1960; De Loof, 1972; Brown, 1976; Yin and Chippandale, 1976). During diapause, diapausing insects accumulate these proteins in haemolymph or these proteins are present in haemolymph throughout the diapause (Brown and Chippandale, 1978; Peferoen et al., 1982; Dilwith et al., 1985; Joplin et al., 1990). The two protein bands, one each of less mobility and higher mobility which have been observed in the present study, gradually started disappearing in late pupal stage and disappeared completely in adult stages of *A. mylitta* (Table 4.80; Fig. 4.20). These observations are in conformity with the observations made by Nijhout, 1994. Joplin et al. (1990), while working on the disappearing proteins of *Sarcophaga crassipalpis* demonstrated that (i) fewer proteins are synthesised by the brain of diapausing pupae, (ii) the number of brain proteins synthesised was greater in older diapausing pupae, (iii) a set of diapause specific brain proteins were seen and (iv) there are no apparent differences in expression of brain proteins during photosensitive stages. A wide range of possibilities exists for protein functions during diapause (Chippandale, 1988), but has been little evidence thus far for their role in diapause regulation (Joplin et al., 1990). Diapause proteins remain in high concentration in haemolymph during diapause, but rapidly disappear after diapause termination (Lefevere et al., 1989). In non-diapausing individuals, those proteins are either not detected or are found in much lower concentration (Peferoen et al., 1982; Salama and Miller, 1992 and De Kort, 1996). These views are also supported by the observations made during present study in *A. mylitta*.

The female specific proteins or vitellogenins are essential for ovarian development and yolk formation in insects. The female specific proteins observed in the haemolymph of A. mylitta starting from V larval instars to adults. Siaktos (1960) observed that transformation of less mobile lipoprotein to lipoproteins of high electrophoretic mobilities occur in plasma of *P. americana*. Haemolymph of insect larvae contains several high molecular weight proteins like arylphorins (Telfer et al., 1983) and lipophorins (Shapiro et al., 1988) in addition to the storage proteins. All these proteins are synthesised in the fat body and released into the haemolymph (Kunkel and Lawer, 1974) to be incorporated later into various organs, especially ovaries (Rohrkasten and Ferienze, 1985; Valle, 1993). Vitellogenic female proteins isolated from a variety of insect species consists of a polypeptide chain, which varies in number from one in *Apis mellifera* to four or more in *Tenebrio molitor, Lecuophaea maderae, Periplaneta americana, L. migratoria* and *Rhodnius prolixus* (Harnish and White, 1982; Kunkel and Nordin, 1985). Micro-vitellogenin is reported to be another group of female-specific proteins (Telfer et al., 1984) and is known as 'relcutin'. These are reported to occur in both the sexes, but are more pre-dominant in female, and are deposited in eggs as reported in *Hyalophora cercopia* (Telfer and Kulakosky, 1984), and in *Manduca sexta* (Kawooya and Law, 1988).

Since the synthesis of protein is determined by organisms genome and these proteins are the interaction of development leading to ontogenetic processes, these results into presence of a stage-specific and sex-specific protein patterns. In the present investigation, a comparison has been made based on their mobility without direct identification/molecular weight determination. However, the present study may facilitate to further proceed on antigenic analysis, which allows positive identification of specific proteins in the tissues. Possible immunological approach in future may help in understanding the utilisation of specific proteins by tissues during their development. The nature of disappearance of diapausing proteins in *A. mylitta* and generated ecological data on their survival in nature, once married together, may pave the way for best possible exploitation of this commercially important sericigenous insect for the humankind.

5.9.6. Total free amino acids

Insect haemolymph contains usually high concentration of free amino acids in solution. Compared with human haemolymph, the concentration of free amino acids in insect haemolymph is from 20 to 100 times higher (Auclair, 1953 a & b). A special characteristic of insects, especially holometabolous insects is the high titre of amino acid present in their haemolymph (Tsuji, 1909 - cited by Wyatt, 1961; Wyatt, 1961; Florkin, 1954; Price, 1961; Chen, 1962, 1966; Schoffeniels and Gilles, 1970; Evans and Crossley, 1974; Firling, 1977). It has also been demonstrated by many authors that the amino acids, in addition to their function as protein constituents, they enter into diverse metabolic pathways and participate in many other physiological activities. A large part of our knowledge about amino acids in insects derives from studies dealing with larval development (Chen, 1966).

However, in the present study, the ontogenetic availability of quantitative free amino acids was estimated in both of its' non-diapausing and diapausing generations. The results indicated that the quantitative level of free amino acids in the haemolymph of A. mylitta showed an increasing trend with development of instars both in NDD and DD generations. The amino acids quantum came down during the late fifth instar and spinning larval stage. Thereafter, the level of amino acids was almost static through-out the pupal stage with little variation. In adult females and males of the generations, the quantitative free amino acids were the highest. In entire larval stages i. e from I to V instars, the level of amino acids was higher in diapause destined generation. Though, there was a quick fall in the level of amino acids in spinning stage. It was significantly higher ($P < 0.01$) in DD larvae. However, NDD late pupae and adults had significantly higher free amino acids in their haemolymph than DD generation (Table 4.81; Fig. 5.29).

Between the sexes up to V instar males had significantly higher amino acids than females. The adult males too had significantly higher amino acids than females in NDD generation (Table 4.82). In case of sexes of DD generation the males had higher amino acids than females in all the stages except in spinning and pre pupal stage (Table 4.83; Fig. 5.29).

Thus quantum of free amino acids was recorded to be highest in adult males than females and the level consistently increased from I instar to adults barring a deflection in spinning and late fifth larval stage (Tables 4.82 & 4.83; Fig. 5.29).

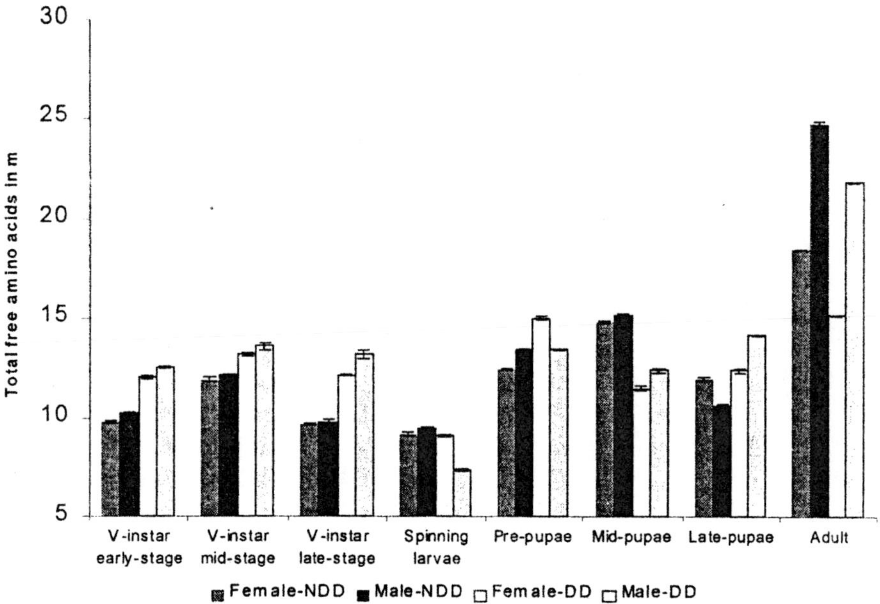

Fig. 5.29. Comparison between mean values of total free amino acids (FAA) of non-diapausing & diapausing generations and females & males of *A. mylitta* (I = ±SE).

In general, all amino acids which were commonly contained in proteins have been identified, either in tissues extract or haemolymph of insects (Chen, 1962). Among these, the aliphatic amino acids have dominant part. Both qualitative and quantitative changes in the content of free amino acids have been followed during post embryonic development of a large variety of insect species. The more extensive information are those on *Galleria mellonella* (Auclair and Dubreuil, 1952); *Bombyx mori* (Florkin, 1937a; 1959; Sarlet et al., 1952; Wyatt and Kalf, 1956); in *Prodenia eridania* (Levenbook, 1962); in *Corcyra cephalonoca* (Ganti and Shanmugasundaram, 1963); in *Culex pipens* (Chen, 1958); in haemocytes and plasma of *Calliphora vicina* (Evans and Crossley, 1974); in *Antheraea mylitta*; larval and pupal haemolymph (Jolly et al., 1972; Agarwal and Jolly, 1981; Sinha et al., 1988). Free amino acids were also estimated in III instar larvae of *Calliphora vicina* wherein it was observed that the amino acids found in haemocyte fraction only accounted for 6 % of the total amino acids of the haemolymph (Evans and Crossley. 1974). High percentage of dicarboxylic amino acids, glutamate and aspartate, 62 to 69 %, respectively, appeared to be sequestered in haemocyte fraction at 72 hour prior to pupation. Jolly et al. (1972) have observed the presence of, thirteen amino acids viz., cisteic acid, aspartic acid, proline, valine, leucine/isoleucine, cystine, glutamic acid, serine, glycine, threonine, lysine, histidne and a-alanine in haemolymph. Interestingly, b-alanine, glutamine, asparagines and arginine were absent in haemolymph protein but were present in free state (Jolly et al., 1972). Contrarily, hydroxyl proline, phenyl alanine, thyroxine and methionine absent in free state were present in haemolymph protein. Eighteen amino acids were also found by Sinha et al. (1988) in the haemolymph during larval instars and nineteen in pupal stages of *A. mylitta*. They observed that there were variation in the concentration of amino acids in both NDD and DD generations. Qualitative difference was not observed in number of amino acids in larvae and pupae of *A. mylitta* (Sinha et al., 1988). Concentration of amino acids has been reported to decrease towards the end of pupal stage. The reasons for absence of b-alanine and tyrosine in adult may be attributed to their utilization in formation of adult structure in *A. mylitta* (Sinha et al., 1988; Agrell, 1949). Cystine appeared in fifth instar and was observed to present till adult emergence (Vanderzant and Reiser, 1956; Golberg and Meillon, 1948; Sinha et al., 1988). There are also reports that during diapause under natural conditions, proline, serine and alanine increases in haemolymph of *Diatraea grandiosella; O. nubialis, Pectinophora gossy-piella, Anthraea pernyi, Heliothis armigera* (Mansingh, 1967; Morgan and Chippandale, 1983; Rostom et al., 1972; Boctor, 1981). Mitsuhashi (1978) reported high titre of serine in the haemolymph of diapausing larvae of a pyralid moth *Chilo supressalis* and a high titre of serine in the haemolymph of post-diapausing pupae of *Mamestra brassicae*. Comparative biochemical study on quantitative presence of total free amino acids in bivoltine and trivoltine types of *A. mylitta* gradually increased as soon as adult emergence was nearing (Mohanty and Mitra, 1988). Similar trend in *Philosamia ricini* have been observed (Pant, 1984). In the present study too, the increased level of quantitative free amino acids was observed in late pupal stage and highest in adults of *A. mylitta* and supports the views of above authors. The amino acid concentrations are shown to increase with increase in instars except in late V instar in *Pericallia ricini* (Jeyakumar et al., 1995). The increase or decrease in level of free amino acids is reported to be governed by the season of rearing too in *A. assama* (Sharma et al., 1995). Changes in free amino acid at various developmental stages in *Achaea janata* L. are reported (Chahnde, 1998). More so, several functions are attributed to free amino acids in insect haemolymph such as (i) osmo-regulation (Beadle and Shaw, 1950), (ii) protein synthesis (Buck, 1953), (iii) energy production for flight (Sactor, 1965), cocoon spinning (Fukuda and Matuda, 1953; Wyatt, 1961). Amino acid pool in *A. mylitta* might be useful for the above activities and also in regulation of diapause stage which needs further studies in due course of time or may be considered as future research strategy.

5.9.7. Uric acid (mg/ml)

Haemolymph contains high amount of end products of the nitrogen metabolism such as uric acid, allantoin, allantoic acid, urea and ammonia. Uric acid is often very concentrated, sometimes near saturation and crystals are commonly found in haemolymph (Florkin and Jeuniaux, 1974). In *A. mylitta* ontogenetic availability of uric acid was estimated in both of its non-diapausing and diapausing generations in the present study. Uric acid was generally higher in DD larvae than NDD larvae from I instar to IV instar (I-NDD 0.0098 mg/l, IV-NDD - 0.019 mg/ml & I-DD 0.0300mg/l, IV- DD Larvae - 0.0574 mg/ml). In different aged V instar larvae of both the sexes, similar trend was not observed and it fluctuated because sometimes females or males of NDD generation had higher uric acid than their respective counterparts of DD generation. However, female pupal stages and adults of NDD generation had higher uric acid than males (Table 4.82 to 4.84; Fig. 5.30).

The haemolymph uric acid concentration have been reported in several Lepidopteran insects such as in *Antheraea*, it was 1.9 mg/ml (Leifert, 1935), in *Bombyx mori* adult - 1.35

to 1.9 mg/ml (Florkin, 1936a & b), larvae - 6.9, 0.8 to 1.6 (Jucci and Dieana, 1930); in *Prodenia* - 1.5 to mg/ml (Babers, 1938). Similarly, uric acid ranges from 5.3 to 10.7 in the larvae of *Apis* (Bogojawlensky, 1932), 5.3 to 14.5 in *Hydrophilus* (Gerould, 1929); to 20.0 in the larva of *Celerio* (Heller and Macklowska, 1930). In silkworm *B. mori*, it is reported to be about 1.0 mg/ml at other times (Kuwana, 1937) but it may go up to 1.6 mg/ml in mature larva when waste products are discharged into haemolymph (Jucci and Dienna, 1930). Uric acid in haemolymph of L. dispar fifth instar was reported to be 0.2 mM/l (Pannabacker et al., 1992). The trend was observed to be characteristically higher in the solitary phase of locusts than in active gregarious phase (Matthee, 1945). It may occur absorbed in protein colloid (Drilhon and Florence, 1946).

Ammonia is also reported to be an excretory product in both aquatic and terrestrial insects (Levenbook, 1966; Cocharan, 1977; Firling, 1977). Uric acid is generally regarded as a major nitrogenous end-product, which is produced in fat body and is eliminated after transportation through haemolymph (Mullin, 1985). Its concentration levels are generally low because urates are insoluble. Uric acid concentration is influenced by feeding (Barrett and Friend, 1975; Ramakrishna and Pawar, 1975), photoperiodicity (Hillard and Butz, 1969), development (Buckner and Caldwill, 1980) and parasitisation (Condon and Gordon, 1977). Uric acid in insects is also found as uric acid ribosides in *G. mellonella* (Krazyzanowska and Niemierko, 1979, 1980) which facilitates transport of citric acid and these are gradually stored in fat body in the form of ptridine (Mullin, 1985) and prior to adult emergence through haemolymph they are stored in wings (Harmsen, 1966).

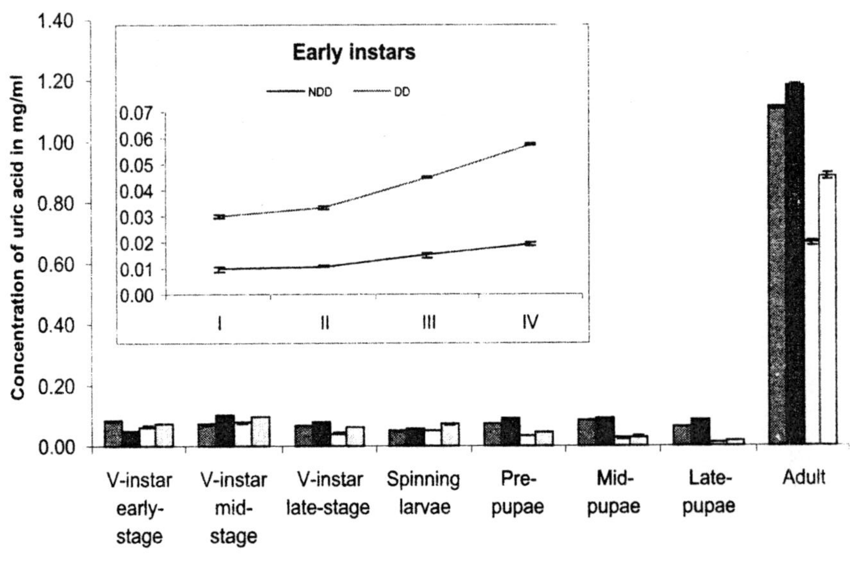

Fig. 5.30. Comparison between mean values of Uric Acid (I = ±SE) of non-diapausing & diapausing generations and females & males of *A. mylitta*

5.9.8. Haemolymph Cholesterol

Lipids are generally defined as compounds which are poorly soluble in water and are soluble in organic solvents (Mullin, 1985). Cholesterol appears to be transported in haemolymph plasma, predominantly in a non-esterified form (Downer and Chino, 1979). In the present study, the cholesterol content was observed to be higher in the non-diapausing larvae of A. mylitta from second to spinning larval stage than diapausing larvae (II-NDD - 4.6588 mg/ml & II-DD - 2.8626 mg/ml; SLF-NDD - 1.8612 mg/ml & SLF-DD - 0.4752 mg/ml; SLM-NDD - 1.2429 mg/ml & SLM-DD - 0.3564 mg/ml). Though this trend was observed to be fluctuating in pupal stage as sometimes content was higher in diapausing generation stages and sometimes in non-diapausing generation stages. The adults of non-diapausing generation had higher cholesterol than diapausing generation (Table 4.85; Fig. 5.31).

The difference in haemolymph cholesterol level was generally higher in females from mid-aged V instar to the emergence of adults in non-diapausing generation and the same trend was observed in larval and pupal stages of diapausing generation, though the trend reversed in diapausing generation adults (AF-DD - 0.3388 mg/ml & AM-DD - 1.1368 mg/ml) - (Tables 4.86 & 4.87; Fig. 5.31).

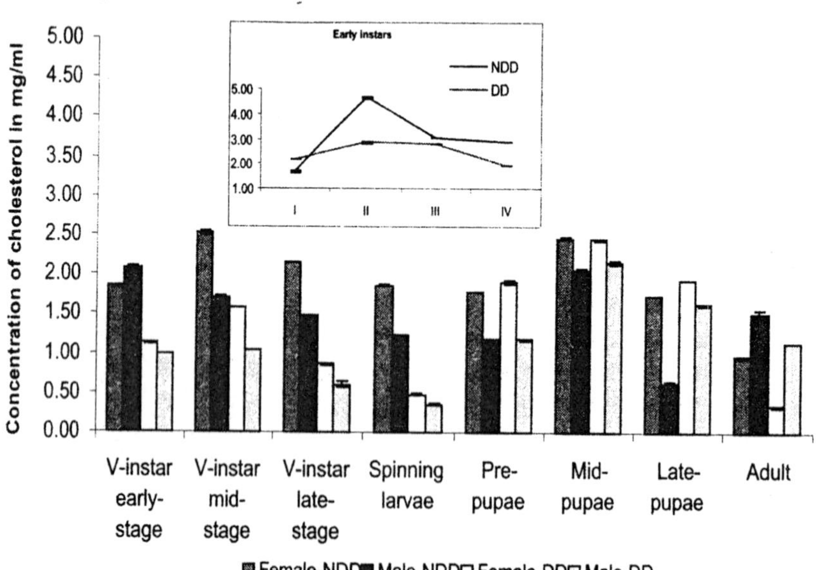

Fig. 5.31. Comparison between mean values of cholesterol (I = ±SE) of non-diapausing & diapausing generations and females & males of *A. mylitta*

All species of arthropods investigated so far are incapable of the de novo synthesis of sterols from small organic precursors (Chino and Gilbert, 1971; Grienisen, 1994). Sterol is needed for all life processes, including ecdysteroid bio-synthesis and therefore, it is derived from the diet. Carnivorous insects ingest cholesterol (c), while phytophagous insects ingest mainly 24-alkylated sterols (Grienisen, 1994) and some are able to synthesise 14C sterol from 2-14C acetate (Saito et al., 1963). Certain insects selectively take up cholesterol from dietary mixture of sterols (Thompson et al., 1963). Levels of cholesterol have also been investigated in different insect species during development (Gilbert, 1964; Goodfellow and Gilbert, 1967; Pant, 1984). This has also been observed in the haemolymph of bi-voltine and tri-voltine *A. mylitta* (Mohanty and Mitra, 1988; Sinha et al., 1992). The trend observed by them was similar to the present study, though they did not indicate about the presence of cholesterol in different sexes. The depletion in cholesterol level during late pupal development is perhaps due to it's active uptake and utilisation during histogenesis in membrane assembly in a general and by stereogenic tissues in particular for which it serves as a pre-cursor (Chapman, 1980). The findings on cholesterol in *P. ricini* (Pant and Nautiyal, 1974), in *A. mylitta* (Mohanty and Mitra, 1988; Sinha et al., 1992) are in agreement with the present observation. Cholesterol is also reported to be essential for the sustained spermatogenetic activity and the maturation of spermatozoa in the testicular tissues of male insects (Blum, 1970). Zaidi and Khan (1974) have reported the variation in cholesterol content in *Dysdercus cingulatus* and Hurakadli et al. (1989) in *Philosamia ricini*. The higher lipid content in males is most likely correlated with mating behaviour (Nakasone and Ito, 1967). The females have been known to possess low amount of lipid and fats than that of males (Niemierko et al., 1956). High concentration of lipid content in V instar mature and spinning larvae are also reported in *A. mylitta* (Agrawal et al., 1981). Lower lipid level in females is due to their utilisation in egg development (Agrawal et al., 1981; Akao, 1931; Ozaki, 1936). During the period preceding the emergence, a decrease in level of cholesterol has been observed (Chino and Gilbert, 1971; Sinha et al., 1992). Males have higher level of cholesterol as evidenced in the present study, because it is used in the form of ecdysteroid hormones, which is involved in the maturation of spermatocytes (Blum and Meckensen, 1968 cited by Blum, 1970; Schmidt and William, 1953 cited by Dumser, 1980) and is necessary for development of testes of several saturniid moths (Kambysellis and Williams, 1971a,b), *Rhodnius prolixus* (Dumser and Davey, 1974), in *Periplaneta americana* (Ishi et al., 1963) and in P. ricini (Hurkadli et al., 1989). Thus, the need for cholesterol in late pupal stage in *A. mylitta* may mobilise cholesterol from haemolymph and other tissues for proper development of reproductive organs.

5.10. Conclusion

- Thus, it can be concluded that there are five types of haemocytes in *Antheraea mylitta* Drury, namely, Prohaemocytes, Plasmatocytes, Granulocytes, Spherulocytes and Oenocytoids. Many fold increases were recorded in pre-pupal stages of both non-diapausing and diapausing generations. This may be due to their requirement for removal of histolysed tissues during pupation.

- An increase in number of haemocytes with increase in larval instars suggests that the growing larval age requires increased nutrients, transport of nutrient, phagocytosis and metabolism to meet its future demand such as silk synthesis, metamorphosis and enduring adverse climatic conditions (in diapausing generation). However, this needs confirmation through specific studies in *A. mylitta* in future.

- The present study was restricted to localisation of general protein, bound lipids, nucleic acids, glycogens (carbohydrates), alkaline phosphatases activity in the haemocytes of larvae, pupae and adults of *A. mylitta*. Reaction of these chemicals in varying degrees in different ontogenetic stages of *A. mylitta* in its NDD and DD generation in different categories of haemocytes confirms their role in transportation, storage and supply of nutritional material for proper survival, growth and reproduction.

- Specific gravity was a bit higher in non-diapausing generation (1.0159 to 1.0182) than diapausing generation (1.0047 to 1.0093) in larval forms. As the diapause generation has to pass long summer, they may accumulate more water, which brings down the specific gravity.
- In the present study, mean range of hydrogen ion concentration in *A. mylitta* was observed to range in between 6.87 to 6.93 in non-diapausing generation and 6.71 to 6.91 in diapausing generation. This study confirms the presence of hydrogen ion towards acidic side of neutrality as reported in other Lepidoptera. Probably buffering capacity of haemolymph of *A. mylitta* falls within this range.
- The levels of Na^+ was lowest in *A. mylitta*, whereas the concentration of K^+, Mg^{++} and to some extent Ca^{++} was high in all the life-stages. The higher level of Mg^{++} and K^+ found in A. mylitta may be attributed to its evolution along the Angiosperms or its progressive adaptations to plant diets (high K^+ and Mg^{++}).
- Fluctuation in the level of trehalose reflects upon the energy related physiological state of insects. Increased level of trehalose in diapausing females and males and its gradual decline during diapause termination or during emergence confirms its role in normal development and metamorphosis of *A. mylitta*.
- Increase in haemolymph protein during course of larval development followed by decrease/ sharp decline in pupal stages and further decrease in adults may be due to their utilization during adult morphogenesis and completion of generation. The present study will thus facilitate to proceed further on antigenic analysis, which allows positive identification of specific proteins in the tissues. Possible immunological approach in future may help in understanding the utilisation of specific proteins by tissues during their development in *A. mylitta*. The nature of disappearance of diapausing proteins in *A. mylitta* and generated ecological data on their survival in nature, once married together, may pave the way for best possible exploitation of this commercially important sericigenous insect for the humankind.
- Quantitative level of free amino acids in the haemolymph of *A. mylitta* showed an increasing trend with development of instars both in NDD and DD generations. The amino acids quantum came down during the late fifth instar and spinning larval stage. Changes in the quantum of free amino acids in *A. mylitta* may be attributed to their role in osmo-regulation, protein synthesis, energy production and cocoon spinning etc.
- The uric acid may be the main excretory material in *A. mylitta* like other Lepidopteran and sericigenous insects. A fluctuating trend in uric acid concentration may be attributed to be under the influence of feeding, photoperiodicity, development etc.
- In the present study, the fluctuation of cholesterol content in non-diapausing generation and diapausing generation may be due to their possible role in ecdysteroid bio-synthesis, and depletion in cholesterol level during late pupal development is perhaps due to it's

active uptake and utilisation during histogenesis, in membrane assembly in general and particularly for spermatogenetic activity and the maturation of spermatozoa in the testicular tissues of male *A. mylitta*.

- Thus, the present study entails about the type of haemocytes, their population structure in haemolymph, their possible role in storage of nutritional material, and the changes in level of different inorganic and organic metabolites across different developmental stages. This will further facilitate the exploitation of *A. mylitta* for its best use by the tasar silk industry in varied agro-climatic conditions.

7

Bibliography

A

Abrol, D. P. (1995). Haemocytes of honeybees, *Apis mellifera* L. and *Apis cerana indica* F. (Hymenoptera: Apidae) and their alteration by mite parasitosis. Indian Bee Journal. 57(1): 6?7.

Adedokun, T. A. and Denlinger, D. L. (1985). Metabolic reserves associated with pupal diapause in the flesh fly *Sarcophaga crassipalpis*. 31(3): 229-233.

Agrawal, S. C. and Jolly, M.S. (1981). Protein bound amino acids in the larval and pupal haemolymph of Antheraea mylitta D. Ind. J. Entomol. 43(2): 145-147.

Agrawal, S. C.; Jolly, M. S. and Banerjee, N. D. (1981). Lipid content in silkworm, Antheraea mylitta D. Indian J. Entomol. 43(3): 294-296.

Agrell, I. (1949). Occurrence and metabolism of free amino acids during insect metamorphosis. Acta Physiol. Scand. 18: 247-285.

Ahmad, A. (1988). Free haemocytes in adult Polistes hebraeus Fabr. (Hymenoptera: Vespidae). J. Entomol Res. 12: 28-35.

Ahmad, A. (1992). Study of haemocytes of two coleopterous insects, *Aulacophora foveicollis* Lucas (Chrysomelidae) and *Mylabris pustulata* Thunberg (Cantharidae). J. Animal Morphol. Physiol. 17: 1655-1676.

Ahmad, A. (1995). Changes in haemocyte counts following topical application of Becdysone and Makisterone-A on fifth instar nymphs of *Dysdercus cingulatus* Fabr. (Hemiptera: Pyrrhocoridae). Entomologia Croatica. 1: 41?48.

Akai, H. and Sato, H. (1971). An ultrastructure of the haemopoietic organs of the silkworm, *Bombyx mori*. J. Insect Physiol. 17: 1665-1676.

Akai, H. and Sato. S. (1973). Ultrastructure of the larval haemocytes of the silkworm, *Bombyx mori* (Lepidoptera: Bombycidae). Int. J. Insect Morph. Embryol. 2(3): 207-231.

Akao, A. (1931). On the metabolism of protein, fats and carbohydrates in pupae of silkworms. J. Chosenmed. Ass. 21: 928-933.

Akao, A. (1935). Zinc and reproduction. Keijo Jour. Med. 6: 49-60.

Ambrose, D. P. and George, P. J. E. (1994). Total and differential count of haemocytes in the life stages and adult haemocyte morphology in Catamiarus brevipennis Serville (Insecta:

Heteroptera: Reduviidae). Environment and Ecology. 12(4): 860-864.

Ambrose, D. P. and George, P. J. E. (1996b). Total and differential haemocyte diversity in three morphs of Rhynocoris marginatus (Insecta: Heteroptera: Reduviidae). Fresenius Environmental Bulletin. 5(3-4): 202?206.

Ambrose, D. P.; Maran, S. P. M. and Rajan, K. (1999). Haemogram changes in the life stages of the assassin bug, Rhynocoris marginatus Fabricius (Heteroptera: Reduviidae). Shashpa. 6(1): 23-28.

Anderson, R. S.; Day, N. K. B. and Good, R. A. (1972). Specific hemagglutinin and a modulator of complement in cockroach haemolymph. Infect. Immun. 5: 55-59.

Andrew, W. (1965). Comparative Haematology. Grune and Stratton, New York.**

Annonymous (1958). B.D.H. Biological Stains and Staining Methods. British Drug House Ltd., Poole, England. pp. 1-34.

Annonymous (2004). Sericulture and Silk Industry Statistics - 2003. Published by Central Sik Board, Munistry of Textiles, Governmet of India. pp. 1-7.

Arnold, J. W. (1952). The haemocytes of the Mediterranean flour moth, *Ephestia kühniella* Zell. (Lepidoptera: Pyralidae). Can. J. Zool. 30: 352-364.

Arnold, J. W. (1961). Further observations on amoeboid haemocytes in Blaberus giganteus L. (Orthoptera: Blattidae). Can. J. Zool. 37: 371-375.

Arnold, J. W. (1972). Haemocytology in insect biosystematics: The Prospect. Can. Entomol. 104: 655-659.

Arnold, J. W. (1974). The haemocytes of the insects. In The Physiology of Insecta (M. Rockstein, ed.). Vol. 5. Second edition. Academic Press, New York. pp. 201-254.

Arnold, J. W. (1979a). Controversies about haemocyte types in insects. In Insect Haemocytes (A. P. Gupta, ed.). Cambridge University Press, Cambridge. pp. 231-258.

Arnold, J. W. (1979b). Biosystematics of the genus Euxoa (Lepidoptera: Noctuidae). XIII. Further observations on haemocytological distinctions between species. Can. Entomol. 111: 771-775.

Arnold, J. W. and Hinks, C. F. (1975). Biosynthesis of the genus Euxoa (Lepidoptera: Noctuidae). II. Haemocytological distinction between two closely related species, Euxoa campestris (Grote) and E. declarata Walker Can. Entomol. 107: 1095-1100.

Arnold, J. W. and Hinks, C. F. (1976). Haemopoiesis in Lepidoptera. I. The multiplication of circulating haemocytes. Can. J. Zool. 54(6): 1003-1012.

Arnold, J. W. and Hinks, C. F. (1979). Insects haemocytes under light microscopy: techniques. In Insects Hemocytes, Development, Forms, Functions, and Techniques. Edited by A. P. Gupta, Cambridge University Press, New York. pp-531-558.

Arnold, J. W. and Salked, E. H. (1967). Morphology of the haemocytes of the giant cockroach, *Blaberus giganteus*, with histochemical tests. Can. Entomol. 99: 1138-1145.

Arnold, J. W. and Sohi, S. S. (1974). Haemocytes of Malacosoma disstria Hübner. (Lepidoptera: Lasiocampidae): Morphology of the cells in fresh blood and after cultivation in vitro. Can. J. Zool. 52(4): 481-485.

Arora, G. S. and Gupta, J. J. (1979). Memoirs of Zoological Survey of India. 17(1): 25-28.

Arvy, L. (1953). Données histologiques sur les leuconpoiése. C. R. Acad. Sci. 235: 1539-1541.

Arvy, L. (1956). Données histophysiologiques sur les centres leucopoietiques de quelques insectes holométaboles. 14th Int. Cong. Zool., Copenhagen, 478-479.

Arvy, L. and Lhoste, J. (1946). Les variation du lecucogramme au cours de la métamorphose cher Forficula auricularia L. Bull. Soc. Zool. France. 70: 144-148.

Arvy, L.; Gabe, M. and Lhoste, J. (1948). Contributions á l'étude morphologique du sang de Chrysomela decimlineata Say. Bull. Biol. 82: 37-60.

Arvy, L.; Gabe, M. and Lhoste, J. (1949). Contribution s' le'tude morphologique des sang des Mantidae. Rev. Canad. Bid. 8: 184-200.

Ashhurst, D. E. (1979). Hemocytes and connective tissue: a critical assessment. In Insect Hemocytes. Edited by A. P. Gupta. Pages 319-330. Cambridge University Press, Cambridge.

Ashhurst, D. E. and Richards, A. G. (1964). Some histochemical observations on the blood cells of the wax moth, *Galleria mellonella* L. J. Morph.114: 247-253.

Ashida, M. and Dhoke, K. (1980). Activation of prophenoloxidase by the activating enzyme of the silkworm, *Bombyx mori.* Insect Biochem. 10: 37-47.

Auclair, J. I. and Dubrenil, R. (1953a). Can. J. Zool. 31: 30. **

Auclair, J. L. (1953b). The qualitative and quantitative changes in free amino acid concentration of haemolymph of the wax moth, *Galleria mellonella* (Lepidoptera). Can. J. Entomol. 85: 63.

Auclair, J. L. and Dubrenil, R. (1952). A simple ultra-micro method for the quantitative estimation of amino acids by paper partition chromatography. Can. J. Zool. 30: 109-113.

Azambuza, P. D.; Gracia E. S. and Ratcliffe, N. A. (1991). Aspects of classification of Hemiptera haemocytes from six triatomine species. Mem. de Instit. Os waldo Cruz. 86: 1-10.

B

Babers, F. H. (1938). Blood, Prodenia (Lepidoptera). J. Agric.Res. 57: 697-106.

Babers, F. H. (1941). Glycogen in Prodenia (Lepidoptera). J. Agric. Res. 62: 509-530.

Bade, M. L. and Wyatt, G. R. (1962). Metabolic conversions during pupation of Cecropia silkworm 1. deposition and utilization of nutrient reserves. Biochem J. 83: 470-478.

Baerwald, R. J. and Bousch, G. M. (1970). Fine structures of the haemocytes of Periplaneta americana (Orthoptera: Blattidae) with particular reference to marginal bundles. J. Ultrastruct. Res. 31: 151-156.

Balavenkatasubbaiah, M.; Natraju, B.; Thiagrajan, V. and Datta, R. K. (2001). Haemocyte counts in different breeds of silkworm, *Bombyx mori* L. and their changes during the progressive infection of BmNPV. Indian J. Seric. 40(2): 158-162.

Baldwin, E. (1962). "The nature of biochemistry". Cambridge Unuiversity Press, London and Newyork.

Bandopadhyaya, N. (1970). Effect of Malathion on the haemolymph and malpighian tubules of Periplaneta americana. D. Phil. Thesis. Calcutta University, Calcutta, India.**

Bardoloi, S. and Hazarika, L. K. (1992). Seasonal variations of body weight, lipid reserves, blood volumes and haemocyte population of *Antheraea assama* Westwood (Lepidoptera: Saturniidae). Environ. Entomol. 21(60): 1398-1403.

Bardoloi, S. and Hazarika, L. K. (1995). Variation in haemocyte population during different larval instars of Antheraea assama Westwood (Lepidoptera: Saturniidae) and their roles in the defence mechanism of the insect. Journal of the Assam Science Society. 37(2): 96?102.

Barduco, M. C.; Gregorio, E. A. and Toledo, L. A. (1988). Haemocytes of Diatraea saccharalis (Lepidoptera: Pyralidae) during the larval period: morphological and quantitative study. Revista Brasileira de Biologia. 48(4): 925-932.

Barrat, J. O. and Arnold, G. (1910). Blood beetles. Quart. Jour. Microscop. Sci. 56: 149-165. **

Barrett, F. M. and Friend, W. G. (1975). Differences in the concentration of free amino acids in the haemolymph of adult male and female Rhodnius prolixus. Comp. Biochem. Physiol. 52B: 427-431.

Bartninkaite, I. (1995). Comparative morphology of the haemocytes of larvae of four species of ermine moths of the family Yponomeutidae. Ekologija. 3: 3?10.

Bauer, E.; Trenczek, T. and Dorn, S. (1998). Instar-dependent haemocyte changes in Pieris brassicae after parasitisation by Cotesia glomerata. Entomologia Experimentalis et Applicata. 88(1): 49?58.

Baust, J. G. and Miller, L. K. (1970). Variations in glycerol content and its influence on cold hardiness in Alaskan carabid beetle, Pterostichus brevicornis. J. Insect Physiol. 16: 979-995.

Bayram, S. and Kilincer, N. (1987). Some changes in the blood cells of larvae of Agrotis segetum Den.-Schiff. (Lepidoptera: Noctuidae) parasitized by Periscepsia carbonaria Panz. (Diptera: Tachinidae). Turkiye I. Entomoloji Kongresi Bildirileri, 13-16 Ekim, Ege Universitesi, Bornova, Izmir. pp. 437-446.

Beadle, L. C. (1939). Regulation of haemolymph, mosquito. Jour. Exp, Biol. 19: 346-362.

Beadle, L. C. and Shaw, J. (1950). The retention of salt and regulation of the non-protein fraction in the blood of aquatic larva, *Sialis lutaria*. J. Exp. Biol. 27: 96-100.

Beaulaton, J. (1979). Haemocytes and hemocytopoiesis in silkworm. Biochimie. 61: 157-164.

Beaulaton, J. and Monpeyssin, M. (1976). Ultrastructure and cytochemistry of haemocytes of Antheraea pernyi Guer. (Lepidoptera: Attacidae) during the fifth larvál stage. I. prohaemocytes, plasmatocytes and granulocytes. J. Ultrastucture Res. 55: 143-146.

Beaulaton, J. and Monpeyssin, M. (1977). Ultrastructure et cytochimie des haemocytes d'Antheraea pernyi Guer. (Lepidoptera: Attacidae). II. Cellules a spherules et oenocytoides. Biol. Cell. 28(1): 13-18.

Beck, G.; Cardinale, S.; Wang Lan; Reiner, M.; Sugumaran, M. and Wang, L. (1996). Characterization of a defense complex consisting of interleukin 1 and phenol oxidase from the haemolymph of the tobacco hornworm, Manduca sexta. Journal of Biological Chemistry. 271(19): 11035?11038.

Beck, S. D. and Hanec, W. (1960). Diapause in European corn borer, *Pyrausta nubilalis* (Hbn). J. Insect. Physiol. 4:304-318.

Begum, R.; Gohain, R. and Hazarika, L. K. (1992a). Age related changes in the haemocytes of the fifth instar Philosamia ricini Boisd. J. Enviro. Biol. 17(1): 149?155.

Begum, R.; Gohain, R. and Hazarika, L. K. (1992b). Humidity induced degeneration of haemocytes of eri worm, *Philosamia ricini* Boisd. (Lepidoptera: Saturniidae). Bulletin of Entomology, New Delhi (India). 33(1?2): 146?154.

Begum, R.; Gohain, R. and Hazarika, L. K. (1994). Age related changes in the haemocytes of

Philosamia ricini Boisd. (Lepidoptera: Saturniidae). Proceedings of National Seminar on Life Sciences. Dibrugarh University. pp. 113-118.

Begum, R.; Gohain, R. and Hazarika, L. K. (1998). Detoxication of deltamethrin by the haemocytes of Philosamia ricini Boisd. (Lepidoptera: Saturniidae). Indian Journal of Sericulture. 37(2): 142?147.

Benassi, C. A.; Colombo, G. and Peretti, G. (1959). Experientia. 15: 457-458.**

Benassi, C. A.; Colombo, G.; Peretti, G. and Allergi, G. (1961). Biochem. J. 80: 332.**

Berenbaum, N. C. (1958). The Histochemistry of bound lipids. Quar. J. Micro. Sci. 99: 231-242.

Bernays, E. A. and Chapman, R. F. (1974) Changes in haemolymph osmotic pressure in Locusta migratoria larvae in relation to feeding. J. Ent. 48A: 149-155.

Best, F. (1906). In Histochemistry Theoretical and AppliedII Edition. Pearse, A. G. E. (Editor).1961.J & A Churchill, London, pp 1-998.

Berry, S. J.; Krishnakumaran, A. and Schneiderman, H. A. (1964). Control of synthesis of RNA and protein in diapausing and injured Cecropia pupae. Science (Washington, D. C.). 146: 928-940.

Bhadur, J. and Pathak, J. P. N. (1971). Changes in the total haemocyte counts of the bug, Halys dentata under certain specific conditions. J. Insect Physiol. 17: 29-34.

Bhoumik, S. K. (1972). Bacteria in the haemocoelic fluid of the cockroach, *Periplaneta americana* L. and effects of their exhibition. D. Phil. Thesis, Calcutta University, Calcutta, India. **

Bialaszewicz, K. and Landan, C. (1938). Blood and heart changes during growth and metamorphosis, silkworm. Acta. Biol. Exptl. 12: 307-320.

Binachi, De A. G. and Terra, W. R. (1976). Haemolymph protein patterns during the spinning stage and metamorphosis of Rhychosiara americana. J. Insect Physiol. 22: 535-540.

Bindokas, V. P. and Adams, M. E. (1988). Haemolymph composition of the tobacco budworm, Heliothis virescens F. (Lepidoptera: Noctuidae). Comp. Biochem. Physiol. 90A: 151-155.

Bischof C. and Ortel, J. (1996). The effects of parasitism by Glyptapanteles liparidis (Braconidae: Hymenoptera) on the haemolymph and total body composition of gypsy moth larvae, Lymantria dispar (Lymantriidae: Lepidoptera). Parasitology Research. 82(8): 687?692.

Blaich, R. (1969). Zool. Jabrb. Anat. 86: 576-614.**

Blum, M. S. (1970). Invertebrate testis in "The Testis" (Johnson, A. D.; Gomes, W. R. and Vandermark, N. L. eds.). Academic Press, New York and London, Vol. II. pp. 393-431.

Boctor, I. Z. (1981). Changes in the free amino acids of the haemolymph of diapause and non-diapause pupae of the cotton ball worm, *Heliothis armigera* Hbn. (Lepidoptera: Noctuidae). Experintia 37: 125-126.

Bodnaryk, R. P. and Morrison, P. E. (1966). J. Insect Physiol. 12: 963-976.

Bogdan, C; Rollinghoff, M. and Diefenbach, A. (2000). Reactive oxygen and reactive nitrogen intermediates in innate and specific immunity. Curr. Opin. Immunol. 12: 64-76.

Bogojawlensky, K. S. (1932). Haemocytes. Arch. Anat. Hist. and Embryol. 11: 361-386.

Bone, G. J. (1945). Sodium to potassium ratio and diet. Ann. Soc. Roy. Zool. Belgique. 75: 123-132.

Bonhog, P. F. (1955). Histochemical studies on the overies nurse tissues and oocytes of the milkwed bug, Oncopeltus fasciatus. Dall. J. Morph. 96: 381-411

Brecher, L. (1925). Physical chemistry of blood, caterpillar. Zeitcher. Vergleich. Physiol. 2: 691-713.

Brehlin, M. M. (1972). Etude du mecnisme de la coagulation de I'hemolymph d'unacriduien, *Locusta migratoria* migratoroides (R and F). Acrida I: 161-175

Brehelin, J. A.; Matz, G. and Porte, A. (1975). Encapsulation of implanted foreign bodies by haemocytes in Locusta migratoria and Melonontha melonontha, Cell Tissuue Res. 160: 283-289.

Brehelin, M. and Zachary, D. (1986). Insect haemocytes: A new classification to rule out the controversy. In "Immunity in Invertbrates" (M. Brehlin, ed.). Springer-Verlag, Berlin. pp 36-48.

Bronskill, J. F. (1959). Embryology of Pimpla turionellae L. (Hymenoptera: Ichneumonidae). Can. J. Zool. 37: 655-688.

Bronskill, J. F. (1960). The capsule and its relation to the embryogenesis of the icheneumonid parasitoid Mesoleius tenthridinus Morl. In thc larch saw fly Pristiphora erichsonii Htg. (Mymenoptera: Tenthridinidae). Can. J. Zool. 38: 769-775.

Brown, J. J. (1976). Camperative study of the ultrastructure and metabolism of the fat body of diapausing and non-diapausing larvae of the southwestern corn borer, *Diatraea grandiosella.* Ph.D.dissertatation, Unversity of Missouri. 95 pages. Unversity microfilms. Ann Arbor. Mich. No. 77-5590.**

Brown, J. J. (1980). Haemolymph protein reserves of diapausing and non-diapausing codling moth larvae, *Cydia pomonella* L. (Lepidoptera: Tortricidae). J. Insect Physiol. 26: 487-491.

Brown, J. J. and Chippendale, G. M. (1978). Juvenile hormone and a protein associated with the larval diapause of the South Western corn-borer, *Diatraea grandiosella*. J. Insect Physiol. 16: 1057-1069.

Buck, T. B. (1953). Physiological properties and chemical composition of insect blood. In Insect Physiology (K. D. Roeder, ed.). John Wiley, NewYork pp 147-190.

Buckner, J. S. and Caldwill, J. M. (1980). Uric acid levels during the last larval instar of Manduca sexta, an abrupt transition from excretion to storage in fat body. J. Insect Physiol. 26: 27-32.

Bühlmann, G. (1974). Viltellogenin in adulten Weibchon der Schabe Nauphoeta cinerea. Immunologische Unter suchungen über Herkunft und Einbau. Rev. Suisse Zool. 81: 642-647.

Burton, R. C.; Hopper, D. G.; Sauer, J. R. and Frick, J. H. (1972). Properties of the haemolymph of the corn eat worm. Comp. Biochem. Physiol. 42B: 713-716.

Butt, T. M. and Shield, K. S. (1996). The structure and behaviour of gypsy moth (*Lymantria dispar)* haemocytes. Journal of Invertebrate Pathology. 68(1): 1?14.

Buxton, P. A. (1930). Water loss and humidity, mealworm. Proc. Roy. Soc., London. B106: 560-577.

C

Carey, F. G. and Wyatt, G. R. (1960). Uridninediphosphate derivatives in the tissues and haemolymph of insects. Biochem. Biophys. Acta. 41: 178-179.

Cebesoy, S. and Ayvali, C. (1996). Some histochemical observations on the haemocytes of Agrotis segetum (Dennis and Schiff.) (Lepidoptera: Noctuidae). Turkish Journal of Zoology. 20(3): 231?239.

Chahnde, R. D. (1998). Changes in free amino acids at various developmental stages in the Achaea janata L. Geobios. 25: 288-290.

Chang, C. K.; Liu, F. and Feng, H. (1964). Metabolism of eri-silkworm during metamorphosis. II. The properties of trehalose and its effects on metabolism. Acta. Ent. Sin. 13: 494-502.

Chapman, R. F. (1980). The Insects, Structure and Function. ELBS Edition, Great Britain. 819 pp.

Chaudhury, A.; Kapila, M. L.; Dubey, O. P.; Sinha, S. S. and Medda, A. K. (1993). Turn over of carbohydrates in tissues of tasar silkworms Antherea mylitta Drury (Lepidoptera: Saturniidae) during late stage of diapause. Ind. J. Physiol. and Allied Sci. 47(3): 128-135.

Chen, G. M. (1962). Free amino acids in insects. In "Amino Acid Pools" Elsevier Publishing Co. Amsterdam. pp. 115-135.

Chen, P. S. (1956). Electrophoretische Bestimmung des Proteingehaltes in Blut normaler und letaler (ltr) larven von *Drosophilla melanogaster*. Rev. Suisse Zool. 63: 216-229.

Chen, P. S. (1958). Studies on the protein metabolism of Culex pipens L. II. Quantitative differences in free amino acid during larval and pupal development. J. Insect. Physiol. 2: 38-51.

Chen, P. S. (1959). Trennug der Blutproteine von Drosphila von Culex - Larven mittels Starke-Gel-Elektrophorese. Rev. Suisee Zool. 66: 280-289.

Chen, P. S. (1966) Amino acid and protein metabolism in insect development. In Advances in insect Physiology. Edited by J. W. L. Beament, J. E Trehen and V. B. Wigglesworth 3: 53-132

Chen, P. S. (1971). Biochemical aspects of insect development. S. Krager, Basel. pp. 55-56.

Chen, P. S. and Hadorn, E. (1954). Vergleidende Unter suchugen iiber die freien Aminos a uren in der larvalen haemolympheon Drosophilla, *Ephestia* and *Corethra*. Revue Suisse Zool. 61: 437-451.

Chen, P. S. and Levenbook, L. (1966a). Studies on haemolymph proteins of the blow fly Phormia regina. I. Changes in ontogenetic pattern. J. Insect. Physiol. 12: 1595-1609.

Chen, P. S. and Levenbook, L. (1966b). Studies on haemolymph of proteins of the blow fly Phormia regina. II: Synthesis and breakdown as revealed by isotopic labelling. J. Insect. Physiol. 12: 1611-1625.

Chino, H. (1957). Conversion of glycogen to sorbitol and glycerol in the diapausing egg of the Bombyx silkworm. Nature. 180: 606-607.

Chino, H. (1958). Carbohydrate metabolism in the diapausing egg of silkworm, *Bombyx mori*. II. Conversion of glycogen into sorbitol and glycerol during diapause. J. Insect Physiol. 2: 1-12.

Chino, H. and Gilbert, L. I. (1971). The uptake and transport of cholesterol by haemolymph lipoprotein. Insect biochem. J. 1: 337-347.

Chippanldale, G. M. (1988). Role of proteins in insect diapause. Endocrinological Frontiers in Physiological Insect Ecolgy (F. Sehanel, A. Zabza and D. L. Denlinger, eds.). Wroclaw. Technical Univ. Press, Wroclaw. pp 331-346.

Chippandale, G. M. (1970a). Metamorphic changes in the haemolymph and midgut proteins of the south western corn borer Diatraea grandiosella. J. Insect Physiol. 16: 1909-1920.

Chippandale, G. M. (1970b). Metabolic changes in fat body proteins of the South Western corn borer, *Diatraea grandiosella*. J. Insect Physiol. 16: 1057-1068.

Chippandale, G. M. and Beck, S. D. (1966). Haemolymph proteins of Ostrinia nubialis during diapause and pre pupal differentiation. J. Insect Physiol. 12: 1629-1638.

Chippandale, G. M. and Kilby, B. A. (1969). Relationship between the proteins of haemolymph and fat body during development of Pieris brassicae. J. Insect. Physiol. 15: 905-926.

Clark, E. W. and Chadbourne, D. S. (1960). The haemocytes of non-diapause and diapause larvae and pupae of the pink bollworm. Ann. Entomol. Soc. Am. 55: 682-685.

Clark, R. M. and Harvey, W. R. (1964). Cellular membrane formation by Plasmatocytes of diapausing cercopia pupae. J. Insect Physiol. 11: 161-175.

Clegg, T. S. and Evans, D. R. (1961). Blood trehalose and flight metabolism in the blow fly. Science. 134: 54-55.

Cocharan, D. G. (1977). Excretion in insects. In Insect Biochemistry and Function. (D. J. Candy and B. A. Kilby, eds.). Chapman and Hall, London. pp. 177-281.

Cohen, A. C. and Patana, R. (1982). Ontogenic and stress related changes in haemolymph chemistry of bee armyworm. Comp. Biochem. Physiol. 71A: 193-198.

Coles, G. C. (1965). The haemolymph and moulting to Rhodnius prolixus Stal. J. Insect Physiol. 11 : 1317-1323.

Colhoun, E. H. (1959). Physiological effects in organophosphorous poisonging. Can. J. Biochem. hysiol. 37: 1137-1144.

Collins, M. M. and West, R. D. (1961). Wild Silk moths of United states, Cedar Rapids, Iowa, Collins Radio Co. pp. 50-55.

Condon, W. J. and Gordon, R. (1977). Effects of the mermithid nematode, Mermis nigrescens on the levels of haemolymph and faecal uric acids in it's hosts: the migratory locust, Locusta migratoria. Cand. J. Zool. 55: 690-692.

Costin, N. M. (1975). Histochemical observation of the haemocytes of Locusta migratoria. Histochemical J. 7: 21-43.

Craig, R. and Clark, K. (1938). Buffers and pH of blood, caterpillars. Jour. Econ. Entomol. 31: 51-54.

Crossley, A. C. (1968). The fine structure and mechanism of breakdown of larval intersegmental muscles in the blow fly Calliphora erythrocephala. J. Insect Physiol. 14: 1389-1407.

Crossley, A. C. (1979). Biochemical and ultrastructural aspects of synthesis, storage and secretion in haemocytes. In Insect Haemocytes (A. P. Gupta, ed.). Cambridge University Press, Cambridge. pp 423-473.

Crossley, A. C. S. (1975). The cytophysiology of insect blood. Advances in Insect Physiology. 11: 117-122.

Crotch, W. J. B. (1956). A silkmoth rearers hand book. Amateur Entomologists Society, London.

Cuénot, L. (1896). Etudes physiologiques sur les Orthopteres. Arch. Biol. 14: 293-341.

Dadd, R. H. (1985). Nutrition organisms. In Comprehensive Physiology, biochemistry and Pharmacology (G. A. Kerkut and L. I. Gilbert, eds.). Vol. 4: 313-390. Pergamon Press, Oxford.
Dahlman, D. L. (1974). Haemolymph characteristics of developing tobacco horn worm reared as larvae on tobacco leaf or synthetic diet. Comp. Biochem. Physiol. 49A: 369-375.
Dajajakusumah, T. and Miles, P. W. (1966). Aust. J. Bio. Sci. 19: 1081-1094.
Danilevsky, A. S. (1965). Photoperiodism and seasonal development of insects (English Trans.). London: Oliver & Boyd. pp 282.
Datta, R. K.; Pramanik, H. And Chatterji, S. N. (1980). Electrophoretic studies on the haemolymph proteins of different breeds of mulberry silkworm Bombyx mori L. Proc. Cell. Biol. Conference, Calcutta: p23.
avis, B. J. (1964). Disc electrophoresis II. Methods and pplication to human serum proteins. Ann. N. Y. Acad. Sci. 121: 404-427.
vis, R. P. and Schneiderman, A. (1960). A autiradiographic study of wound healing in diapausing silkworm pupae. Anat. Rec. 131: 348**.
Kort, C. A. D. (1996). Cosmic influences on the expression of a specific gene in Colorado potato beetle: the diapause protein 1 gene . Archives of insect biochemistry and Physiology. 32: 567-573.
oof A. (1972). Diapause pheromone in non-diapausing last instar larvae, pupae and pharate dults of the Colorado beetle. J. Insect. Physiol. 18: 1039-1047.
oof, A. and De Wilde, J. (1970). The relationship between haemolymph proteins and itellogenesis in the Colorado potato beetle, Leptinotarsa decemlineata. J. Insect Physiol. 5: 1455-1466.
a, C.; Dunphy, G. B. and Rau, M. E. (2000). Interaction of haemocytes and prophenoloxidase stem of fifth instar nymph of Acheta domesticus with bacteria. Dev. Compar. Immunol. 367-379.
le, J. (1960). Diapause in the adult Colorado beetle (Leptinotarsa decemlineata) as an ocrine deficiency syndrome of the corpora allatta. The ontogeny of insects. (Acta Symposii utione Insectarum, Praha) (L. Hardy ed.) pp 226-230. academic press.
A. (1960). Flame Photometry. McGraw Hill Book Co. Inc. pp-295.
owski, S. R.; Galzowa and Rhoshdestwenska, W. (1932). Haemolymph pH and bolism, silkworm. Biochem. Zeitscler. 275: 445-463.
. F. (1958). General Cytochemical Methods. Vol. I. Academic press, New York, pp 423-443.
, D. L. (1985). Hormonal control of diapause. In Comprehensive Insect Physiology mistry and Pharnacology. Vol. 8:353-412. (Eds, Kerkut, G. A. and Gilbert, L. I.), non Press, Oxford.
(1947). Puparium formation, Sarcophaga (Diptera). Proc. Roy. Soc., London. B231:
.
1. (1958). Zonenelektrophoretische Untersuchungen der Hamolymph - Proteine von 1 in verdschiedenen Stadien der Larvenenontwicklung. Z. Naturforsch. 13b: 215-

Dilwith, J. W.; Lenz, C. J.; Chippendale, G. M. (1985). An immunological study of the diapaus associated protein of the southwestern corn borer, *Diatraea grandiosella* Insect biochen 15: 711-722.

Dinamarka, M. L. and Levenbook, L. (1966). Oxidation, utilization and incorporation into protei of alanine and lysine during metamorohosis of the blow fly, *Phormia regina* (Meign.). Arch Biochem. Biophys. 76: 71-78.

Doira, H. (1968). Development and sexual differences of blood proteins in the silkworm, *Bomb mori* . Sci. Bull. Fac. Agric. Kyushu Univ. 23: 205-214.

Dortland, J. F. and de Korte, C. A. D. (1978). Protein synthesis and storage in the fat body of colardo potato beetle, *Leptinotarsa decemlineata*. Insect Biochem. 8: 93-98.

Downer, R. G. H. and Chino, H. (1979). Cholesterol and cholesterol esters in haemolymph American cockroach, *Periplanata americana* L. Can. J. Zool. 57: 1333-1336.

Drilhon, A. (1934). Internal environment, Lepidoptera. Comp. Rend. Soc. Biol. 115: 1194-11

Drilhon, A. and Florence, G. (1946). Biochemistry of blood, Lepidoptera. Bull. Soc. Chim. I 28: 160-167.

Ducceschi, V. (1902). Blood, silkworm larvae. Atti. Reale Accad. Georogofili. Firenze. 80: 382.

Duchateau, Gh.; Florkin, M. and Leclercq, J. (1953). Arch. Int. physiol. Biochem. 67: 518-5

Duchateau-Bosson, G.; Jeuniaux, Ch. and Florkin, M. (1962). Effect of moulting, diet, hist on the amino acids of silkworm. Arch. Int. Physiol. Biochem. 70: 287-291.

Duchateau-Bosson, G.; Jeuniaux, Ch. and Florkin, M. (1963). Contributiens á la biochim ver á soie XXVII. Trehalose, trehaleise et mue. Arch. Int. Physiol. Biochem. 71: 566-

Dumser, J. B. (1980). The regulation of spermatogenesis in insects. Ann. Rev. Entomol. 25 369.

Dumser, J. B. and Davey, K. G. (1974). Endocrinological and other factors influencing development in Rhodnius prolixus. Can J. Zool. 52: 1011-1022.

E

Edney, E. B. (1977). Water Balance in Land Arthropods. Springer-Verlag, Berlin.

Egorova, T. A. (1963). Trehalose in the tissues of Antheraea pernyi. Natuch. Dokl. Vys Biol. Nauki 1: 88-91.

Ehlers, D.; Quast, M. and Mohrig, W. (1989). Adhesiveness of haemocytes of the larvae of mellonella L. Zoologische Jahrbucher. 93(3): 319-326.

El Ghar, G. E. S. A.; Khalil, M. E. and Eid, T. M. (1996). Some biochemical effects extracts in the black cutworm, *Agrotis ipsilon* (Hufnagel) (Lepidoptera: Noctuidae) of Applied Entomology. 120(8): 477?482.

El Mandarawy, M. B. R. (1997). Effects of insect diapause and parasitisation of a Bracon brevicornis Wesm. on the haemolymph of its host *Sesamia cretica* Led. J the Egyptian Society of Parasitology. 27(3): 805?815.

El-Ibarashy, M. T. (1965). A comparative study of metabolic effects of the corpus allat adults coleoptera in relation to diapause.

Meded. Landb. Hogesch. Wagninen. 65(11): 65.

Engelmann, F. (19790. Insect Vitellogenin: Identification, biosynthesis and role in vitellogenesis. Advances In Insect Physiology. 14: 49-107.

Epstein, W. (1930). Origin of voltinism, Sericaria (Lepidoptera). Russki Zool. Zhur 10: 77-89.

Eslin, P. and Prevost, G. (1996). Variation in Drosophila concentration of haemocytes associated with different ability to encapsulate Asobara tabida larval parasitoid. Journal of Insect Physiology. 42(6): 549?555.

Evans, P. D. and Crossely, A. C. (1974). Free amino acids in the haemocytes and plasma of the larva of Calliphora vicinia. J. Exp. Biol. 61: 463-472.

F

Feir, D. (1964a). Haemocyte counts of the large milkweed bug Oncopeltus fasciatus Nature. 202: 1136-37.

Feir, D. (1964b). Liquid nitrogen fixation: A new method for haemocyte counts and mitotic indices in tissues section. Ann. Entom. Soc. Am. 62(1): 246-247.

Feir, D. (1979). Multiplication of haemocytes. In Insect Haemocytes (A. P. Gupta ed.). Cambridge University Press, Cambridge. pp. 67-82.

Feir, D. and O'Connor, M. (1965). Mitotic activity in the haemocytes of Oncopeltus fasciatus Dall. Experimental cell research 39: 637-642.

Feir, D. and Drazywda, I. (1969). Comp. Biochem. Physiol. 31:197-201.**

Fenoglio, C.; Bernardini, P. and Gervaso, M. V. (1993). Cytochemical characterization of the haemocytes of Leucophaea maderae (Dictyoptera. Blaberoidea). Journal of Morphology. 218(2): 115?126.

Feulgen, R. and Rossenbeck, H. (1924). In Histochemistry Theorotical and applied by Pearse, A. G. E. II Edition, Publ. J. & A.Churchill. London. 1961. pp.1-998.

Fink, D. E. (1925). Metabolism during development. Jour. Gen. Physiol. 7: 527-545.

Firling, C. (1977). Amino acid and protein changes in the haemolymph of developing fourth instar Chironomous tentans.J. Insect Physiol. 23: 17-22.

Fittingoff, C. M. and Riddiford, I.M. (1990). Heat sensitivity and protein synthesis during heat shock in tobacco horn worm, *Manduca sexta*. J. Comp. Physiol. B160: 349-356.

Florkin, M. (1936a). Protein content blood. Comp. Rend. Soc. Biol. 123: 1024-1026.

Florkin, M. (1936b). Degree of urema. Comp. Rend. Soc. Biol. 123: 1247-1249.

Florkin, M. (1937a). Contributions á létude du plasma sanguine des insects. Mém. Acad. R. Méd. Belg. 16: 1-69.

Florkin, M. (1937b). Variations de composition du plasma sanguin au cours de la metamorphose du ver á soie. Arch. Int. Physiol. 60 : 17-31.

Florkin, M. (1954). Aspects zooliogiques es pools d'ácids aminés non protéiques. Bull. Soc. Zool. Fr. 79: 369-407.

Florkin, M. (1959). The free amino acids of insect haemolymph. Proc. VI Int. Congr. Biochem. 12: 63-73.

Florkin, M. and Jeuniaux, C. (1974). Haemolymph composition. In Physiology of Insecta (M. Rockstein, ed.) Vol. V., 2nd edition. pp. 255-307.

Fox, F. R. and Mills, R. R. (1969). Changes in haemolymph and cuticle proteins during the moulting process in the American cockroach. Comp. Biochem. Physiol. 29: 1187-1195.

Fraenkel, G. and Blewett, M. (1944). The utilisation of metabolic water in insects. Bull. Ent. Res. 35: 127-139.

Francois, J. (1974). Etude ultrastructurale des hémocytes du Thysanoure Thermobia domestica (Insecta: Aptérygoté). Peclobiologia. 14: 157-162.

Francois, J. (1975a). Haemocytes and the haemopoietic organs of Thermobia domestica (Thysanura: Lepismatidae). Int. J. Insect Morphol. Embryol. 4: 477-494.

Francois, J. (1975b). L én capsulation hemocytaire experimentale chez le lépisme Thermobia domestica. J. Insect Physiol. 21: 1535-1546.

Fukuda, T. and Matuda, M. (1953). Nippon Sanshigaka Zasshi (In Japanese, English abstract). 22: 243.

Fukuda, S. (1952). Function of the pupal brain and suboesophageal ganglion in production of non-daipause and diapause eggs. Proc. Japan Acad. 29: 582-586.

Fukuda, T. and Hayashi, T. (1958). J. Biochem. Japan 45: 469-474.**

Furusawa, T.; Narutaki, A.; Mitsuda, K. (1993). Changes in amino acid pools during embryonic development of the Japanese oak silkworm, Antheraea yamamai (Lepidoptera: Saturniidae). Applied Entomology and Zoology. 28(2): 234-237.

Fyhn, H. J. and Saether, T. (1970). Regulation of the haemolymph osmolarity during metamorphosis in the oak Silkmoth, *Antheraea pernyi*. J. Insect Physiol. 16: 263-269.

G

Gaffinet, G. and Gregoire, C. H. (1975). Coagulocyte alterations in clotting haemolymph of Carausius moronus L. Arch. Inst. Physiol. Biochem. 83(4): 707-722.

Ganti, Y. and Shanmugasundaram, E. R. B. (1963). A study on the free amino acids during growth and metamorphosis of Corcyra cephalonica St. J. Exp. Zool. 152: 1-4.

Gardiner, E. M. M. and Strand, M. R. (1999). Monoclonal antibodies bind distinct classes of haemocytes in the moth Pseudoplusia includens. J. Insect Physiol. 45(2): 113?126.

Geiger, J. G.; Krolak, J. M. and Mills, R. R. (1977). Possible involvement of cockroach haemocytes in the synthesis and storage of cuticle protein. J. Insect Physiol. 23: 227-230.

Gerould. J. H. (1929). History, reversal of heat action. Biol. Bull. 64: 424-431.

Gese, P. K. (1950). Concentration of inorganic constituents in the blood, *Samia* (Lepidoptera) pupa. Physiol. Zool. 23: 109-113.

Giannotti, E. and Caetano, F. H. (1991). Morphological characterization of the haemocytes of Polistes lanio lanio (Hymenoptera: Vespidae) during post? embryonic development. Revista Brasileira de Biologia. 51(1): 179?184.

Gilbert, L. I. (1964). Physiology of Insects, Acad. Press Inc. Ltd., New York. Vol. 1, pp. 149.

Gillespie, J. P.; Kanost, M. R. and Trenczek, T. (1997). Biological mediators of insect immunity. Annual Review of Entomology. 42: 611?643.

Gilliam, M. and Shimanuki, H. (1967). In vitro phagocytosis of Nosema apis spores by honey bee haemocytes. J. INvertbrate Pathology, 9:387-389.

Gillott, C. (1995). Entomology, 2nd edition. Plenum Press, New York. p. 639.

Glaser, R. W. (1925). pH of haemolymph. Jour. Gen. Physiol. 7: 599-602.

Goffinet, G. and Grégoire, C. H. (1975). Coagulocyte alterations in clotting haemolymph of Carausius moronus L. Arch. Int. Physiol. Biochem. 83(4): 702-722.

Golberg, I. and Meillon, B. (1948). The nutrition of the larva of Aedes aegypti L. 4. Protein and amino acid requirementa . Biochem. J. 43: 379-387.

Goodfellow, R. and Gilbert, L. I. (1967). Advances in Insect Physiology, Acad. Press Inc., New York. 4: pp. 69.

Gregoire, Ch. (1971). Haemolymph coagulation in arthropods. In chemical Zoology. Edited by M. Florkin, Vol. 6, Pages 145-186.

Grienisen, L. M. (1994). Recent advances in our knowledge of ecdysteroid biosynthesis in insects and crustaceans. Insect Biochem. Molec. Biol. 24 (2): 115-132.

Grimstone, A. V.; Rotherman, S. and Salt, G. (1967). An electron microscope study of capsule formation by insect blood cells. J. Cell. Sci. 2: 281-292.

Gringorten, J. L. and Friend, W. G. (1979). Haemolymph volume changes in Rhodnius prolixux during flight. J. Exp. Biol. 83: 325-333.

Grison, P. and Lee Berre, J. R. (1953). Queques conséquences physiologiques de l'inanition chez límago de Leptinotarsa decemlineata Say. (Col. Chrysomélides). Rev. Path. Vég. Ent. Agric. Fr. 32: 73-86.

Gupta, A. P. (1969). Studies of the blood of Meloidae (Coleptera). I. The haemocytes of Epicauta cinerea (Forster), and synonnymy of haemocyte terminologies. Cytologia. 34(2): 300-344.

Gupta, A. P. (1979a). Haemocyte types: their structures, synonymies, interrelationships and taxonomic significance. In Insect Haemocytes. (A. P. Gupta, ed.). Cambridge University Press, Cambridge. pp. 85-127.

Gupta, A. P. (1979b). Identification key for haemocyte types in hanging drop preparation. In Insect Haemocytes pp. 527-529. (Eds. A. P. Gupta), Cambridge university prees, Cambridge.

Gupta, A. P. (1983). Neurohemal and neurohemal-endocrine organs and their evolution in arthropods. In Neurohemal Organs of Arthropods. (A. P. Gupta, ed.). Charles C. Thomas, Publishers. Springfield, Illions. pp. 17-50.

Gupta, A. P. (1985). Cellular Elements in the haemolymph. In Comprehensive Insect Physiology, Bichemistry and Pharmacology. (G. A. Kerkut and L. I. Gilbert, eds.). Pergamon Press, Oxford, New York, Toronto, Sydney, Paris, Frankfurt. 3: 401-451.

Gupta, A. P. (1986). Haemocytes and Humoral Immunity. John Wiley, New York.

Gupta, A. P. (1991a). Gap cell junctions, cell adhesion molecules, and molecular basis of encapsulation. In Immunology of insects and other arthropods. (A. P. Gupta, ed.). pp. 133?167.

Gupta, A. P. (1991b). Insect immunocytes and other haemocytes: roles in cellular and humoral immunity. Immunology of Insects and Other Arthropods. (A. P. Gupta, ed.). pp. 19?118; 538.

Gupta, A. P. and Sutherland, D. J. (1966). In vitro transformation of the insect plasmatocytes in some insects. J. Insect Physiol. 12: 1369-1375.

Gupta, A. P. and Sutherland, D. J. (1967). Phase contrast and biochemical studies of spherule cells in cockroaches (dictyoptera) Ann. Entomol. Soc. Am. 60: 557-65.

Gupta, A. P. and Sutherland, D. J. (1968). Effect of sublethal doses of chlordane on the haemocytes and midgut epithelium of Periplanata americana. Ann. Entomol. Soc. Am. 61: 910-918.

H

Han, SungSik; Lee, MinHo; Kim, WooKap; Wago, HaruHisa; Yoe, SungMoon; Han, S. S.; Lee, M. H.; Kim, W. K.; Wago, H. H. and Yoe, S. M. (1998). Haemocytic differentiation in haemopoietic organ of Bombyx mori larvae. Zoological?Science. 15(3): 371?379.

Harmsen, R. (1966). A quantitative study of the pteridines in Piresis brassicae. I. During post embryonic development. J. Insect Physiol. 12: 9-22.

Harnish, D. G. and White, B. N. (1982). Insect Vitellins: Identification, purification and charcterisation from eight orders. J. Exp. Biol. 220: 1-10

Harpaz, F.; Kislev, N. And Zelcer (1969). I. Electron microscopic studies on haemocytes of the Egyptian, Spodoptera littoralis (Boisdual) infected with a neuclear polyhedrosis virus, as compared to noninfected haemocytes. J. Invertbr. Pathol. 14: 175-185.

Harvey, W. R. (1962). Metabolic aspects of insect diapause. Ann. Rev. Entomol. 7: 57-80.

Hasegawa, K. (1952). Studies on the voltinism of silkworm, Bombyx mori L. with special reference to organs concerning determination of voltinism. J. Fac. Agric. Tottori Univ. 1: 83-124.

Hawk, P. B. (1979). Hawk's Physiological Chemistry. Edited by Bernard L. Oser. Forteenth Edition. Publisher by Tata McGRAW-HILL Publishing Company Ltd. New Delhi, India. pp. 1046-1047; 1067; 1239-1241.

Hayakawa, Y and Chino, H. (1982a). Phosphofructokinase as a possible key enzyme regulating glycerol or trehalose accumulation in diapausing insects. Insect Biochem. 12: 639-642.

Hayakawa, Y. and Chino, H. (1982b). Temperature dependent activation of glycogen phosphorylase and synthatase of fat body of silkworm Philosamia cynthia. The possible mechanism of the temperature-dependent interconversion between glucogen and trehalose. Insect Biochem. 12: 361-368.

Hazarika, L. K. and Gupta, A. P. (1987). Variations in haemocyte populations during various developmental stages of Blattella germanica L. (Dictyoptera: Blattellidae). Zool. Sci. 4: 307-313.

Hazarika, L. K. and Gupta, A. P. (1997). Effect of juvenile hormone on haemocyte counts of Blattella germanica. Pesticide Res. J. 9(2): 252?255.

Hazarika, L. K.; Bardoloi, S. and Kakoti, A. (1994). Effects of host plants on haemocyte population and blood volume of Antheraea assama Westwood (Lepidoptera: Saturniidae). Sericologia. 34(2): 301-306.

Hegazi, E. M.; El?Shazli A. and Abd?El?Aziz, G. M. (1998). Effect of superparasitism of Microplitis rufiventris parasitoid on the total and differential haemocyte counts of its host, Spodoptera littoralis. Alexandria Journal of Agricultural Research. 43(2): 89?102.

Heller, J. (1932). Role of haemolymph in metabolism and metamorphosis, Lepidoptera. 255: 205-221.

Heller, J. and Moklowska, A (1930). Composition and changes in metamorphosis, blood, *Drosophila* (Lepidoptera). Biochem. Zeitchr. 219: 473-489.

Hernandez, S.; Lanz, H.; Rodriquez, M. H.; Torres, J. A.; Martinez, P. A. and Tsutsumi. (1999). Morphological and cytological characterisation of female Anopheles albimanus (Diptera: Culicidae) haemocytes. Cambridge University Press, Cambidge. pp. 29-66.

Hillard, S. D. and Butz, A. (1969). Daily fluctuations in the concentration of total sugars and uric acids in the haemolymph of Periplaneta americana. Ann. Ent. Soc. Amer. 62: 71-74.

Hinks, C. F. and Arnold, J. W. (1977). Haemopoiesis in Lepidoptera. II. The role of the haemopoietic organs. Canadian J. Zool. 55(10): 1740-1755.

Hirano, M. and Yamashita, O. (1983). Developmental changes in trehalose synthesis in fat body of the silkworm, Bombyx mori: Trehalsoe synthetase related to regulation of haemolymph trehalose during metamorphosis. Insect Biochem. 13(6): 593-599.

Hirano, M. and Yamashita, O. (1980). Changes in utilisation and turn over rate of haemolymph trehalose during the metamorphosis of Bombyx mori. J. Seri. Sci. Jap. 49: 509-511.

Hoffmann, J. A. (1967a). Etude de la récupération hémoctaire aprés hemoraggies experimantales cher l'orthoptera,. *Locusta migratoria* L. J. Insect Physiol. 15: 1375-1384.

Hoffmann, J. A. (1967b). Etude des oenocytoids chez Locusta migratoria L. (Orthoptera). Archiv. Zool. Exp. Gen. 108: 251-291.

Hoffmann, J. A. (1969). Etude de la recuperation hémocytaire après hémorrages expérimentales chez lórthoptére, Locusta migratoria. J. Insect Physiol. 15: 1375-84.

Hoffmann, J. A. (1970). Regulations endocrines de la production et de la differenciation des hemocytes cher. Un insecte Orthoptera: *Locusta migratoria* L. Gen Comp. Endocr. 15: 198-219.

Hoffmann, D.; Brehelin, M. and Hoffman, J. A. (1977). Premiers resultants sur les reactions de defense antibacteriennes de larves de Locusta migratoria. Ann. Parasitol. Hum. Comp. 52 : 87-88.

Hollande, A. C. (1909). Contribution á létude du sang des Coleopteres. Arch. Zool. Exp. Gén (Ser. 5) 2: 271-294.

Hollande, A. C. (1911). Etudes histologiques comparée du sang des insectes á hémorrhee et des insectes sans hemorrhée et des insectes. Arch. Zool. Exp. Gén. (Ser. 5) 6: 283-323.

Hopf, H. S. (1940). Phosphorus distribution at high temperature, haemolymph. Biochem. J. 34: 1396-1401.

Horhov, D. W. and Dunn, P. E. (1982). Changes in circulation haemocyte population of Manduca sexta larvae following injection of Bacteria. J. Invertbr. Pathol. 40: 327-339.

Horie, Y. (1961). Physiological studies on the alimentary canal of the silkworm, Bombyx mori III. Absorption and utilization of carbohydrates. Bull. Seri. Expt. Stn. 16, 287-309.

Hotkiss, R. D. (1948). In Histochemistry Theoretical and Applied. Edited by A. G. Pearse. 2nd Edition. J. & A. Churchill Ltd., London. 1961. p.1-998.

Howden, C. F. and Kilby, B. A.(1960). Biochemical studies on insect haemolymph. I. variations in reducing power with age and the effect of diet. J. Insect Physiol. 4: 258-269.

Hrdy, I. (1958). The blood picture of imago of the cricket, Gryllus domesticus L. Acta. Soc. Entomol. Cechoslov. 54: 305-311.

Hurkadli, H.K.; Hodi, M. A. and Nadkarni, V. B. (1989). Cholesterol content in the developing and adult testes of eri silkworm, *Philosamia ricini* (Hutt.). Sericologia 29(1): 121-124.

I

Ibrahim, S. M.; El?Maasarawy, S. A. S.; El?Sheikh, M. A. K. (1993). Haemocyte pattern in Agrotis ipsilon F1 progeny of gamma?irradiated pupae. Bulletin of the Entomological Society of Egypt, Economic?Series. 20: 141?150.

Ishi, S.; Kapilanis, T. N. and Robins, W. E. (1963). Distribution and fate of 4-C14 cholesterol in the adult male American cockroach. Ann. Entomol. Soc. Am. 56: 115.

Ivanova-Kasas, O. M. (1959). Die embryonale Entwicklung der Blattwespe Pontonia cabreae L. (Hymenoptera: Tenthredinidae). Zool. Jahrb. Anat. Ontog. Tiere. 77: 193-228.

Iwama, R. and Rashida, M. (1986). Biosynthesis of prophenoloxidase in haemocytes of larval haemolymph of the silkworm, Bombyx mori. Insect Biochem. 16: 547-555.

J

Jankovic Hladni, M.; Chen, A. C.; Ivanovic, J.; Djordjevic, S.; Stanic, V.; Peric, V. and Frusic, M. (1992). Effects of diet and temperature on Morimus funereus larval haemolymph cation concentrations. Archives of Insect Biochemistry and Physiology (USA). 20(3): 205-214.

Jeuniaux, C. (1958) Resorption du liquide exuvial chez la vér a soie *(Bombyx mori, L.)*. Arch. Int. Physiol. Biochem. 66: 121-122.

Jeuniaux, Ch. (1961). Arch. Int. Physiol. Biochem. 69: 750-751.**

Jeyakumar, A.; Prakash, D. S. and Kannan, S. (1995). Impact of varying biochemical profiles of Ricinus communis Linn. on the haemodynamics of *Pericallia ricini* Fabr. (Arctiidae: Lepidoptera). Entomon. 20(3?4): 169-173.

Jiang, H.; Wang, Y.; Ma, C. and Kanost, M. R. (1997). Subunit composition of Pro-phenoloxidase from Manduca sexta : Molecular cloning of subunit pro-PO P1. Insect Biochem.mol. Biol. 27: 835-850.

Jolly, M. S.; Sinha, A. K. and Agrawal, S. C. (1972). Free amino acids in the larval and pupal haemolymph of Antherea mylitta D. reared on *Terminalia tomentosa.* Indian J. Seri., 11: 63-67.

Jones, J. C. (1956). The haemocytes of Sarcophaga bullata Parker. J. Morph. 99(2): 233-257.

Jones, J. C. (1959). A phase contrast study of the blood cells inProdenia eridania (Order: Lepidoptera). Quart. J. Mic. Sci. 100(10):17-23

Jones, J. C. (1962). Current concepts concerning insect haemocytes. Amm. Zoologist. 2: 209-246.

Jones, J. C. (1964). The circulatory system of insects. In The Physiology of Insecta (M. Rockestein, ed.). Vol. 3. Academic Press, New York. pp. 23-94.

Jones, J. C. (1965). The haemocytes of Rhodnius prolixus (Stal.) Biol.Bull. (Woods hole). 129: 282-294.

Jones, J. C. (1967a). Changes in the haemocyte picture of Galleria mellonella. Biol.Bull. (Woods hole). 132: 211-221.

Jones, J. C. (1967b). Estimated changes within the haemocyte population during the last larval and early pupal stages of Sarcophaga bullata Parker.

J. Insect Physiol. 13: 645-646.

Jones, J. C. (1967d). Effect of repeated haemolymph withdrawls and ligaturing the head on differential counts of Rhodnius prolixus Stal. J. Insect Physiol. 13: 1351-1360.

Jones, J. C. (1970). Haemocytopoeisis in insect. In Regulation of Hematopoiesis (A. S. Gordon, ed.). Appleton - Century - Crofts, New York. pp. 7-65.

Jones, J. C. (1979). Pathways and pitfalls in the classification and study of insect haemocytes. In Insect Haemocytes (A. P. Gupta, ed.). Cambridge University Press, Cambridge. pp. 279-300.

Jones, J. C. and Tauber, O. E. (1951). Normal total haemocyte counts of Tenebrio molitor. Ann. Entomological Society of America. 44: 539-543.

Joplin, K. H.; George, D. Y. and Denlinger, D. L. (1990). Diapause specific proteins expressed by the brain during the pupal diapause of the flesh fly, Sarcophaga crassipalpis. J. Insect Physiol. 36(4): 775-783.

Joshi, P. A. and Lambdin, P. L. (1996). The ultrastructure of haemocytes in Dactylopius confusus (Cockerel), and the role of granulocytes in the synthesis of cochineal dye. Protoplasma. 192(3?4): 199?216.

Joshua, H.; Fischi, J.; Henia. E.; Ishay, J. and Gitter, S. (1973). Cytological, biochemical and bacteriological data of the haemolymph of Vespa orientalis. Comp. Biochem. Physiol. 45B: 167-175.

Jucci, C. and Deiana, G. (1930). Uric acid in blood, silkworm. Boll. Soc. Ital. Biol. Sper. 5: 167-170.

Jung Hee N.; Kim E. S.; Choi Chung Sik; Lee InHee; Kang ChangSoo; Kim HakRyul; Nah, J. H.; Kim, E. S.; Choi, C. S.; Lee, I. H.; Kang, C. S. and Kim, H. R. (1996). Purification and characterization of storage protein of Hyphantria cunea. Korean Journal of Entomology. 26(3): 289-297.

Jungreis, A. M.; Jatlow, P and Wyatt, G. R. (1973). Inorganic ion composition of haemolymph of the cecropia silk moth, changes with diet and ontogeny. J. Insect Physiol. 19: 225-233.

Jungries, A. M. (1974). Physiology and composition of moulting fluid and mid gut lumen contents in Silkmoth, Hylophora cecropia. J. Comp. Physiol. 88: 113-127.

Jungries, A. M. (1978). The composition of larval-pupal moulting fluid in the tobacco horn worm, Manduca sexta. J. Insect Physiol. 24: 65-73.

Jungries, A. M. (1979). Physiology of moulting in insects. Advances in Insect Physiology. 14: 109-182.

Jungries, A. M. and Wyatt, G. R. (1972). Sugar release and penetration in insect fat body: Relations to regulation of haemolymph trehalose in developing stages of Hyalophora cecropia. Biol. Bull. Mar. Biol. Lab., Woods Hole. 143: 367-391.

K

Kambysellis, M. F. and Williams (1971a). In vitro development of insect tissues I. A macromolecular factor pre-requisite for silkworm spermatogenesis. Biol. Bull. 141: 527-540.

Kamby sellis, M. F. and Williams (1971b). In vitro development of insect tissues II. The role of ecdysone in the spermatogenesis of the silkworm. Biol. Bull. 141: 541-552.

Kar, P.K.; Srivastava, P.P.; Sinha, R.K. and B. R. R. Pd. (1994). Protein concentration in the pupal haemolymph of different races and F1 of top cross of Antheraea mylitta D. Indian J. seri. 33(2): 174-175.

Karaki, M. (1969). Electrophoretic studies on blood proteins of Samia cynthia ricini during development. Biol. J. 18: 32-34.

Kares, E. A. (1994). Effect of parasitism by Zele nigricornis (Walk) on the larval haemolymph of Spodoptera littoralis (Boisd.). Annals of Agricultural Science, Moshtohor., 32(4): 2139?2147.

Kawaguchi, Y. and Doira, H. (1965). Gene controlled incorporation of haemolymph protein in the silkworm. Proc. Seric. Sci. Kyushu Univ. 1:51**.

Kawaguchi, Y.; Akagi, S.; Banno, Y.; Koga, K.; Kuwano, E. and Doira, H. (1997). Protein profiles of larval haemolymph of Bombyx mori in artificially induced tri-moulting larvae using a 1,5; disubstituted imidazole. Journal of the Faculty of Agriculture, Kyushu University. 42(1?2): 203?209.

Kawooya, J. K. and Law, J. H. (1988). Role of lipophorins in lipid transport to the insect. J. Biol. Chem. 263: 8748-8753.

Keun Song Jeong ; Nah JungHee; Lee InHee; Kang ChangSoo; Kim HakRyul; Song J. K.; Nah, J. N.; Lee I. H.; Kang, C. S. and Kim, H. R. (1995). Comparative analysis of storage proteins of fall webworm (Hyphantria cunea Drury). Entomological Research Bulletin. 21: 25?33.

Kim, C. S.; Yoon, I. B. and Kim, W. K. (1990a). Ultrastructure of the haemocyte during metamorphosis in Lymantria dispar L. Korean Journal of Entomology. 20(3): 189?196.

Kim, C. S.; Yoon, I. B. and Kim, W. K. (1990b). Ultrastructure of the haemocytes in Lymantria dispar L. Korean Journal of Entomology. 20(4): 223-230.

Kinnear, J. F. and Thomson, J. A. (1975). Nature, origin and fate of haemolymph proteins in Calliphora. Insect biochem. 5: 531-552.

Koeppe, J. K. and Gilbert, L. I. (1973). Immunological evidence for the transport of haemolymph protein into the cuticle of Manduca sexta. Biol.Rev. 19: 615-624.

Koeppe, J. K.; Mills, R. and Brunet, P. C. J. (1972). Cuticle sclerotisation by the Americam cockroach: High molecular weight carriers of phenolics befa glucosides. Biol. Rev. 48: 333-375.

Koopmanschap, B.; Lammers, H. and Kort, S. de. (1992). Storage proteins are present in the haemolymph from larvae and adults of the Colorado potato beetle. Archives of Insect Biochemistry and Physiology (USA). 20(2): 119-133.

Kramer, K. J. ; Spiers, R. D. and Childs, C. N. (1978a). Insuline like glucagons-like peptides in insects (Manduca sexta) haemolymph. J. Biol. Chem. 10: 179-182.

Kramer, K. J.; Spiers, R. D. and Childs, C. N. (1978b). A method for separation of trehalose from insects. Anal. Biochem. 86: 692-696.

Krzyzanowska, M and Niemierko, W. (1979). Accumulation, transport and excretion of uric acid ribosides by ligated larvae of Galleria mellonella L. Insect. Biochem. 9: 11-18.

Krzyzanowska, M and Niemierko, W. (1980). Purine and uric acid ribosides in the ligated larvae of Galleria mellonella L. Insect. Biochem. 10: 323-330.

Kukra, M. And Weiser, J. (1973). Alanine aminotransferase, alkaline phosphatase and protease activity in Barathera brassiacae during microsporidian infection. J. Invertebr. Pathol. 21: 121-122.

Kunkel, J. G. and Lawer, D. M. (1974). Larval specific serum proteins in the order Dyctyoptera. I. Immunological characterization in larval *Blatella germanica* and cross reaction through out the order. Comp. Biochem. Physiol. B47:696-710.

Kulkarni, A. P. and Mehrotra, K. N. (1970). Amino acid nitrogen and proteins in the haemolymph of adult desert locusts Schistocerca gregaria. J. Insect Physiol. 16: 2181-2189.

Kunkel, J. G. and Nordin, J. H. (1985). Yolk proteins. Comp. Insect. Physiol. Biochem. Pharma. 1: 83-109.

Kurihara, Y.; Shimazu, T. and Wago, H. (1992a). Classification of haemocytes in the common cutworm, Spodoptera litura (Lepidoptera: Noctuidae). I. Phase microscopic study. Applied Entomology and Zoology. 27(2): 225?235.

Kurihara, Y.; Shimazu, T. and Wago, H. (1992b). Classification of haemocytes in the common cutworm, Spodoptera litura (Lepidoptera: Noctuidae). II. Possible roles of granular plasmatocytes and oenocytoids in the cellular defense reactions. Applied Entomology and Zoology. 27(2): 237?242.

Kurnick, N. B. (1955). In Histochemistry Theorotical and applied by Pearse, A. G. E. Publ. J. & A.Churchill. London. 1961 edition. pp.1-998.

Kuthiala, A. and Chippendale, G. M. (1989). Relationship between the fatty acids of fat body triacylglycerol and lipophorin diacylglycerol of non-diapause and diapause larvae of the south-western corn borer, *Diatraea grandiosella*. Archives of Insect Biochemistry and Physiology. 12(2): 123-131.

Kuwana, Z. (1937). Uric acid excretion in silkworm. Japan Jour. of Zool. 7: 305-309.

L

Lackie, A. M. (1988a). Haemocyte behaviour. Adv. Insect Physiol. 21: 85-178.

Lackie, A. M. (1988b). Immune mechanisms in insects. Parasitology?Today. 4(4): 98?105.

Lackie, A. M.; Takle, G.; Tetley, L. (1985). Haemocyte emcapsulation in the locust Schistocerca gtregaria (Orthoptera) and in the cockroach Periplaneta americana (Dictyoptera). Cell Tissue Res.240:343-351.

Lai Fook, J. (1973). The structure of the haemocytes of the Calpodes ethilus (Lepidoptera). J. Morph. 139: 79-104.

Laigo, F. M. and Paschke, J. D. (1966). Variation in total haemocyte counts as induced by Nosemosis in the cabbage looper, Trichoplusia ni. J. Invetbr. Pathol. 8: 175-179.

Laufer, H. (1959). Identification and characterizaton of some blood protein in the development of giant Silkmoth. Annat. Record. 134:597.

Laufer, H. (1960). Blood protein in insect development. Ann. N.Y. Acad. Sci. 89: 490-515.

Laufer, H. (1961). Forms of enzymes in insect development. Ann. N.Y.Acad. Sci. 94:825-835.

Lavine, M. D. and Strand, M. R. (2001). Surface characteristic of foreign targets that elicit an encapsulation response by the moth Pseudoplusia includens. J. Insect Physiol. 47: 965-974.

Lavine, M. D. and Strand, M. R. (2002). Insect haemocytes and their role in immunity. Insect Biochem. Mol. Biol. 32: 1295-1309.

Lea, M. S. (1964). A study of the haemocytes of the silkworm Hyalophora cecropia, Ph. D. dissertation, North Western University, Illions (Original not seen). Multiplication of haemocytes. In Insect Haemocytes (A. P. Gupta, ed.). Cambidge University Press, Cambridge.

Lea, M. S. (1986). A Sericesthis iridescent virus infection of the haemocytes of the wax moth, Galleria mellonella: effects on total and differential counts and haemocyte ontogeny. J. Invertbr. Pathol. 48: 42-51.

Lea, M. S. and Gilbert, L. I. (1961). Cell division in diapausing silkworm pupae. Amer. Zool. 1: 368-369.

Lea, M. S. and Gilbert, L. I. (1966). The haemocytes of Hyalophora cecropia (Lepidoptera). J. Morph. 118(2): 197-216.

Lee, H. S.; Chang, B. S. and Yoe, S. M. (1992). A biochemical study on trehalase in the haemolymph of the cabbage butterfly, *Pieris rapae*. Korean Journal of Entomology. 22(2): 113?118.

Lee, HaengYeun; Lee, YongHo; Kang, SeongHoon; Kim, WooKap; Kim, H. R.; Lee, H. Y.; Lee, Y. H.; Kang, S. H. and Kim, W. K. (1998). Purification and characterization of a male-specific protein in the haemolymph of the wax moth, Galleria mellonella L. Archives of Insect Biochemistry and Physiology. 37(4): 257?268.

Lee, R. M. (1961). The variation of blood volume with age in desert locust (Schistocerca gregaria Forsk.). J. Insect Physiol. 6: 35-61.

Lees, A. D. (1955). The Physiology of Diapause in Arthropods. Cambridge University Press, New York, pp. 151.

Lefevere, K. S.; Kopmanchap, A. B. and De Korte, C. A. D. (1989). Chamges in concentration of metabolites in haemolymph during and after diapause in female Colorado potato beetle Leptinotarsa decemlineata. J. Insect Physiol. 35(2): 121-128.

Leifert, H. (1935). Nitrogen excretion, Antheraea (Lepidoptera). Zool. Jabrb. Zool.Physiol. 55: 131-190.

Levenbook, L and Bauer, A. (1984). The fate of larval storage protein caliphorin during adult development of Calliphora vicinia. Insect Bioche. 14: 77-86.

Levenbook, L. (1950). Composition of blood, Gastrophilus (Diptera) larvae. Biochem. Jour. 477: 336-346.

Levenbook, L. (1962). The distribution of free amino acids, glutamine and glutamate in the Southern armyworm Prodenia eridania. J. Insect Physiol. 8: 559-567.

Levenbook, L. (1966). Haemolymph amino acids and peptide during larval growth of the blowfly, Phoremia regina. Comp. Biochem. Physiol. 18: 341-351.

Levenbook, L. (1985). Insect storage proteins. In Comprehensive Insect Physiology, Biochemistry and Pharmacology (G. Kerkut, and L. I. Gilbert, eds.). Pergamon Press, Oxford. Vol 10. pp. 307-346.

Lillie, R. O. (1954). Histopathological Techniques and Practical Histochemistry. Publishres-Blackiston Company, New York.

Linquist, S. (1981). The heat shock response. A. Rev. Biochem. (1981). 55: 1151-1191

Linquist, S. (1986). Relation of protein synthesis during heat shock. Nature. 293: 311-314.

Liu, F. and Feng (1965). The metabolism of blood sugars of the armyworm. Acta et. Sin. 14: 432-440.

Lockey, D. and Ourth, D. D. (1996). Formation of pores in Escherichia coli cell membranes by a cecropin isolated from hemolymph of Heliothis virescens larvae. European Journal of Biochemistry. 236(1): 263?271.

Lockwood, A P. M. and Croghan, P. C. (1959). Osmotic pressure in Pterobius. Nature. 184: 370-371.

Loret, S. M. and Strand, M. R. (1998). Follow-up of protein release from Pseudoplusia includens haemocytes: a first step toward identification of factors mediating encapsulation in insects. European Journal of Cell Biology., 76(2): 146?155.

Loughton, B. G. (1965). The development and distribution of protein in Lepidoptera. J. Insect. Physiol. 11: 1651-1661.

Loughton, B. G. and Tobe, S. S. (1969). Blood volume in the African migratory locust. Can. J. Zool. 47: 1333-1336.

Loughton, B. G. and West, A. S. (1965). The development and distribution of haemolymph proteins in Lepidoptera. J. Inscet Physiol. 11: 919-932.

Lowenberger, C. (2001). Innate Immune response of Aedes aegypti. Insect Biochem. Mol. Biol. 31: 219-299.

Lowry, O. H.; Rosebrough, N. J.; Farr, A. L. and Randall, R. S. (1951). Protein measurement with the Folin Phenol reagent. J. Biol. Chem. I. 93: 265-275.

Luckhart, S.; Cupp, M. S. and Cupp, E. W. (1992). Morphological and functional classification of the haemocytes of adult female Simulium vittatum (Diptera: Simuliidae). Journal of Medical Entomology. 29(3): 457?466.

Ludwig, D. (1954). Changes in distribution of nitrogen in blood of Japanese beetle Papillio japonica during growth and metamorphosis. Physiol. Zool. 27: 325-334.

Lue, P. F. and Dixon, S. E. (1967). Can. J. Zool. 45: 205-214.**

M

Mack, S. R. and Vanderberg, J. P. (1978). Haemolymph of Anopheles stephensi from non-infected and Plasmodium berghei-infected mosquitoes. I. Collection Procedure and Physical Characteristics. J. Parasitol. 64: 918-923.

Mahalingam, V. and Muralirangan, M. C. (1996). Haemocytic responses of Atractomorpha crenulata (Fabricius) (Orthoptera: Pyrgomorphidae) to fungal (Aspergillus flavus) infection. Entomologist. 115(2): 97?101.

Mansingh, A. (1967). Changes in the free amino acids of the haemolymph of Antheraea pernyi during induction and termination of diapause. J. Insect Physiol. 13: 1645-1655.

Marcuzzi, G. (1956). L'osmoregolazione nel Tenebrio molitor L. Col. Tenebrionidae. Atti Accad. Naz. Lincei rc. 20: 492-500.

Martin, J. S. (1969). Control of haemolymph lipid concentration during locust flight: An adipokinetic hormone from corpora cardiaca. J. Insect Physiol. 15: 2319-2344.

Martin, M. D.; Kinnear, J. F. and Thompson, J. A. (1969). Developmental changes in the late larva of Calliphora stygia II. Protein synthesis. Aust. J. Biol. Sci. 22: 935-946.

Masler, E. P. and Kovaleva, E. S. (1997). Aminopeptidase like activity in haemolymph plasma from larvae of the gypsy moth, *Lymantria dispar* (Lepidoptera: Lymantriidae). Comparative Biochemistry and Physiology. B: Biochemistry and Molecular Biology. 116(1): 11?18.

Mason, L. J.; Johnson, S. J. and Woodring, J. P. (1990). Influence of age and season on whole-body lipid content of Plathypena scabra (Lepidoptera: Noctuidae). Environmental Entomology. 19(5): 1259-1262.

Matthee, J. J. (1945). Biochemical differences, phases locusts and noctuids. Bull. Ent. Res. 36: 343-371.

Mayer, R. J. and Candy, D. J. (1969) Control of haemolymph lipid concentration during locust flight: An adipokinetic hormone from corpora cardiaca. J. Insect Physiol. 15: 611-629.

Meister, M.; Hetru, C. and Hoffman, J. A. (2000). The antimicrobial host defence of Drosophila. In Origin and Evolution of Vertebrate Immune System (L. Du Pasquer and G. W. Litman, eds.). Current Topics in Microbiology. Vol. 248. Springer-Verlag, Berlin. pp. 17-36.

Mellanby, K. (1939). The functions of insect blood, Biol. Rev. 14: 243-260.

Millara, P. (1947). Contribution á l'étude cytologique et physiologique des leucocytes d'insectes. Bull. Biol. France et Belg. 81: 129-153.

Miller, S. G. and Silhack, D. (1982). Identification and purification of storage proteins in tissues of the greater wax moth Galleria mellonella (L.). Insect Biochem. 12: 277-292.

Mitsuhashi, T. D. (1978). Free amino acids in the haemolymph of some lepidopterous insects. Appl. Ento. Zool. 13: 318-320.

Mohanty, A.K. and Mitra, A.C, (1988) A comparative biochemical study of the haemolymph in bivoltine and trivoltine pupae of the tasar silkworm, *Antheraea mylitta* Drury during development. Sericologia 28(1): 125-132.

Monpeyssin, M. and Beaulaton, J. (1978). Haemocytopoiesis in the Oak silkworm, Antheraea pernyi and some of the Lepidoptera. J. Ultrastruct. Res. 64: 33-45.

Moore, S. and Stein, W. H. (1948). In Method in Enzymology. Vol. III. (Colowick, S. P. and Kaplan, N. O., eds). Academic Press. New York.

Moran, D. T. (1971). The fine structure of insect blood cells.Tissue Cell 3: 413-422.

More, N. K. and Sonawane, S. Y. (1990). (record 1996). Insect haemocytes: a comparative study. Bulletin of Entomology New Delhi. 31(2): 164?174.

Morgan, T. D. and Chippendale, G. M. (1983). Free amino acids of the haemolymph of the Southern corn borer and the European corn borer in relation to diapause. J. Insect. Physiol. 29(10): 735-740.

Mullin, D. E. (1985). Chemistry and physiology of the haemolymph. In Comprehensive Insect Physiology and Pharmacolgy, (G. A. Kerkut and L. I. Gilbert eds.), Pergamon Press. 3: 355-400.

Munson, S. C. and Yeager, J. F. (1944). Fat inclusions in the blood cells of the southern army worm Prodenia eridania (Cram.). Ann. Entomol. Soc. Am. 37: 369-400.

Muta, T. and Iwanga, S. (1996). The role of haemolymph and coagulation in innate immunity. Curr. Opin. Immunol. 8: 41-47.

N

Nagata, M. and Yoshitake, N. (1988). Protein metabolism in the larval development of the silkworm, Bombyx mori: Protein reutilization at fourth moult. J. Ser. Sci. Jpn. 58(3): 221-233.

Nakasone, S. and Ito, T. (1967). Fatty acid composition of the silkworm, *B. mori* L. J. Insect Physiol. 13: 1237-1246.

Nakasone, S. and Kobayashi, M. (1965). Acrulamide gel electrophoresis of blood proteins during the moulting and metamorphosis in the silkworm, *Bombyx mori.* Japan J. Seric. 34: 257-262.

Nanavty, M. (1965). Silk from grub to glamour. Paramount Publishing House, Bombay pp.192-209.

Nappi, A. J. (1974). Insect haemocytes and problem of host recognition of foreignness. In Contemporary Topics in Immunology (E. L. Copper, ed.). Vol. 4. Plenum Press, New York. pp. 207-224.

Nappi, A. J. (1975). Parasites' encapsulation in insects. In Invertebrate Immunity (K. Maramorosch and R. E. Shape, eds.). Academic Press, New York. pp. 293-326.

Nath, J. and Butler, L. (1973). Alkaline phosphatase during development of black carpet beetle. Ann. Rev. Ent. Soc. Am. 66: 280-284.

Natochin, Yu. V.; Parnova, R. G.; Gurevich, V. V.; Didian. S. E.; Peznik, L. V. and Shakhmatova, F. I. (1988). Chemical Composition of haemolymph and regulation of ionic structure of cells in the caterpillat Pieris brassicae. Z. Ecol. Biok. Fiziol. 24: 149- 156.

Neill and Neely. (1956). J. Clin. Pathol. 9: 162 In Hawk's Physiological Cemistry (1979). Chapter 29: Blood Analysis. pp. 975-1152. (P. B. Hawk, ed.) McGraw-Hill, Inc., New York.

Neimierko, S.; Wlodawer, P. and Wojtezak, A. F. (1956). Lipid and phosphorus metabolism growth of the silkworm, B. mori L. Acta Biol. Exp. Vars., 17: 255-276.

Nelson, D. R.; Terranova, D. C. and Sukkestad, D. R. (1967). Comp. Biochem. Physiol. 20: 907-918.**

Neuwirth, M. (1973). The structure of the haemocytes of Galleria mellonella (Lepidoptera). J. Morph. 139(1): 105-124.

Nicolson, S. W. (1980). Water balance and osmoregulation in Onymacris plana, a tenebrionid beetle from the Namile desert. J. Insect. Physiol. 26: 315-320.

Nicolson, S.; Horesfield, P. M.; Gardiner, B. O. C. and Maddrell, S. H. P. (1974). Effects of starvation and dehydration on osmotic and ionic balance in Carausius morosus. J. Insect. Physiol. 20: 2061-2069.

Nijhout, H. F. (1994). Insect Hormones. Chapter Seven, Daipause. Princeton University press, Princeton, N. J. pp 160-175.

Nittono, Y. (1960). Studies on the blood cells in the silkworm, *Bombyx mori* L. Bull. Seri. Exp. Stn. Tokyo. 16: 171-266 (In Japanese, English Summary).

Nowosielski, J. W. and Patton, R. L. (1965). Variation in the haemolymph protein, amino acid and lipid levels in adult house criket, *Acheta domesticus* L. of different stages.J. Insect Physiol. 11: 263-270.

O

Ogel, S. (1955). A contribution to the study of blood cells in Orthoptera. Comm. De la Fac.ctes Sci. Univ. (Sci). 1 : 21-25.

Ojha, N. G. ; Sharan, S.K.; Srivastava, P. P. ; Singh, B. M. K. and Sinha, P. S. (1997). Determination of susceptible age for cold trearment for the extension of pupal diapause in tropical tasar silkworm, *Antherea mylitta* Drury Proc. XX Int. Cong. Of Zool. , Aug. 25th to 31st, Fierenze, Italy.

Ortel, J. (1995). Effects of metals on the total lipid content in gypsy moth (Lymantria dispar, Lymantriidae, Lepidoptera) and its haemolymph. Bulletin of Environmental Contamination and Toxicology. 55(2): 216?221.

Osanai, M. and Yonezawa, Y. (1984). Age related changes in amino acid pool sizes in the adult Silkmoth Bombyx mori, reared at low and high temperatures. A biochemical examination of the rate of living theory and urea accumulation when reared at high temperatures. Exptl. Geront. 19: 37-51.

Ozaki, J. (1936). Studies on the mineral matter contents of prepupae and pupae and sexual differences in pupae oils of silkworm moth, *B. mori*. Res. Bull. Imp. Tokyo. Seri. Coll. 1: 121-155.

P

Pannabecker, T. L.; Andrews, F and Beyenbach, K. W. (1992). A quantitative analysis of the osmolytes in the haemolymph of larval Gypsy moth, Lymantria dispar. J. Insect Physiol. 38(11): 823-830.

Pant, R and Nautiyal, G. C. (1974). Cholesterol turnover in Philosamia ricini during development. Indian J. Biochem. Biophy. 11: 156-158.

Pant, R. (1984). Some biochemical aspect of the eri silkworm, *Philosamia ricini.* Sericologia. 24(1):53-91.

Pant, R. and Agrawal, H. C. (1964). Free amino acids of the haemolymph of some insects. J. Insect Physiol. 10: 443-446.

Pant, R. and Agrawal, H. C. (1965). Changes observed in the free amino acids content of Philosamia ricini pupal haemolymph during metamorphosis. Indian J. Exptl. Biol. 3: 133-136.

Patel, N. G. (1971). Protein synthesis during insect development. Insect Biochem. 1:391-427.

Pathak, J. P. N. (1983). Effect of endocrine gland on the unfixed total haemocyte counts of the bug, *Halys dentata.* J. Insect Physiol. 29: 91-94.

Pathak, J. P. N. (1986). Haemogram and its endocrine control in insects, In: Immunity in Invertebrates (M. Brehelin, ed.) Springer-Verlag, Berlin, Berlin. pp 49-54.

Pathak, J. P. N. (1991). Effect of endocrine extracts on the blood volume and population of haemocytes in Halys dentata (Pentatomidae: Heteroptera). Entomon. 16(4): 251-255.

Pathak, S. C. and Kulshreshtha, V. (1993). Variation in haemocyte types with reference to reproductive activity in Blattella germanica L. (Dictyoptera: Blattellidae) and the occurrence of un-described haemocyte types in some adult stages. Entomon., 18(3-4): 119-125.

Pathak, S. C. and Saxena, N. (1994). Haemocytes in the fifth instar larvae and pupae of Plusia orichalcea Fabr. (Lepidoptera: Noctuidae). Indian Journal of Entomology. 56(1): 87-92.

Pearse, A. G. E. (1961). Histochemistry Theoretical and Applied 2nd Edition. J & A Churchill, London, pp 1-998.

Pech, L. L. and Strand, M. R. (1996). Granular cells are required for encapsulation of foreign targets by insect haemocytes. Journal?of?Cell?Science. 109(8): 2053-2060.

Pech, L. L.; Trudeau, D. and Strand, M. R. (1995). Effect of basement membranes on the behaviour of haemocytes from Pseudoplusia includens: development of an in vitro encapsulation assay. J. Insect Physiol. 14: 801-807.

Peferoen, M. Stynen, D. and De Loaf, A. (1982). A re-examination of the protein pattern of the haemolymph of Leptinotarsa decemlineata with special reference to vitellogenins and diapausing proteins. Comp. Biochem. Physiol. 728: 345-351.

Peterson, N. S. and Mithel, H. K. (1985). Heat shock proteins. In Comprehensive Insect Physiology nad biochemistry and pharnacolgy Edited by G. A. Kerkut and L. I. gilbert Vol. 10: 347-366.

Pichon, Y. (1970). Ionic content of haemolymph in the cockroach Periplaneta americana. J. Exp. Zool. 210: 361-367.

Pilmer, R. H. A. (1944). Van Slyke's copper sulphate method for measuring specific gravity of whole blood, plasma and serum. Proc. Biochem. Soc. 232nd meeting of Biochemical Society. pp. vii.

Pionar, G. O. Jr. (1974). Insect immunity to parasite nematodes. In Contemporary Topics in Immunology (E. L. Copper, ed.). Plenum Press, New York. 4: 167-178.

Poonia, F. S. and Mishra S. D. (1975). Quantitative changes in the level of carbohydrates in the food plant, haemolymph and excreta in the tasar silkworm, *Antheraea mylitta* D. (Lepidoptera:Saturniidae) during the post embryonic stages. Indian J. Seri. 14: 31-34.

Poisson, R. and Pesson, P. (1939). Contribution a letudr du sang des coccodes (Hemiptera: Sternorhyncha). Le sang de Pulvinaria mesembryan thermi Vallot. Arch. Zool. Exper. Et. Gen. 81: 23-32.

Prasad, C. S. and Nath, G. (1985). Qualitative and quantitative changes in haemolymph proteins of Spodoptera litura Fab. during larval and prepupal development. Indian J. Entomol. 47(1): 71-72.

Prat, J. J. Jr. (1950). Analysis, amino acids in blood. Ann. Ento. Soc. Amer. 43: 573.**

Price, G. M. (1961). Some aspects of amino acid metabolism in the adult house fly, *Musca domestica*. Biocem. J. 2: 175-185.

Pullin, A. S. (1992). Diapause metabolism and changes in carbohydrates related to cryoprotection in Pieris brassicae. Journal of Insect Physiology. 38(5): 319-327.

Pullin, A. S. and Bale, J. S. (1989). Effects of ecdysone, juvenile hormone and haemolymph transfer on cryoprotectant metabolism in diapausing and non-diapausing pupae of Pieris brassicae. Journal of Insect Physiology. 35(12): 911-918.

R

Racioppi, J. V. and Dahlman, D. L. (1980). Effects of L-canavanine on Manduca sexta (Sphingidae: Lepidoptera) larval haemolymph solutes. Comp. Biochem. Physiol. 67C: 35-39.

Raichoudhury, D. P. and Sen Gupta, K. (1959). Studies on the larval blood of silkworm, *Bombyx mori* L. Ind. J. Entomol. 21(1): 6-9.

Raina, A. K. (1976). Ultrastructure of the larval haemocytes of pink bollworm, Pectinophora gossypiella (Lepidoptera: Gelechiidae). Int. J. Insect Morphol. Embryol. 5(3): 187-195.

Raina, A. K. and Bell, R. A. (1974). Haemocytes of pink bollworm, Pectinophora gossypiella during larval development and diapause. J. Insect Physiol. 20: 2171-2180.

Ramakrishnan, N and Pawar, V. M. (1975). Effect of diet on the haemolymph protein of the larvae of Spodoptera litura. Ind. J. Entomol. 37: 172-178.

Ramsay, J. A. (1953). J. Exp. Biol. 30: 358-369.**

Raper, R. and Shaw J. (1948). Amino acids in haemolymph, dragon fly nymph. Nature (London), 162: 999.**

Ratcliffe, N. A. (1993). Cellular defence responses of insects: unresolved problems. In Parasites and Pathogens of Insects (N. E. Beckage; S. N. Thompson and B. A. Federici, eds.). Vol. 1. Academic Press, San Diego, C.A. pp. 267-304.

Ratcliffe, N. A. and Rowly, A. F. (1979). Role of haemocytes in defence against biological agents. In Insect Haemocytes (A. P. Gupta, ed.). Cambridge University Press, Cambidge. pp. 331-414.

Ravindranath, M. H. (1978). The individuality of Plasmatocytes and granular haemocytes of arthropods. A Review. Dev. Comp. Immunol. 2: 581-594.

Ribeiro, C.; Simoes, N. and Brehelin, M. (1996). Insect immunity: the haemocytes of the armyworm Mythimna unipuncta (Lepidoptera: Noctuidae) and their role in defence reactions. In vivo and in vitro studies. Journal of Insect Physiology. 42(9): 815?822.

Richards, O. W. and Davies, R. G. (1977). Imms' General Textbook of Entomolgy. Vol. I. Chapman and hall, London, John amd wiley and sons, Newyork. 418pp.

Richardson, C. H.; Burdette, R. C. and Eagelson, C. W. (1931). The determination of the blood volume of insect larvae. Ann. Entomol. Soc. Amer. 24: 503-507.

Ritter, H. Jr. (1965). Blood of cockroach: Unusual cellular behaiour. Science (Wash. D.C.). 147: 518-519.

Rimoldi, O. J.; Soulages, J. L.; Gonzalez, M. S.; Peluffo, R. O. and Brenner, R. R. (1990). Biochemistry of the evolution of Triatoma infestans. XI. Haemolymph lipophorin. Acta Physiologica et Pharmacologica Latinoamericana. 40(2): 239?255.

Roberts, D. and Brock, H. (1981a). The moajor serum proteins of Diptera larvae. Experientia 37: 1-3-110.

Roberts, D. and Brock, H. (1981b). The fate of larval storage proteins caliphorin during adult development of Calliphora vicinia. Insect Biochem. 14: 77-86.

Roe, M. J.; Kim, J. H.; Yu, S. C.; Kim, K. S. and Kim, W. K. (1993). Haemocyte types and granular composition of Lucilia illustris. Korean Journal of Entomology. 23(1): 31?39.

Rohrkasten, A. and Ferenze, H. J. (1985). In vitro study of selective endocytosis of vitellogenin by locust oocytes. Boux's Arch. Dev. Biol. 194: 411-416.

Rosenberger, C. R. and Jones, J. C. (1960). Studies on the total blood cell counts of the southern armyworm larva, Prodenia eridania (Lepidoptera). Ann. Entomol. Soc. Amer. 44: 351-355.

Rostom, Z. M. F., Algauhari, A. M. I. and Ab. Del Fattah, M. M. (1972). Free amino acids and total protein in the haemolymph of Pectinophora gossypiella during induction and termination of diapause. Proc. Egypt. Acad. Sci. 25: 71-79.

Rowley, A. F. (1977). The Role of haemocytes of Clitumnus extradentetus in haemolymph coagulation. Cell Tissue Res. 182 (4): 513-524.

Ruh, M. F. and Willis, J. H. (1974). Synthesis of blood and cuticular proteins in late pharate adults of the cecropia Silkmoth. J. Insect Physiol. 20:1277-1285.

Ruh, M. F.; Willis, J. H. and Hollowwell, M. P. (1972). Blood protein synthesis in pupae of the silkmoth, Hyalophora cecropia. J. Insect. Physiol. 18: 151-160.

Russo, J. Allo.; Nenaon, J. P. And Brehlin, M. (1993). The haemocytes of mealy bugs Phenacoccus manihoti and Planococcus citri (Insecta: Homoptera) and their role in capsule formation. Can. J. Zool. 72: 252-258.

S

Sactor, B. (1965). The physiology of insects Vol.II Academic Press, New York. **

Saito, H. (1998). Purification and characterization of two insecticyanin?type proteins from the larval haemolymph of the Erisilkworm, *Samia cynthia ricini*. Biochimica et Biophysica Acta: General Subjects., 1380(1): 141?150.

Saito, S. (1963). Trehalose in the body fluid of the silkworm, *Bombyx mori*. J. Insect Physiol. 9:509-519.

Sakamoto, E. and Horie, Y. (1979). Qualitative changes of phosphorous compound in haemolymph during development of silkworm Bombyx mori. J. Seri. Sci. Jpn. 49: 509-511.

Salama, M. S. and Miller, T. A. (1992). A diapause associated protein of the pink bollworm Pectinophora gossypiella Saunders. Archives of Insect Biochemistry and Physiology (USA). 21(1): 1-11.

Salama, M. S.; Schouest, L. P. Jr. and Miller, T. A. (1992). Effect of diet on the esterase patterns in the haemolymph of the corn earworm and the tobacco budworm (Lepidoptera: Noctuidae). Journal of Economic Entomology. 85(4): 1079?1087.

Salt, G. (1970). The Cellular Defence Reactions of Insects. Cambridge Monogr. in experimental Biology. No. 16. Cambridge University Press, Cambridge.

Salt, R. W. (1957). Natural occurrence of glycerol in inects and its relation to their ability to survive freezing. Can. Entomol. 89: 491-494.

Salt, R. W. (1959). Role of glycerol in the cold-hardiness of Bracon cephi (Gahan), Can. J. Zool. 6: 55-74.

Sanjayan, K. P.; Ravikumar, T. and Albert, S. (1996). Changes in the haemocyte profile of Spilostethus hospes (Fab) (Heteroptera: Lygaeidae) in relation to eclosion, sex and mating. Journal of Biosciences. 21(6): 781?788.

Sarlet, H., Ducháteu, G. and Florkin, M. (1952). Les acides aminés due milieu intérieur du verá soie au cours du pilage. Arch. Insect Physiol. pp. 126-127.

Sass, M.; Kiss, A.; Locke, M. (1994). Integument and haemocyte peptides. J. Insect Physiol. 40: 407-421.

Sato, S.; Akai, H. and Sawela, H. (1976). An ultrastructural study of capsule formation by Bombyx haemocytes. Ann. Zool. Japan. 49: 177-178.

Saxena, B. P. (1992). Comparative study of haemocytes of three lepidopterans by light and scanning electron microscopy. Acta Entomologica Bohemoslovaca. 89(5): 323?329.

Saxena, B. P. and Tikku, K. (1990). Effect of plumbagin on haemocytes of Dysdercus koenigii F. Proceedings of the Indian Academy of Sciences. Animal Sciences. 99(2): 119?124.

Saxena, B. P.; Sharma, P. R. and Tikku, K. (1988). Scanning electron microscopical studies of the haemocytes of Spodoptera litura Fabr. Cytologia. 53(2): 385-391.

Scapigliati, G. and Mazzini, M. (1992). Morphological characterization of the haemocytes of Bacillus rossius (Rossi) (Phasmatodea: Bacillidae). Redia. 75(1): 233?240.

Schin, K. and Moore, R. D. (1977). Cation concentration in the haemolymph of the fly Chironomus thummi during development. J. Insect Physiol. 23: 723-729.

Schmidt, A. R. and Ratcliffe, N. A. (1977). The encapsulation of foreign tissue implants in Galleria mellonella larvae. J. Insect Physiol. 23: 175-184.

Schmidt, A. R. and Ratcliffe, N. A. (1978). The encapsulation of oraldite implants recognition of foreignness in Clitumnus extradentatus. J. Insect Physiol. 24: 511-521.

Schmidt, G. and Schwanski, W. (1975). Changes in haemolymph proteins during the metamorphosis of both sexes and castes of polygynous Formica rufa. Comp. Biochem. Physiol. B52: 365-380.

Schoffeniels, E. and Gilles, R. (1970). Nitrogeneous constituents and nitrogen metabolism in arthropods. In Chemical Zoology, Vol. V. (eds. M. Florkin and B. T. Scheer), Chap. 7. pp. 199-227. Academic Press. New York.

Seitz, A. (1913). The macrolepidoptera of the world. Palaearctic fauna. Stuttgart, Alfred Kernen. 2: 215-218.

Seitz, A. (1933). The macrolepidoptera of the world. Indo - Australian bomyces and sphinges. Stuttgart. Alfred Kernen. 10: 509-516.

Sengupta, A. K.; Sinha, A. K. ; Bajpeyi, C. M. ; Sinha, B. R. R. P. and Sinha, S. S. (1995). Effect of changing temperature on the haemolymph protein pattern of diapausing pupae in Antherea mylitta D. Perspectives in Cytology and Genetics (Eds. G.K.Manna and S.C., Roy) , 601-607.

Sha, C. Y. and Xie, Q. J. (1992). Changes of haemocyte number, protein and esterase in the haemolymph of Mythimna separata (Walker) infected with Bacillus thuringiensis Berliner. Entomological Knowledge. 29(4): 215?217.

Shapiro, J. P. (1968). Changes in haemocyte population of the wax moth, Galleria mellonella, during wound healing. J. Insect Physiol. 14: 1725-1733.

Shapiro, J. P.; Silhacek, D. L. and Niedz, R. P. (1992). Storage proteins of the larval root weevil Diaprepes abbreviatus (Coleoptera: Curculionidae): riboflavin binding and subunit isolation. Archives of Insect Biochemistry and Physiology (USA). 20(4): 315-331.

Shapiro, J. P.; Wells, M. A. and law, J. H. (1988). Lipid transport in insects. Ann, Rev. Entomol. 33: 297.

Shapiro, M.; Stock, R. D. and Ignoffo, C. M. (1969). Haemocyte changes in larva of the ball worm, Heliothis zea infected with the nucleo-polyhydrosis virus. J. Invertebr. Pathol. 14: 28-30.

Shapiro, M. (1979a). Changes in haemocyte populations. In Insect haemocytes (A. P. Gupta, ed.). Cambridge University Press, Cambridge. pp. 475-523.

Shapiro, M. (1979b). Techniques for total and differential haemocyte counts and blood volume and mitotic index. In Insect haemocytes (A. P. Gupta, ed.). Cambridge University Press, Cambridge. pp. 549-561.

Sharan, S. K.; Singh, M. K.; Ojha, N. G. and Sinha, S. S. (1994). Regulation of Voltinism in Antheraea mylitta by Manipulation of Rearing Period of Larval Stages. Int. J. Wild Silkmoth & Silk. 1(2): 191-194.

Sharan, S. K.; Mishra, P. K.; Kumar, D.; Singh, B. M. K.; Sinha, B. R. R. P.; Rai, S. and Panday, P. N. (2002). Effect on total and differential haemocyte counts due to infection of Nosema species in Antheraea mylitta Drury (Lepidoptera: Saturniidae) larvae. XIX Cong. of Int. Seri. Commission, (21st-25th Sept.), pp. 308-312.

Sharan, S. K.; Mishra, P. K. and Panday, P. N. (2004). Total haemocyte counts in the ontogeny of Tasar silkworm, Antheraea mylitta Drury (Lepidoptera: Saturniidae) in non-diapausing and diapausing generations. Proc. Zool. Soc. India. 3(1): 91-101.

Sharan, S. K.; Mishra, P. K. and Panday, P. N. (2004). Absolute number of haemocytes in relation with blood volume and body weight of Tasar silkworm Antheraea mylitta Drury (Lepidoptera: Saturniidae) in non-diapause and diapause destined generations. Proc. Zool. Soc. India. In Press.

Sharma, D. K.; Dipali Devi and Margherita Turchetto (1995). Seasonal variations in the haemolymph of free amino acids of Antheraea assama Westwood. Indian J. Seri. 34 (2): 122-126.

Sharma, P. R.; Kalpana?Tikku; Saxena, B. P. and Tikku, K. (1998). A light and electron microscopic study of the haemocytes of adult red cotton bug Dysdercus koenigii. Biologia Bratislava. 53(6): 759?764.

Shimizu, I. (1982). Variation of cation concentrations in the haemolymph of the silkworm, Bombyx mori with diet and larval-pupal development. Comp. Biochem. Physiol. 71A: 445-447.

Shimizu, I. (1992). Comparison of fatty acid compositions in lipids of diapause and non-diapause eggs of Bombyx mori (Lepidoptera:Bombycidae). Comparative Biochemistry and Physiology. 102(4): 713-716.

Shrivastava, S. C. and Richards, A. G. (1965). An autoradiographic study of the relation between haemocytes and connective tissues in the wax moth, *Galleria mellonella* Biol. Bull. (Woods Hole). 128: 337-345.

Shu ZhiQun; Xu XuShi; Zhu ZhiMin; Long QiXin; Xu Lin; Shu, Z. Q.; Xu, X. S.; Zhu, Z. M.; Long, Q. X. and Xu, L. (1997). The haemolymph pathological biochemistry changes of Argyrogramma agnata after heterologous NPV infection. Chinese Journal of Biological Control. 13(3): 118?121.

Siaktos, A. W. (1960). The conjugated plasma proteins of the american cockroach. II. Changes during moulting and clotting processes. J. Gen. Physiol. 43: 1015-1030.

Siegert, K. J. (1986). The effects of chilling and integumentary injury on carbohydrate and lipid metabolism in diapause and non-diapause pupae of Manduca sexta. Comparative Biochemistry and Physiology. 85(2): 257-262.

Siegert, K. J. (1995). Carbohydrate metabolism during the pupal moult of the tobacco hornworm, Manduca sexta. Archives of Insect Biochemistry and Physiology. 28(1): 63?78.

Siegert, K. J.; Speakman, J. R. and Reynolds, S. E. (1993). Carbohydrate and lipid metabolism during the last larval moult of the tobacco hornworm, Manduca sexta. Physiological Entomology. 18(4): 404?408.

Singh, B. M. K. and A. K. Srivastava (1997). Ecoraces of Antheraea mylitta Drury and exploration strategy. Curr. Tech. Seminar, CTR&TI, Ranchi, Base paper. 6:1-39.

Sinha, A. K.; Chaudhary, S. K. and Sengupta, K. (1985). Changes in protein content in the larval and pupal haemolymph of Antherea mylitta D. Sericologia, 25(1): 39-43.

Sinha, A.K., Chauhary, S.K. and Sengupta, K. (1988). Changes in free amino acids in the larval and pupal haemolymph of Antherea mylitta Drury reared on Terminalia arjuna and Terminalia tomentosa. Indian J. Seri. 27(2): 95-108.

Sinha, U. S. P., Sinha, A. K. and Goel, R.K. (1992). Changes in the concentration of cholesterol in the larval and pupal haemolymph of tropical tasar silkworm, *Antherea mylitta* D. Indian J. Seri. 31(1): 73-75.

Sinha, U. S. P.; Sinha, A. K.; Bramhchari, B. N. and Sinha, S. S. (1998). Seasonal variations in the carbohydrate content during different larval stages of tasar silkworm, *Antheraea mylitta* Drury, Sericologia, 38(2), 385-388.

Sinha, U.S.P. and Sinha, A.K. (1994). Changes in concentration of proteins and carbohydtares in developing embryos and larval haemolymph of temperate tasar silkworm, *Antheraea proylei*, Jolly. Indian J. seri., 33(1): 84-85.

Sissakjan, N. M. and Kuvajeva, E. B. (1957). On the peculiarities of protein synthesis in the coelomic fluid of the silkworm, *Bombyx mori*. Biochimija, 22: 686-694.

Slama, K. (1960). Metabolism during diapause development in sawfly metamorphosis. In the Ontogeny of Insects, pp 195-201. Academic Press, London.

Smith, H. W. (1938). The blood of the cockroach Periplanata americana L. Cell structure and degeneration, and cell counts. Studies of contact insecticides. New Hempshire. Agric. Exptl. Stn. (Durhan) Techn. Bull. 71: 1-23.

Smith, V. J. and Ratcliffe, N. A. (1978). Host defence reactions of the shore crab, Carcinus maenas (L.), in vitro. J. Morphol. Biol. Assn., U.K. 58: 367-379.

Smolin, A. N. (1960). Trehalose in haemolymph of oak silkworm. Mosko State Ped. Inst. Studies from chair of organic and biological chemistry. 10: 12-16.

Somme, L. (1965). Further observation on glycerol and cold hardiness in insects Can. J. Zool. 43: 765-770.

Somme, L. (1967). The effect of temperature and anoxia on haemolymph composition and supercooling in three overwintering insects. J. insect. Physiol. 13: 805-814.

Sonawane, Y. S. and More, N. K. (1993). The circulating haemocytes of the bed bug, Cimex rotundatus Sign. (Heteroptera: Cimicidae). J. Animal. Morphol. Physiol. 40: 79-86.

Song Jeong Keun; Nah JungHee; Lee InHee; Kang ChangSoo; Kim HakRyul; Song J. K.; Nah, J. N.; Lee I. H.; Kang, C. S. and Kim, H. R. (1995). Comparative analysis of storage proteins of fall webworm (Hyphantria cunea Drury). Entomological? Research?Bulletin. 21: 25?33.

Sowri, D. M. K. and Sarangi, S. K. (2002). A comparative study on the trehalose level in different varieties of the silkworm, Bombyx mori, during fifth instar larval development. Entomon, 27(1): 57-61.

Srivastava, B. K. and Mathur, L. M. L. (1966). On the hydrogen ion concentration in the alimantry canal and blood of phytophagous larval Lepidoptera. Indian L. Entomol. 28: 423-426.

Stadt, D. (1964). Chemical methods for Medical investigations. E. Merck, A.G. p. 61.

Steinhauer, A. L. and Stephen, W. P. (1959). Changes in blood protein during the development of American cockroach, Periplaneta americana. Ann. Rev. Entomol. Sco. Ammer. 52: 733-738.

Stephen, W. P. and Steinhauer, A. L. (1957). Physiol. Zool. 30: 114-120. **

Stevenson, E. and Wyatt, G. R. (1962). The metabolism of silkmoth tissues. I. Incorporation of leucine into protein. Arch. Biochem. Biophy. 99: 65-71.

Story, K. B.; Mc Donald, D. G.; Booth, C. E. (1986). Effects of temperature acclimation as haemolymph composition in the freeze tolerant larvae of Eurosta solidaginis. J. Insect Physiol. 32: 897-902

Strand, M. R. and Noda, T. (1991). Alterations in the haemocytes of Pseudoplusia includens after parasitism by Microplitis demolitor. Journal of Insect Physiology. 37(11): 839-850.

Strand, M. R. and Pech, L. L. (1995). Immunological basis for compatibility in parasitoid-host relationships. Annual Review of Entomology. 40: 31?56.

T

Takada, M. and Kitano, H. (1971). Studies on the larval haemocytes in the cabbage white butterfly, Pieris rapae crucicora Boisduval, with reference to haemocyte classification, phagocytic activity and encapsulative activity. Kontyii. 39: 385-394.

Tanaka, T. (1987). Morphological changes in haemocytes of the host, Pseudaletia separata, parasitized by Microplitis mediator or Apanteles kariyai. Developmental and Comparative Immunology. 11(1): 57-67.

Tanno, K. (1964). Higher sugar levels in the solitary bee, Ceratina. Low Temp. Sci. Ser. B 22: 51-77.

Tauber, O. E. (1936). Mitosis of circulating cells in the haemolymph of the roach, *Blatta orientalis*. Iowa State Coll J. Sci. 10: 431-439.

Tauber, O. E. and Yeager, J. E. (1934). On the total blood (haemolymph) cell count of the field cricket, Coryllus assimilis pennysyslvanicus Burn. Iowa State Cell J. Sci. 9: 13-24.

Tauber, O. E. and Yeager, J. E. (1935). On the total haemolymph (blood) cell count of insects II, Orthoptera, Odonata, Hemiptera and Homoptera. Ann. Entomol. Soc. Am. 28: 229-240.

Tauber, O. E. and Yeager, J. F. (1936). On the total haemolymph (blood cells) counts of insects II. Neuroptera, Coleoptera, Lepidoptera and Hymenoptera. Ann. Entomol. Soc. Am. 28: 135-145.

Tauber, M. J.; Tauber, C. A.; and sasaki, M. (1986). Seasonal adaptations of insects, Oxford University Press, London.

Taylor, I. R.; Birnie, J. H.; Mitchell, P. H. and Solinger, J. L. (1934). pH changes in metamorphosis, *Galleria* (Lepidoptera). Physiol. Zool. 7: 593-599.

Telfer, W. H. (1954). Immunological studies of insect metamorphosis II. The role of sex limited blood proteins in egg formation by the cecropia silkworm. J. Gen. Physiol. 37: 539-558.

Telfer, W. H.; Keim, P.S. and Law, J. H. (1983). Arylphorin, a new protein from haemolymph fat body and ovary of a phytophagous bug, Dysdercus koenigii (Heteroptera: Pyrrochoriade). Biochem. Arch. 10(4): 297-303.

Telfer, W. H. and Kulakosky, P. G. (1984). Isolation of haemolymph proteins as probes of endocytotic yolk formation in Hyalophora cecropia. In Adv. Invert. Repro. Vol 3. (W. Engels, ed.). Elsevier Science Publishers. Amsterdam.pp 81-86.

Telfer, W. H. and Kunkel, G. (1991). The function and evolution of insect storage hexameres. Ann. Rev. Entomol. 36: 205-228.

Telfer, W. H. and Williams, C. M. (1953). Immunological studies of insect metamorphosis I. Qualitative and quantitative description of the blood antigens of cecropia silkworm J. Gen. Physiol. 36: 389-413.

Telfer, W. H. and Williams, C. M. (1953). Immunolodical studies of insect metamorphosis II. The role of sex limited blood protein in egg formation of the blood antigen of the cecropia silkworm. J. Gen. Physiol. 37: 539-558.

Telfer, W. H. and Williams, C. M. (1960). The effects of diapause, development and injury on the incorporation of radio active glycine into the blood proteins of the cecropia silkworm. J. Insect. Physiol. 5: 61-72.

Terra, W. R.; De Bianche, A. G. and Lara, F. J. (1974). Physical properties chemical composition of the haemolymph of Rhynchosciara americana (Diptera) larvae. Comp. Biochem. Physiol. 47B: 117-129.

Thomson, J. A. (1975). Major protein patterns of gene activity during development in holometabolous insects. In Advances in Insect Physiology (J. E. Trehene, M. J. Berridge and V. B. Wigglesworth, eds.). 11: 321-398. Academic Press, London.

Thompson, M. J.; louloudes, S. J. ; Robbins, W. E. ; Watres, J. A. ; Steefe, J. A. and Mosetting, E. (1963). The identity of major sterol from house flies reared by the CSMA procedure. J. Insect Physiol. 9: 615-622.

Tobias, J. M. (1948b). Potassium, sodium and low sodium in tissues, changes on pupation, Silkworm. Jour. Cellular Comp. Physiol. 31: 143-148.

Tojo, S.; Betchaku, T.; Ziccardi, V. J. and Wyatt, G. R. (1978). Fat body protein granules and srorage proteins in the silk moth, *Hyalophora cecropia*. J. Cell. Boil. 78: 823-838.

Tojo, S.; Morita, M. and Hiruma, K. (1985). Effevt of some juvenile hormone on phase characteristics in the common cutworm, *Spodoptera litura*. J. Insect Physiol. . 31: 243-249.

Tojo, S.; Nagata, M. and Kobayashi, M. (1980). Storage proteins in the silkworm Bombyx mori. Insect biochem. 10: 289-303.

Tonka, T. 1997. Haemocyte count in Galleria mellonella injected with the microsporidian. Sbornik Vairimorpha ephestiae. Jihoceska Univerzita Zemedelska Fakulta, Ceske Budejovice. Fytotechnicka Rada. 14(2): 45?48.

Tsuji, C. (1909). Sanji Hokoku. 35: 1-24. Cited in Wyatt (1961). **

Turunen, S. and Chippandale, G. M. (1979). Possible function of juvenile hormone dependent protein in larval insect diapause. Nature 280: 836-838.

Turunen, S. and Chippandale, G. M. (1980). Proteins of the fat body of non-diapausing and diapausing larvae of the south western corn borer, *Diatraea grandiosella*: Effect of juvenile hormone. J. Insect Physiol. 26: 163-169.

V

Valle, D. (1993). Vitellogenins in insects and other groups- A Review. Mem. Inst. Oswalda Cruz. **: 1-26.

Vanderzant, E. S. and Reiser, R. (1956). J. Econ. Entomol. 49: 454.**

Vaney, C. and Maigon, F. (1905). Variation subies par le glucose. Le glucogine, la graisse et les albumines solubles au cours des metamorphosesdever a soie. C. R. Acad. Sci. (Paris). 140: 1192-1195.

Vass, E. and Nappi, A. J. (2001). Fruit fly immunity. Bio. Essays. 51: 529-535.

Vasuki, K. and Dikshit, T. S. S. (1968). Alkaline and acid phosphatases in the grass hopper, *Poecilocerus pictus* and the effect of endrine. Enzymol. Acta. Biocatal. 34: 277-280.

Vercauteren, R. E. and Aerts, F. (1958). On the cytochemistry of the haemocytes of Galleria mellonella with special reference to Polyphenoloixidase. Enzymol. 20: 167-172.

Vilcinskas, A.; Kopacek, P.; Jegorov, A.; Vey, A. and Matha, V. (1997). Detection of lipophorin as the major cyclosporin-binding protein in the haemolymph of the greater wax moth Galleria mellonella. Comparative Biochemistry and Physiology. C: Pharmacology, Toxicology and Endocrinology. 117(1): 41?45.

Vincent, M. J.; Miranpuri, G. S. and Khachatourians, G. G. (1993). Acid phosphatase activity in haemolymph of the migratory grasshopper, Melanoplus sanguinipes, during Beauveria bassiana infection. Entomologia Experimentalis et Applicata. 67(2): 161-166.

W

Wago H. and Ichikawa (1979). Changes in the phagocytic rate during the larval development and manner of the haemocyte reactions to foreign cells in Bombyx mori Appl. Ent. Zool. 14(4): 397-403.

Wang, C. M. and Patton, R. L. (1969).Lipid in the haemolymph of the cricket Acheta domesticus. J. Insect Physiol. 15: 851-860.

Webley, D. P. (1951). Bold cell counts in the African migratory locust (Locusta migratoria migratorioidaes Reiche and Fairmairee). Proc. Roy. Entomo. Soc. (London) A. 26: 25-37.

Wharton, D. R. A.; Wharton, M. L. and Lola, J. (1965). Blood volume and water content of male American Cockroach, Periplaneta americana L. methods and the influence of age and starvation. J. Insect Physiol. 11: 391-404.

Wheeler, R. E. (1961). Studies on the total haemocyte counts in Periplaneta americana L. with special reference to moulting cycle. M. S. Thesis, University of Maryland. 71 p.**

Wheeler, R. E. (1963). Studies on the total haemocyte count and haemolymph volume in Periplaneta americana L. with special reference to last moulting cycle. J. Insect Physiol. 9: 223-235.

Whitcomb, R. F.; Sharpio, M. and Granados, R. R. (1974). Insect defence mechanism against microorganisms and parasitoids. In The Physiology of Insecta (M. Rockestein, ed.). Vol 5 (2nd edition). Academic Press, New York. pp. 447-536.

Whitmore, E. and Gilbert, L. I. (1974). Haemilymph proteins and lipoproteins in lepidoptera- a comparative electrophorectic study. Comp. Biochem. Physiol. , 47b: 63-78.

Whittaker, J. R. and West, A. S. (1962). A starch gel electrophoretic study of insect haemolymph protein. Can. J. Zool. 40: 655-671.

Wiegand, C.; Levin, D.; Gillespie, J. P.; Willott, E.; Kanost, M. R. ; Trenczek, T. (2000). Monoclonal antibody M13 identifies a plasmatocyte membrane protein and inhibits encapsulation and spreading reactions of Manduca sexta haemocytes. Arch. Insect .biochem. physiol. 45: 95-108.

Wigglesworth, V. B. (1939). The Principals of Insect Physiology. 1st Edition. Methuen. London.

Wigglesworth, V. B. (1955). The role of haemocytes in the growth and moulting of an insect Rhodnius prolixus. (Hemiptera) J. Exp. Biol. 32: 649-663.

Wigglesworth, V. B. (1956a). The function of the amoebocytes during moulting in Rhodnius. Ann. Des Sci. Nat. Zool. II Ser. 18: 139-144.

Wigglesworth, V. B. (1956b). The haemocytes and connective tissue formation in an insect Rhodnius prolixus (Hemiptera). J. Exp.Biol. 32: 649-663.

Wigglesworth, V. B. (1959). Insect blood cells. Ann. Rev. Entomol. 4: 1-16.

Wigglesworth, V. B. (1972). The Principals of Insect Physiology. 7th Edition. English Language Book Society and Chapman and Hall, London. p. 827.

Wigglesworth, V. B. (1979). Haemocytes and growth in insects. In Insect Haemocytes (A. P. Gupta, ed.). Cambridge University Press, Cambridge. pp. 303-318.

Wirtz, R. A. and Hopkins, T. L. (1974). Tyrosine and phenylealanine concentrations in haemolymph and tissues of the American cockroarch, Periplaneta americana during metamorphosis. J. Insect Physiol. 20: 1143-1154.

Witting, G. (1968). Electron microscopic characterization of insect haemocytes. Proc. of 26th Annu. Meet. EMSA, 1968, 68-69.

Wlodawer, P. and Wisniewska, A. (1965). Lipids in the wax moth larvae during starvation. J.Insect Physiol. 11: 11-20.

Wootton, I. D. P. (1964). Microanalysis in Medical Biochemistry. 4th edition. pp 71-74.

Wyatt, G. R. (1961). The Biochemistry of insect haemolymph. Ann. Rev. Entomol. 6: 75-102.

Wyatt, G. R. (1963). Biochemistry of diapause, development and injury in silkmoth pupae. In Insect Physiology (V. J. Brooks, ed.). pp. 23-41.

Wyatt, G. R. (1967). The biochemistry of sugars and polysachharides in insects. Advances in Insect Physiology (J. W. L. Beament, J. E. Treherne and V. B. Wiggelsworth, eds.). Vol 4. Academic Press, Londn. pp. 287-360.

Wyatt, G. R. (1980). The fat body as a protein factory. In Insect Biology in Future. (M. Locke and D. S. Smith, eds.). Academic Press, New York. pp. 201-225.

Wyatt, G. R. and Kalf, G. F. (1956). Trehalose in insects. Fed. Proc. 15: 188.

Wyatt, G. R. and Kalf, G. F. (1957). The chemistry of insect haemolymph.

I. Trehalose and other carbohydrates. J. Gen. Physiol. 40: 833-847.

Wyatt, G. R.; Lougheed, T. C. and Wyatt, S. S. (1956). The chemistry of insect haemolymph: Organic components of the haemolymph of the silkworm Bombyx mori and two other species. J. Gen. Physiol. 39: 853-68.

Wyatt, G. R. and Pan, M. L. (1978). Insect Plasma Proteins. Ann. Rev. Biochem. 47: 799-817.

Y

Yafaeva, Z. Sh. (1962). Ufa. 2: 73. **

Ye, B. H.; Zhu, C. L.; Shen, S. B. and Zhu, X. L. (1992). Preliminary analysis of polypeptide patterns of various organs and haemolymph of Anopheles sinensis Wiedemann. Annals of Medical Entomology. 1(1): 1?3.

Yeager, J. F. (1945). The blood picture of southern armyworm (Prodenia eridania). J. Agric. Res. 71: 1-40.

Yeager, J. F. and Fay, R. W. (1935). Method, specific gravity of haemolymph, roach. Proc. Soc. Expt. Biol. And Med. 32: 1667-1669.

Yeager, J. F. and Munson, S. C. (1941). Histochemical detection of glycogen in blood cells of the southern armyworm Prodenia eridania and in other tissues, especially mid gut epithelium. J. agric. Res. 63: 257-294.

Yin, C. M. and Chippendale, G. M. (1976). Hormonal control of larvae diapause and metamorphosis of the southwestern corn borer, *Diatraea grandiosella.* J. Exp. Biol. 64: 303-310.

Yoo, C. M. and Lee, K. R. (1975). Studies in the haemolymph proteins during the metamorphosis in pine moth, Dendrolimus spectabilis. Butler. Korean J. Zool. 17(2): 81-92.

Yoo, C. M. and Lee, R. (1974). Polyacrylamide gel electrophoresis of proteins of the mealworm Ephestia kühniella Zeller. Korean J. Zool. 16(3): 185-193.

Yushkevich, N. L. (1960). Histochemical investigation of the protein tryptophane in insect tissues (Mealworm), *Tenebrio molitor* [In Russian] Doklady, Nauk SSSR 134: 945-946.

Z

Zachary, D. and Hoffmann, J. A. (1973). The haemocytes of Calliphora erythrocephala Mieg. (Diptera). Z. Zellforsch. 141: 55-73.

Zachary, D.; Brehelin, M. and Hoffmann, J. A. (1975). Role of "thrombocytoid" in capsule formation in dipteran Calliphora erythrocephala. Cell Tissue Res. 162: 343-348.

Zahedi, M. (1993). Haemocytes of the mosquito, Armigeres subalbatus. Mosquito Borne Diseases Bulletin. 10(4): 121?127.

Zaidi, Z. S. and Khan, M. A. (1974). Cholesterol concentration in the haemolymph, fat body and gonads of the red cotton bug, Dysdercus cingulatus. In Hurkadli et al., Sericologia (1989). 29: 125-126.

Zeng, F.; Shu, S.; Park, Y. I. and Ramaswamy, S. B. (1997). Vitellogenin and egg production in the moth, Heliothis virescens. Archives of Insect Biochemistry and Physiology. 34(3): 287?300.

Zhu, C. Z.; Sun, Y. and Cheng, Z. H. (1992). Lectin in the haemolymph of Mythimna separata larvae. Acta Entomologica Sinica. 35(4): 399?404.

Zielinska, Z. M. and Wroniszewska, A. (1957). Studies on the biochemistry of wax moth Galleria mellonella 16. Weight of tissues and organs during atarvation of the larvae. Acta. Boil. Exp. 17: 345-9.

Index

A

B

C

O

P

R

S